Lecture Notes
in Control and Information Sciences   357

Editors: M. Thoma, M. Morari

Isabelle Queinnec, Sophie Tarbouriech,
Germain Garcia, Silviu-Iulian Niculescu (Eds.)

Biology and Control Theory: Current Challenges

 Springer

Editors

Isabelle Queinnec
LAAS-CNRS
7 avenue du Colonel Roche
31077 Toulouse cedex 4
France
E-mail: queinnec@laas.fr

Germain Garcia
LAAS-CNRS
7 avenue du Colonel Roche
31077 Toulouse cedex 4
France
E-mail: garcia@laas.fr

Sophie Tarbouriech
LAAS-CNRS
7 avenue du Colonel Roche
31077 Toulouse cedex 4
France
E-mail: tarbour@laas.fr

Silviu-Iulian Niculescu
Laboratoire des Signaux et Systèmes
(L2S, UMR CNRS 8506), CNRS-Supélec,
3, rue Joliot Curie,
91190, Gif-sur-Yvette
France
E-mail: Silviu.Niculescu@lss.supelec.fr

Library of Congress Control Number: 2007925171

ISSN print edition: 0170-8643
ISSN electronic edition: 1610-7411
ISBN 3-540-71987-3 Springer Berlin Heidelberg New York
ISBN 978-3-540-71987-8 Springer Berlin Heidelberg New York

Preface

Bio and Control: Introductory Ideas

Historically speaking, the connection between biology and control feedback theory goes back to the analogy suggested by NORBERT WIENER in the 40s for explaining some human behaviors by using *feedback* mechanisms: brain control of a standard arm mechanical movement. At the same period of time, ERWIN SCHRÖDINGER pointed out some similarities between physics principles and the laws governing alive organisms, and he suggested a physics-based approach for the modeling and the analysis of such organisms by using the *analogy* (electrical, mechanical, and chemical processes). Without being exhaustive on the corresponding methodologies, these simple ideas are at the origin of a large number of models that tried to reproduce the behavior of living organisms. Without any doubts, such models served and helped in defining the first inter- and trans-disciplinary programs between biology and other sciences and/or disciplines from Mathematics to Physics and Computer Sciences in the last decade.

In this context, the control feedback theory has its own place, and we hope that it is able to bring the beginning of answers in the understanding of biology dynamics. The mathematical description of signals and circuits is not only at the origin of the modern control feedback theory, but also contributed significantly at the emergence of *Systems Biology*, as mentioned by P. WELLSTEAD in his essay *Schrödinger's Legacy: Systems and Life* (ETS Walton Lecture, 2005), and this example is far to be an isolated case.

Formalism, Potential Interactions and Some Expectations...

Taking into account the current competences in the field of the classical control feedback theory, namely, analysis, observation and control of dynamical systems (linear or nonlinear, finite-dimensional or not), the *main objective* for the edition of this book is to propose a (potential) "progressive transfer" of some of the "competences" issued from control feedback theory towards the domains of the life sciences, and especially towards all the domains of life sciences in which a *dynamic behavior* can be pointed out (see also some discussions in the report of the panel on future directions in control, dynamics and systems: *Control in an information rich world*, edited by R.M. MURRAY).

In this sense, take an extremely simple example: a same molecule has different meanings and various interests of study for the specialists in the life sciences domain, depending on the type of action considered:

- Action on isolated enzymatic systems (biochemical);
- Action on a dynamical chain of reaction in the alive cell (biology);
- Action on the human organic functions (pharmacology, physiology);
- Human therapeutic actions (medicine),

and each of these actions can be translated by some particular quantitative and qualitative properties of the corresponding dynamical system modeling the considered action.

Roughly speaking, any behavior of a biological system with respect to one (or several) time scaling can be interpreted in a dynamical system framework or context by using appropriate (analysis and control) tools. Indeed, we think that some knowledge in control feedback theory could deserve to have a *better understanding* of different kinds of dynamic evolution provided that some variables of the system are measurable (or observable), even if one cannot control the object under consideration. Furthermore, several tools and methods for a *qualitative and quantitative analysis* of such evolutions hold.

Next, it is worst to note that, in biology, the notion of *structure* has a larger sense than in automatic control theory. More precisely, the structure defines the set of relations existing between the different elements that constitute the object or the set of objects under consideration. Behind each structure, there is some *complexity* (any alive organism is strongly complex) with its *own hierarchy* (any alive organism is highly organized). Thus, the notion of *closed-loop* exists (and is recognized like that since more than an half century) in the context of biological systems, and as emphasized by HENRI LABORIT *"every life evolution from and after the photosynthesis has been regulated by feedback between more ordered and less ordered structures of the environmental device"*.

In summary, in our opinion, the fact to create some links between control theory community and that one of the life sciences could allow addressing the following problems (see also the challenges from biology briefly presented in *New issues in the Mathematics of Control* by R. BROCKET in *Mathematics Unlimited - 2001 and beyond* for some insights in neurobiology, cell biology and psychology):

- To model, observe and have a perception of the alive structures;
- To analyze the dynamic interconnections between biological systems and structures.

Finally, we also believe that this interaction between specialists in biology and control feedback theory will be useful in both directions. More precisely, biology systems need appropriate analysis tools due to their structure and hierarchy, complexity and environment interference, and we believe that these aspects may generate interesting research topics in control area. Indeed, several works, raising the potential impact of control developments to bring some beginning of answers in the context of biological systems, have been published in the recent years (special sessions or workshops at the interface between these communities, see, for instance, the proceedings of CDC 2005 and 2006). The idea of this book was conceived in the context mentioned above.

How to Read the Book?

This book is organized as follows.

- Part 1 is devoted to model selection and consists of chapters 1 through 4: modeling perspective in chronic myelogenous leukemia (first chapter), an aid for an early diagnosis of HIV/AIDS infection (Chapter 2), a multi-scale control oriented model in ovulatory processes (Chapter 3) and finally some robotics insights in modeling visually guided hand movements (Chapter 4).
- Part 2 is devoted to models for system analysis and consists of chapters 5 through 10, as follows: analysis of monotone and near-monotone biochemical network structures (Chapter 5), system and control in understanding the biological signal transduction (Chapter 6), analysis of some piecewise-linear models of genetic regulatory networks (Chapter 7), the modeling and the analysis of cell death signalling (Chapter 8), a Petri Net approach to persistence analysis in chemical reaction networks (Chapter 9), and finally some geometric ideas in stability analysis of various delay models in bioscience (Chapter 10).
- Part 3 is devoted to analysis and control aspects and consists of chapters 11 through 13: modeling and control of anesthetic pharmacodynamics (Chapter 11), a direct adaptive control of some non-negative and compartmental systems with delays (Chapter 12), and finally the analysis and control of dynamics in biological systems in presence of limitations (Chapter 13).

Note that this partition is somewhat arbitrary as most of the chapters are interconnected, and it mainly reflects the editors' biases and interests.

We hope that this volume will help in claiming many of the problems for control researchers, starting discussions and opening interactive debates between the control and biology communities, and, finally, to alert graduate students to the many interesting ideas at the frontier between control feedback theory and biology. There are, of course, many areas which are not represented through a chapter, and therefore we would like to apologize to those whose areas are not profiled.

Acknowledgements

The idea of this edited book inherits from the organization of an International Workshop on the subject in April, 24-25th, 2006, at LAAS-CNRS, Toulouse, France, co-organized with HeuDiaSyC (UMR CNRS 6599), Compiègne, France. Such a meeting was the third one of a series initiated with the 1st CNRS-NSF Workshop on *"Advances on time-delay systems"* (Paris, La Défense, France, January 2003) and continued by the International Workshop on *"Applications of time-delay systems"* (Nantes, France, September 2004).

Foremost, we would like to thank all the contributors of the book. Without their encouragement, enthusiasm, and patience, this book would have not been possible. A list of contributors is provided at the end of the book.

We also wish to thank Springer for agreeing to publish this book. We wish to express our gratitude to Dr. THOMAS DITZINGER (Senior Editor in Engineering) for his careful consideration and helpful suggestions regarding the format and organization of the book.

Toulouse, France, February 2007
Toulouse, France, February 2007
Toulouse, France, February 2007
Gif-sur-Yvette, France, February 2007

ISABELLE QUEINNEC
SOPHIE TARBOURIECH
GERMAIN GARCIA
SILVIU-IULIAN NICULESCU

Contents

List of Contributors

Frank Allgöwer (University of Stuttgart, Germany) - Chapter 8
David Angeli (University of Firenze, Italy) - Chapter 9
Carolyn L. Beck (University of Illinois at Urbana-Champaign, USA) - Chapter 11
Marc Bloom (New York University School of Medicine, USA) - Chapter 11
Eric Bullinger (National University of Ireland, Ireland) - Chapter 6
Simona Celebrini (Centre de Recherche Cerveau et Cognition, France) - Chapter 4
VijaySekhar Chellaboina (University of Tennessee, USA) - Chapter 12
Frédérique Clément (INRIA Rocquencourt, France) - Chapter 3
Hidde de Jong (INRIA Rhône-Alpes, France) - Chapter 7
Patrick De Leenheer (University of Florida, USA) - Chapter 9
Nki Echenim (INRIA Rocquencourt, France) - Chapter 3
Thomas Eißing (University of Stuttgart, Germany) - Chapter 8
Rolf Findeisen (University of Stuttgart, Germany) - Chapter 6
Germain Garcia (LAAS-CNRS, France) - Chapter 13
Jean-Luc Gouzé (INRIA Sophia Antipolis, France) - Chapter 7
Frédéric Grognard (INRIA Sophia Antipolis, France) - Chapter 7
Keqin Gu (Southern Illinois University at Edwarsville, USA) - Chapter 10
Wassim M. Haddad (Georgia Institute of Technology, USA) - Chapter 12
Qing Hui (Georgia Institute of Technology, USA) - Chapter 12
Christophe Jouffrais (IRIT, France) - Chapter 4
Dimitrios Kalamatianos (National University of Ireland, Ireland) - Chapter 6
Peter S. Kim (Stanford University, USA) - Chapter 1
Peter P. Lee (Stanford University, USA) - Chapter 1
Doron Levy (Stanford University, USA) - Chapter 1
Hui-Hung Lin (National Cheng-Kung University, China) - Chapter 11
Claude H. Moog (IRCCyN, France) - Chapter 2
Constantin-Irinel Morarescu (HeuDiaSyC, France) - Chapter 10
Wim Michiels (K.U. Leuven, Belgium) - Chapter 10
Silviu-Iulian Niculescu (L2S, France) - Chapter 10
Djomangan Adama Ouattara (IRCCyN, France) - Chapter 2
Isabelle Queinnec (LAAS-CNRS, France) - Chapter 13
Jayanthy Ramakrishnan (University of Tennessee, USA) - Chapter 12

Eduardo D. Sontag (Rutgers University, USA) - Chapters 5 and 9
Michel Sorine (INRIA Rocquencourt, France) - Chapter 3
Philippe Souères (LAAS-CNRS, France) - Chapter 4
Sophie Tarbouriech (LAAS-CNRS, France) - Chapter 13
Yves Trotter (Centre de Recherche Cerveau et Cognition, France) - Chapter 4
Steffen Waldherr (University of Stuttgart, Germany) - Chapter 8
Peter Wellstead (National University of Ireland, Ireland) - Chapter 6

Part I

Model Selection

Mini-Transplants for Chronic Myelogenous Leukemia: A Modeling Perspective

Peter S. Kim[1], Peter P. Lee[2], and Doron Levy[3]

[1] Department of Mathematics, Stanford University, Stanford, CA 94305-2125
  `pkim@math.stanford.edu`
[2] Division of Hematology, Department of Medicine, Stanford University,
  Stanford, CA, 94305
  `ppl@stanford.edu`
[3] Department of Mathematics, Stanford University, Stanford, CA 94305-2125
  `dlevy@math.stanford.edu`

Summary. We model the immune dynamics between T cells and cancer cells in leukemia patients after a bone-marrow (or a stem-cell) transplant. We use a system of nine delay differential equations that incorporate time delays and account for the progression of cells through different stages. This model is an extension of our earlier model [3]. We conduct a sensitivity analysis of the model parameters with respect to the minimum cancer concentration attained during the first remission and the time until the first relapse. In addition, we examine the effects of varying the initial host cell concentration and the cancer cell concentration on the likelihood of a successful transplant. We observe that higher initial concentrations of general host blood cells increase the chance of success. Such higher initial concentrations can be obtained, e.g., by reducing the amount of chemotherapy that is administered prior to the transplant, a procedure known as a mini-transplant. Our results suggest that mini-transplants may be advantageous over full transplants. We identify the regions of the parameters for which mini-transplants are advantageous using statistical tools.

Keywords: Chronic myelogenous leukemia, stem-cell transplant, bone-marrow transplant, non-myeloablative, mini-transplant, immune response, delay differential equations.

1 Introduction

Allogeneic bone-marrow or stem-cell transplantation (ABMT or ASCT) is currently the only known curative treatment for CML [11]. Prior to ABMT, the patient receives chemotherapy to lessen the disease and to lower the patient's immune cell population. This pre-treatment procedure is performed to reduce immune suppression by leukemia cells and to prevent graft rejection by the host.

Typically, the patient receives large doses of chemotherapy to eliminate almost all leukemia and immune cells. The treatment is called a full (or myeloablative) transplant, because the chemotherapy destroys, or ablates, nearly all the myeloid stem cells, which are the cells that produce new blood.

I. Queinnec et al. (Eds.): Bio. & Ctrl. Theory: Current Challenges, LNCIS 357, pp. 3–20, 2007.
springerlink.com   © Springer-Verlag Berlin Heidelberg 2007

However, in some cases, patients cannot tolerate full chemotherapy, so they are given mini (or non-myeloablative) transplants. In mini-transplants, patients receive milder doses of chemotherapy that do not ablate the myeloid stem cells. As a result, the treatment depends more heavily on the donor immune cells to expand and destroy remaining leukemia cells.

In [3], we modeled the immune dynamics of a full transplant. Our results suggested that the expansion of donor T cells depends more on general host blood cells than on leukemia cells alone. This is because a successful transplant relies on a blood-restricted graft-versus-host disease, in which donor T cells react to antigen that is present on general host blood cells. This less discriminate reactivity results in greater proliferation of immune cells and often proves necessary, because leukemia cells usually do not provide sufficient stimulus.

In this paper we present an extension of the model of [3], in which we made two major changes: First, we assume that all target cells have two possible states: alive and dying. Dying target cells are cells that are in the process of dying due to cyotoxic interactions with T cells. These cells linger for about five minutes, during which they may still stimulate other circulating T cells. In addition, we also introduce discounting factors for cell death rates to prevents from double-counting cell deaths due to the natural decay and the cytotoxic T cell responses.

In this paper, we use the extended model to study the dynamics of mini-transplants and to determine conditions that increase the chance of a successful cure. For full and mini-transplants, chemotherapy indiscriminately kills a large number of non-leukemic host blood cells. Since these cells stimulate donor immune cells, high levels of chemotherapy might reduce the potency of the anti-leukemia immune response. In this study, we seek to understand the trade off between eliminating leukemia and host immune cells and maintaing the stimulus to donor immune cells. Under certain circumstances, we find that mini-transplants may prove more advantageous not only due to its reduced toxicity to the patient but also because it preserves a larger population of host blood cells that provide a high enough stimulation to drive the expansion of donor immune cells.

The paper is organized as follows. In Section 2, we present the state diagrams and the corresponding delay differential equations that govern the dynamics of the various cell populations. In Section 3, we present a summary of the parameter estimates that we used in the model. In Section 4, we summarize a previous result from the original paper, [3]. Then, we discuss a typical solution of the revised model and analyze the sensitivity of the revised model with respect to the parameters, using Latin Hypercube sampling. In particular, we focus on the sensitivity of the model to the initial leukemia and general host cell concentrations. We also demonstrate that minitransplants, in which chemotherapy is administered in weaker doses, may increase the chances of a successful transplant. In Section 5, we discuss our interpretations of our results and outline future directions of study.

2 The Model

We follow our previous work [3], and track the time evolution of six cell populations. From the donor, we consider anti-cancer T cells, anti-host T cells, and all other donor cells (exclusive of the two populations explicitly mentioned). From the host, we consider cancer cells, anti-donor T cells, and general host blood cells, The anti-cancer T cells represent the cells that respond to a cancer-specific antigen and exclusively mediate the GVL effect, while the anti-host T cells represent those that respond to a general blood antigen and mediate blood-restricted GVHD. The cancer population, general donor, and general host populations can each exist in two states: alive or dying (due to previous interaction with cytotoxic T cells). This two-state formulation is a modification that is not present in [3]. As a result, our new model consists of nine delay differential equations (rather than six as in [3]).

In the following subsections we present the state diagrams and the corresponding equations for the cell populations. The various state diagrams follow the notation shown in Figure 1. The rectangles stand for the possible states for each population. The arrows represent transitions between states. The terms next to the arrows denote the rates at which cells move from one state to another. Most transitions have an associated time delay indicated by the values in the circles. The values in the circles represent the time it takes to complete the associated transition. Each circle can be thought of as a gate that holds cells in a given state until the appropriate time elapses.

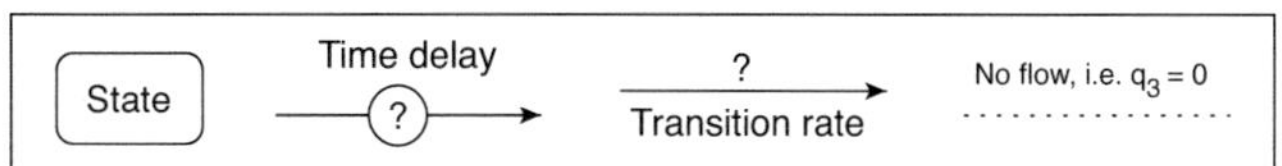

Fig. 1. The symbols used in the state diagrams

The model only explicitly measures population levels in each of the six base states, one for each cell population. Each term in the equations represents the beginning or the termination of a path connected to the base state (located at or near the center of each state diagram, denoted by the population label). The interaction initiation terms contain no delays, and the rates are proportional to the product of the two interacting populations and a mixing coefficient, k, in accordance with the law of mass action [10]. Termination terms contain each rate and delay encountered along their associated paths. Thus, the value of a given population variable, for example T_C, will at times be less than the total number of such cells, because it will not include cells that are within the pipeline of interactions with other populations.

In [3], our model formulation allowed population levels to cross zero. This artifact required us to impose a stopping criterion that forced all populations to remain non-negative by stopping at zero rather than passing through to negative values. In the new model, presented here, the various terms in the model incorporate discounting factors that prevent the populations from crossing zero.

2.1   General Blood Cells

Figure 2(a), presents the state diagram for the alive and dying general donor blood cells, D_A and D_D. These cells provide stimulus for a graft rejection response from T_D. They play a passive role in all interactions – they do not inspect other cells.

Donor blood cells move through the state diagram as a result of interactions with T cells. Alive cells may be shifted to the dying state with probability $p_1^{T_D/D}$. (Here and throughout the paper, we denote probabilities in the form $p_i^{X/Y}$, where X represents the T cell population and Y represents the target cell population involved in the interaction. In other words, X/Y should not be interpreted as an exponent.) Also, stem cells provide a constant flow of new alive cells, and existing cells die at the natural death rate d_D.

Once a cell shifts to the dying state, D_D, it has ρ units of time to live and continue stmulating ambient T cells before death. While in this liminal dying process, cells may also die at the natural death rate d_D. Cells that have died naturally during this liminal period do not undergo a second death due to a previous cytotoxic T cell encounter ρ time units ago. Hence, we adjust the second death rate by the discounting factor $e^{-d_D\rho}$.

The alive and dying general host cells, H_A and H_D, (shown in Figure 2(b)) have an analogous role to the donor cells D_A and D_D, in that they provide stimulus for a blood-restricted GVHD response from T_H, and the population has a similar diagram and DDEs to the donor population.

T cells do not distinguish between alive and dying target cells. They are stimulated equally be both. Hence, we define the collective population variables $D = D_A + D_D$ and $H = H_A + H_D$ to denote the total donor and host cells, respectively.

The corresponding DDEs for Figures 2(a) and 2(b) are

$$\frac{dD_A}{dt} = S_D - d_D D_A - p_1^{T_D/D} k D_A T_D, \tag{1}$$

$$\frac{dD_D}{dt} = -d_D D_D + p_1^{T_D/D} k D_A T_D - e^{-d_D\rho} p_1^{T_D/D} k D_A(t-\rho) T_D(t-\rho)$$

and

$$\frac{dH_A}{dt} = S_H - d_H H_A - p_1^{T_H/H} k H_A T_H, \tag{2}$$

$$\frac{dH_D}{dt} = -d_H H_D + p_1^{T_H/H} k H_A T_H - e^{-d_H\rho} p_1^{T_H/H} k H_A(t-\rho) T_H(t-\rho).$$

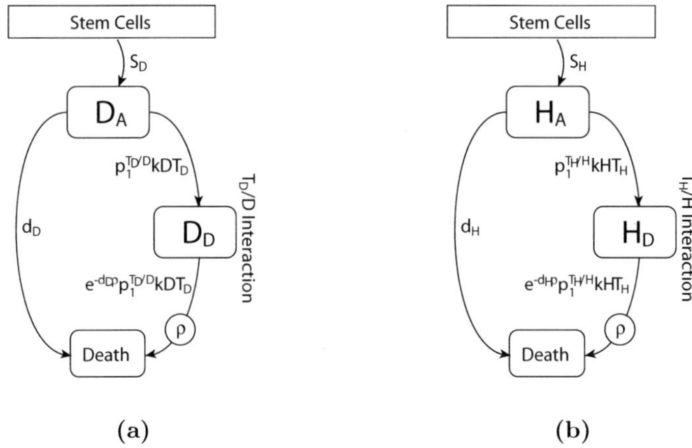

Fig. 2. General donor and host blood cell diagrams

2.2 Anti-cancer T Cells

Anti-Cancer T cells, T_C, interact with two other populations, the cancer cells, C, and the graft-rejecting cells T_D. The state diagram for the anti-cancer T cells is shown in Figure 3. As with the general blood cells, the cancer cells have two states, alive and dying, denoted by C_A and C_D, respectively. Also, the total cancer concentration is denoted by $C = C_A + C_D$. In the T_C/C interaction (the left wing of Figure 3) the T cells examine cancer cells and decide whether to react to the stimulus or to ignore it with probabilities $p_1^{T_C/C}$ and $p_2^{T_C/C}$, respectively. If the T cells ignore the stimulus, they return to the base state after a delay of σ, which represents the time for a nonproductive interaction. If the T cells react, they have a chance of destroying their targets through a cytotoxic response associated with a delay of ρ. After responding, the cells may enter a cycle of proliferation with probability $q_2^{T_C/C}$. Alternatively, they may forego the proliferation plan and simply recover cytotoxic capabilities (involving the replenishing of cytotoxic granulocytes) in preparation for their next encounter, returning them to the pool of active cells after a delay of v.

We assume that T cells divide an average of n times during proliferation, resulting in 2^n times as many cells. Each cycle of division takes τ units of time to complete, hence the entire proliferation cycle requires $n\tau$ units of time. We assume that during this time, all proliferating T cells are unavailable to interact with other cells, and thus they are not included in the measure of T_C. When interacting with cancer cells, there is a probability $q_3^{T_C/C}$ that the T cells will become anergic (i.e. tolerant) from the interaction. This effectively amounts to death in our simulations.

The T_D/T_C interaction (the right wing of Figure 3) represents encounters between anti-cancer cells and anti-donor T cells, where here the T_C cells are only targets. With probability $p_1^{T_D/T_C}$, the T_C cell may perish due to a graft-rejection response from T_D. In addition, some cells die at the natural rate d_{T_C}.

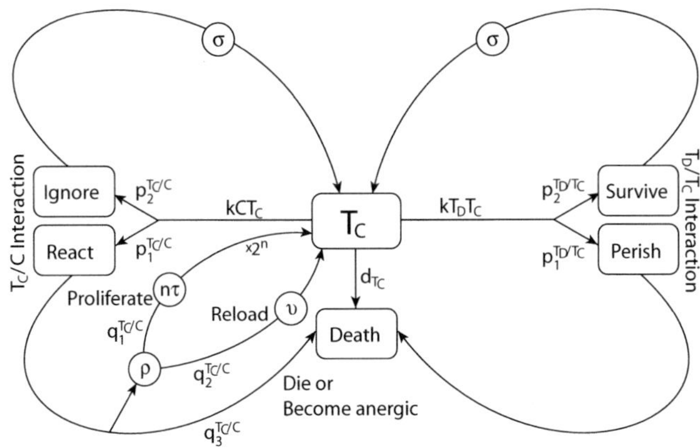

Fig. 3. Anti-cancer T cell diagram

The DDE that corresponds to Figure 3 is:

$$\frac{dT_C}{dt} = -d_{T_C}T_C - kCT_C - kT_DT_C + p_2^{T_C/C}kC(t-\sigma)T_C(t-\sigma)$$

$$+ p_2^{T_D/T_C}kT_D(t-\sigma)T_C(t-\sigma) \tag{3}$$

$$+ 2^n p_1^{T_C/C}q_1^{T_C/C}kC(t-\rho-n\tau)T_C(t-\rho-n\tau)$$

$$+ p_1^{T_C/C}q_2^{T_C/C}kC(t-\rho-\upsilon)T_C(t-\rho-\upsilon).$$

2.3   Anti-host T Cells

Anti-host T cells, T_H, undergo interactions with all host cells C, H, and T_D. The state diagram for the anti-host T cells is given in Figure 4. In the T_H/C interaction (upper left wing of Figure 4), the anti-host T cells react with cancer in the same way as anti-cancer T cells. In other words, they can either respond to the stimulus or ignore it with probabilities $p_1^{T_H/C}$ and $p_2^{T_H/C}$, respectively. After reacting, there is a probability $q_3^{T_H/C}$ that the T cells will become anergic after the encounter with cancer cells.

In the T_H/H interaction (lower left wing of Figure 4), the T cells react in the same way with general host blood cells H, except that the general host blood cells cannot cause anergy.

In the T_H/T_D interaction (right wing of Figure 4), the two types of T cells each have a chance of killing the other. Hence, the probabilities that an anti-host T cell survives the encounter and goes on to react to or ignore its target are $p_2^{T_D/T_H}p_1^{T_H/T_D}$ and $p_2^{T_D/T_H}p_2^{T_H/T_D}$, respectively. The extra factor of $p_2^{T_D/T_H}$ is the probability that the target anti-donor T cell does not kill the anti-host T cells. The remaining anti-host T cells get killed by their interactions with their target anti-donor T cells. Additionally, we assume that although T cells may be killed, they do not become anergic from such interactions. Hence, $q_3^{T_H/T_D} = 0$. Some cells also die at a natural death rate d_{T_H}.

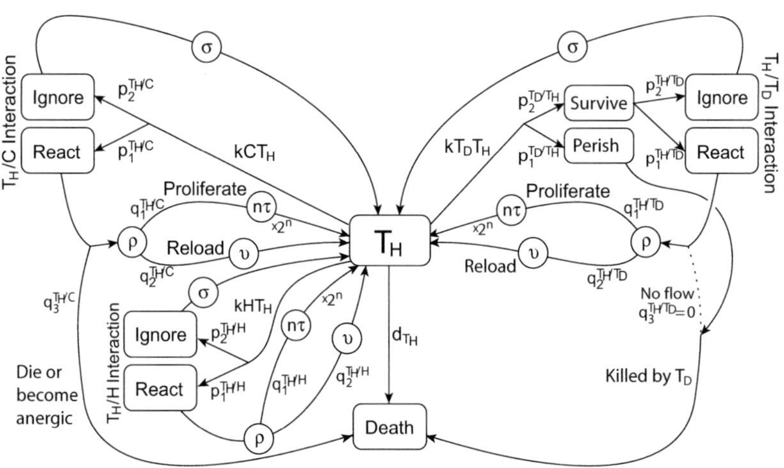

Fig. 4. Anti-host T cell diagram

The DDE that corresponds to Figure 4 is:

$$\frac{dT_H}{dt} = -d_{T_H}T_H - kCT_H - kT_DT_H - kHT_H$$
$$+ p_2^{T_H/C}kC(t-\sigma)T_H(t-\sigma)$$
$$+ p_2^{T_D/T_H}p_2^{T_H/T_D}kT_D(t-\sigma)T_H(t-\sigma)$$
$$+ p_2^{T_H/H}kH(t-\sigma)T_H(t-\sigma)$$
$$+ 2^n p_1^{T_H/C}q_1^{T_H/C}kC(t-\rho-n\tau)T_H(t-\rho-n\tau)$$
$$+ 2^n p_1^{T_H/H}q_1^{T_H/H}kH(t-\rho-n\tau)T_H(t-\rho-n\tau) \qquad (4)$$
$$+ 2^n p_2^{T_D/T_H}p_1^{T_H/T_D}q_1^{T_H/T_D}kT_D(t-\rho-n\tau)T_H(t-\rho-n\tau)$$
$$+ p_1^{T_H/C}q_2^{T_H/C}kC(t-\rho-v)T_H(t-\rho-v)$$
$$+ p_1^{T_H/H}q_2^{T_H/H}kH(t-\rho-v)T_H(t-\rho-v)$$
$$+ p_2^{T_D/T_H}p_1^{T_H/T_D}q_2^{T_H/T_D}kT_D(t-\rho-v)T_H(t-\rho-v).$$

2.4   Cancer Cells

Being target cells, cancer cells, C, have similarly passive roles in the interactions as those of D and H, and thus the value of C represents the number of cells in the base state and the two *Perish* states (i.e. all states other than *Death*). The state diagram for the cancer cells is given in Figure 5.

In addition, cancer multiplies at a logistic growth rate indicated by the closed loop at the top of the diagram. The logistic parameter r_C represents the net growth rate of cancer, which includes its natural death rate, and the parameter m_C represents the carrying capacity of the cancer population.

The C/T_H and C/T_C interactions are analogous to the T_D/D and the T_H/H interactions for the general blood cells in Section 2.1.

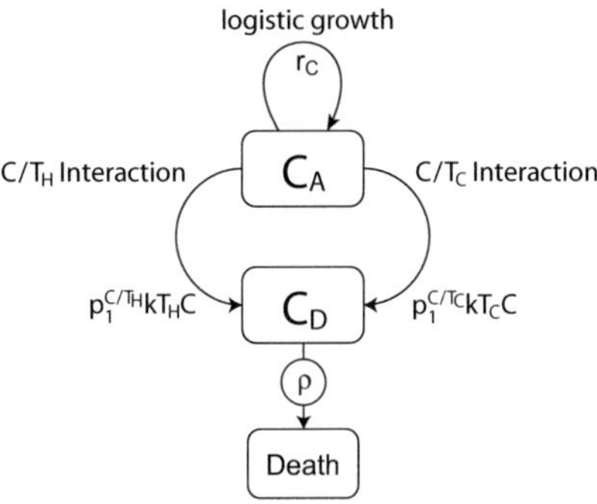

Fig. 5. Cancer diagram

The DDE for the evolution of C, that corresponds to Figure 5 is:

$$\frac{dC_A}{dt} = r_C C(1 - C/m_C) - p_1^{C/T_C}kCT_C - p_1^{C/T_H}kCT_H, \tag{5}$$

$$\frac{dC_D}{dt} = p_1^{C/T_C}kCT_C + p_1^{C/T_H}kCT_H - p_1^{C/T_C}kC(t-\rho)T_C(t-\rho)$$
$$+ p_1^{C/T_H}kC(t-\rho)T_H(t-\rho).$$

2.5 Anti-donor T Cells

The anti-donor T cells, T_D, respond to anti-cancer T cells T_C, general donor blood cells D, and anti-host T cells T_H the same way anti-host T cells respond to C, H, and T_D respectively. Hence, the diagram for anti-donor T cells is analogous to the one for anti-host T cells in Figure 4. This state diagram is given in Figure 6. The only difference is that we assume anti-cancer T cells cannot cause anergy in anti-donor T cells, so $q_3^{T_D/T_C} = 0$ on the lower right wing of Figure 6. Similarly, $q_3^{T_D/T_H} = 0$ on the lower left wing.

The DDE that corresponds to Figure 6 is:

$$\frac{dT_D}{dt} = -d_{T_D}T_D - kT_CT_D - kT_HT_D - kDT_D$$
$$+ p_2^{T_D/T_C}kT_C(t-\sigma)T_D(t-\sigma)$$
$$+ p_2^{T_H/T_D}p_2^{T_D/T_H}kT_H(t-\sigma)T_D(t-\sigma)$$
$$+ p_2^{T_D/D}kD(t-\sigma)T_D(t-\sigma)$$
$$+ 2^n p_1^{T_D/T_C}q_1^{T_D/T_C}kT_C(t-\rho-n\tau)T_D(t-\rho-n\tau) \tag{6}$$
$$+ 2^n p_2^{T_H/T_D}p_1^{T_D/T_H}q_1^{T_D/T_H}kT_H(t-\rho-n\tau)T_D(t-\rho-n\tau)$$
$$+ 2^n p_1^{T_D/D}q_1^{T_D/D}kD(t-\rho-n\tau)T_D(t-\rho-n\tau)$$
$$+ p_1^{T_D/T_C}q_2^{T_D/T_C}kT_C(t-\rho-\upsilon)T_D(t-\rho-\upsilon)$$
$$+ p_2^{T_H/T_D}p_1^{T_D/T_H}q_2^{T_D/T_H}kT_H(t-\rho-\upsilon)T_D(t-\rho-\upsilon)$$
$$+ p_1^{T_D/D}q_2^{T_D/D}kD(t-\rho-\upsilon)T_D(t-\rho-\upsilon).$$

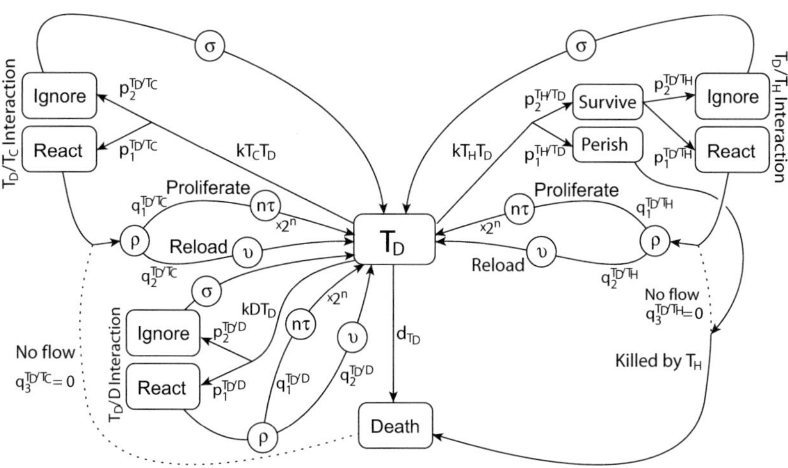

Fig. 6. Anti-donor T cell diagram

3 Parameters

The various parameters and associated references used in our model are summarized in Table 1 and Table 2. Whenever possible, we obtained their values directly from the literature, although at times we had to make do with an

Table 1. Parameters

Param.	Description	Estimate	References
Delays (day)			
ρ	Time for reactive T cell/ antigen interactions	0.0035 (5 min)	[6]
σ	Time for unreactive interactions	0.0007 (1 min)	[6]
τ	Time for cell division	$0.5 - 1.5$	[2, 7]
υ	T cell recovery time after killing another cell	1	[6]
Growth and Death Rates (day^{-1})			
$d_{T_C}, d_{T_H}, d_{T_D}$	T cell death rates	0.23	[4]
d_D, d_H	General death rate	0.1	Estimated
r_C	Net cancer growth rate	$10^{-3} - 10^{-2}$	[5, 10, 12]
m_C	Logistic carrying capacity for cancer	$1.5 - 3.5 \times 10^5$ (cells/μL)	Estimated
n	Avg # of T cell divisions after stimulation	< 8 times	[1]
Proportionality Constant for Mass Action $(\text{cells}/\mu\text{L})^{-1}\text{day}^{-1}$			
k	Kinetic mixing rate	10^{-3}	[8]

Table 2. Transition Probabilities

Param.	Description	Estimate
Probabilities		
$p_1^{X/Y}$	Prob. of a reactive X/Y interaction	$p_1^{X/Y} + p_2^{X/Y} = 1$
$p_2^{X/Y}$	Prob. of an unreactive X/Y interaction	
$q_1^{X/Y}$	Prob. that X proliferates after interaction	$q_1^{X/Y} + q_2^{X/Y} + q_3^{X/Y} = 1$
$q_2^{X/Y}$	Prob. that X keeps probing after interaction	
$q_3^{X/Y}$	Prob. that X becomes anergic after interaction	
$p_1^{T_H/H}, p_1^{T_H/T_D}, p_1^{T_D/D}, p_1^{T_D/T_C}, p_1^{T_D/T_H}$		0.9
$p_1^{T_C/C}, p_1^{T_H/C}$		0.8
$p_1^{C/T_C}, p_1^{C/T_H}$		0.6
$q_i^{T_H/H}, q_i^{T_H/T_D}, q_i^{T_D/D}, q_i^{T_D/T_C}, q_i^{T_D/T_H}$		0.5 $(i = 1, 2)$
$q_i^{T_C/C}, q_i^{T_H/C}$		0.25 $(i = 1, 2)$

Table 3. Initial Concentrations

Param.	Description	Estimate
Initial Concentrations (cells/μL)		
$T_C(0)$	Anti-cancer T cells	$10^{-1} - 1$
$T_H(0)$	Anti-host T cells	$10^{-1} - 1$
$D_A(0)$	General donor cells	$\leq 10^3$
$C_A(0)$	Cancer cells	$< 1.7 \times 10^{-3}$
$T_D(0)$	Anti-donor T cells	$\ll 1.7 \times 10^{-4}$
$H_A(0)$	General host cells	$\sim 10^3$
Stem Cell Supply Rates ((cells/μL) day^{-1})		
S_D	Donor cell resupply rate	$10^{-6} - 10^{-5}$
S_H	Host cell resupply rate	$10^{-8} - 10^{-7}$

estimation when specific information was unavailable. The initial concentrations and stem cell levels are summarized in Table 3. A detailed discussion of the parameters can be found in [3].

4 Numerical Simulations and Results

In delay differential equations, the values of the variables for $t < 0$ must be set as an initial condition. In our case, we handle this constraint by assuming that the host cell populations are at a steady state before the transplant, while the donor-derived cell populations are not present until the time of the transplant (which is at $t = 0$). In other words, the values up to time 0 are denoted by the vector $[T_{C,0}\delta(t), T_{H,0}\delta(t), D_0\delta(t), C_0, T_{D,0}, H_0]$, where the terms $T_{C,0}$, $T_{H,0}$, D_0, C_0, $T_{D,0}$, and H_0 denote the concentrations of the six cell populations at time 0, and

$$\delta(t) = \begin{cases} 0 & \text{if } t \neq 0, \\ 1 & \text{if } t = 0. \end{cases}$$

In all our numerical simulations we use the 'dde23' delay-differential equation solver in MATLAB 6.5. The relative and absolute tolerances are left in their default values of 10^{-3} and 10^{-6}, respectively.

4.1 Parameter Sensitivity Analysis Using LHS

In the original model of [3], we performed a sensitivity analysis that focused on two parameters at a time, i.e. we varried the values of two parameters while keeping all other parameters fixed. Using this approach we determined the relative importance of several parameters in determining the model behavior. Such a result from [3] is shown in Figure 7 which was obtained by varying the average number of T cell divisions per stimuation against the initial anti-host T cell

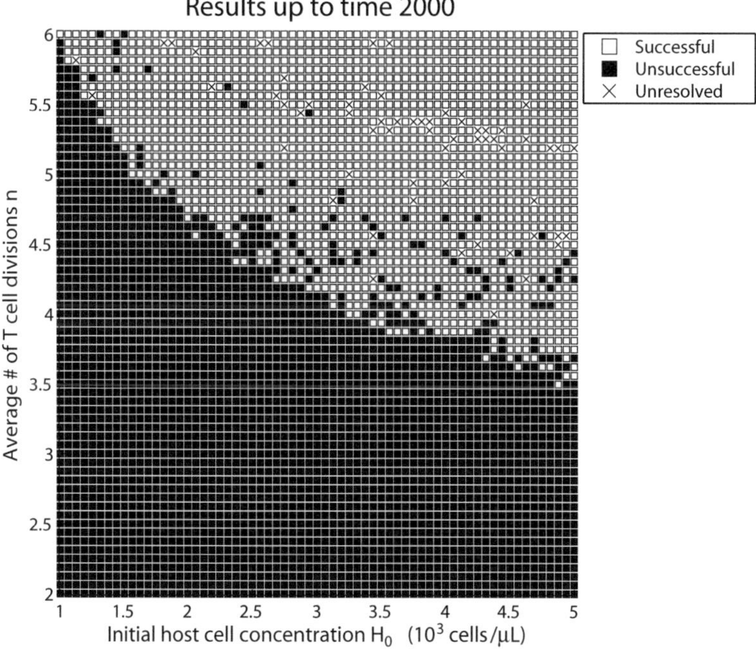

Fig. 7. Outcomes of model simulations with respect to the average number of T cell divisions per stimulation and the initial anti-host T cell oncentration

Table 4. Correlations between parameters (over the given ranges) and minimum cancer concentrations. Note that some parameters have a wider range than others. Correlation coefficients are as follows: Pearson product-moment correlation (PPMC), Spearman rank-order correlation (SROC).

Param.	Description	PPMC	SROC	Range
ρ	Time for productive interaction	-0.0043	-0.0141	$0.0035 \pm 25\%$
σ	Time for nonproductive interaction	-0.0432	-0.0700	$0.0007 \pm 25\%$
τ	Time for one cell division	0.0358	0.0295	$1.0000 \pm 25\%$
υ	Time for T cell recovery after cytotoxic event	0.0053	-0.0002	$1.0000 \pm 25\%$
$p_1^{T_H/H}$	Prob. of anti-host T cell interaction with general host cell	-0.0664	-0.0722	$0.9 \pm 25\%$
$p_1^{T_C/C}$	Prob. of anti-cancer T cell interaction with cancer cell	0.0070	0.0282	$0.8 \pm 25\%$
p_1^{C/T_C}	Prob. of cancer cell interaction with anti-cancer T cell	-0.0339	-0.1427	$0.6 \pm 25\%$
$q_1^{T_H/H}$	Prob. that anti-host T cell divides after interaction	-0.2067	-0.2664	$0.5 \pm 25\%$
$q_1^{T_C/C}$	Prob. that anti-cancer T cell divides after interaction	-0.0836	-0.0633	$0.25 \pm 25\%$
d_{T_C}	Death rate of anti-cancer T cell	0.0818	0.1164	$0.1 \pm 25\%$
d_H	Death rate of host cell	0.0675	0.1320	$0.1 \pm 25\%$
r_C	Growth rate of cancer	0.0587	0.1106	0.01 to 0.05
m_C	Carrying capacity of cancer	-0.0135	0.0032	1 to 3×10^5
n_T	Average number of T cell divisions per stimulation	-0.2384	-0.3699	2 to 4
k	Kinetic mixing coefficient	**-0.3043**	**-0.3940**	0.5 to 2×10^{-3}
$C_A(0)$	Initial cancer concentration	**0.4082**	**0.2885**	0 to 1×10^{-2}
$T_C(0)$	Initial anti-cancer T cell conc.	-0.0813	-0.0770	0 to 10
$T_H(0)$	Initial anti-host T cell conc.	-0.1836	-0.2100	0 to 10
$T_D(0)$	Initial anti-donor T cell conc.	-0.0449	-0.0788	0 to 1×10^{-5}
$H_A(0)$	Initial host cell conc.	**-0.4858**	**-0.5892**	0 to 3000
$D_A(0)$	Initial donor cell conc.	-0.0208	-0.0436	0 to 1000
S_H	Stem cell supply rate of host cells	-0.0326	0.0273	0 to 1×10^{-6}
S_D	Stem cell supply rate of donor cells	-0.0505	-0.0530	0 to 1×10^{-4}

concentration. In the model from [3], cancer populations could reach zero, and hence a successful outcome was defined as cancer elimination. An unsuccessful outcome was defined as the case where cancer relapses to over 1.5×10^5 cells/μL. An unresolved outcome as the case where neither scenario has occurred within 2000 days of transplantation.

Figure 7 suggests that higher initial anti-host T cell concentrations and higher average numbers of T cell division improve the chances of a successful outcome.

Table 5. Correlations between parameters (over the given ranges) and the time to cancer relapse. We define the time of relapse to be the time that the total cancer concentration recovers to over 1000 cells/μL.

Param.	Description	PPMC	SROC	Range
ρ	Time for productive interaction	0.0275	0.0190	$0.0035 \pm 25\%$
σ	Time for nonproductive interaction	0.0739	0.0951	$0.0007 \pm 25\%$
τ	Time for one cell division	0.0226	-0.0058	$1.0000 \pm 25\%$
υ	Time for T cell recovery after cytotoxic event	0.0219	-0.0095	$1.0000 \pm 25\%$
$p_1^{T_H/H}$	Prob. of anti-host T cell interaction with general host cell	0.0283	0.0120	$0.9 \pm 25\%$
$p_1^{T_C/C}$	Prob. of anti-cancer T cell interaction with cancer cell	-0.0805	-0.0479	$0.8 \pm 25\%$
p_1^{C/T_C}	Prob. of cancer cell interaction with anti-cancer T cell	0.1345	0.1398	$0.6 \pm 25\%$
$q_1^{T_H/H}$	Prob. that anti-host T cell divides after interaction	0.1094	0.1148	$0.5 \pm 25\%$
$q_1^{T_C/C}$	Prob. that anti-cancer T cell divides after interaction	-0.0246	-0.0139	$0.25 \pm 25\%$
d_{T_C}	Death rate of anti-cancer T cell	-0.0416	-0.0495	$0.1 \pm 25\%$
d_H	Death rate of host cell	-0.1142	-0.0849	$0.1 \pm 25\%$
r_C	Growth rate of cancer	**-0.8072**	**0.8601**	0.01 to 0.05
m_C	Carrying capacity of cancer	0.0378	0.0505	1 to 3×10^5
n_T	Average number of T cell divisions per stimulation	0.1771	0.1792	2 to 4
k	Kinetic mixing coefficient	0.1818	0.1936	0.5 to 2×10^{-3}
$C_A(0)$	Initial cancer concentration	-0.1061	-0.1171	0 to 1×10^{-2}
$T_C(0)$	Initial anti-cancer T cell conc.	0.0997	0.0766	0 to 10
$T_H(0)$	Initial anti-host T cell conc.	0.0814	0.0878	0 to 10
$T_D(0)$	Initial anti-donor T cell conc.	0.0802	0.0728	0 to 1×10^{-5}
$H_A(0)$	Initial host cell conc.	**0.3160**	**0.3446**	0 to 3000
$D_A(0)$	Initial donor cell conc.	0.0623	0.0678	0 to 1000
S_H	Stem cell supply rate of host cells	-0.0137	0.0082	0 to 1×10^{-6}
S_D	Stem cell supply rate of donor cells	0.0497	0.0488	0 to 1×10^{-4}

In addition, the border line between the predominantly successful and the predominantly unsuccessful region is fairly clear.

To expand on the pairwise sensitivity analysis of [3] and to study the effects of all model parameters on the model behavior, we use Latin hypercube sampling (LHS) [9]. This sensitivity analysis was not conducted for the model in [3]. such a study is useful for statistically determining which parameters are the most influential in affecting the outcome of the model's behavior.

LHS involves solving the system of equations multiple times with randomly sampled parameter values. The samples are chosen such that each parameter is distributed over its range of admissible values. We vary a wide range of parameters and measure their correlations to the minimum cancer concentration attained during the simulation and the time to the first cancer relapse. We define relapse as the time that the cancer population first recovers to 1000 cells/μL after its initial drop. Tables 4 and 5 show a list of varied parameters and correlations to the minimum cancer concentrations and relapse times, respectively.

Over the ranges that parameters were varied, the parameters that most strongly affect the minimum cancer concentration turn out to be the kinetic mixing coefficient k, the initial cancer concentration $C_A(0)$, and the initial host cell concentration $H_A(0)$. The most significant of the three is the initial host cell concentration. This observation implies that it is beneficial to preserve a high population of non-cancerous general host cells before the transplant. As discussed in [3], these general host cells provide stimulus to expand the T cells involved in the blood-restricted graft-versus-host immune response.

The two parameters that most strongly influence the relapse time are the cancer growth rate r_C and the initial host cell concentration. Clearly, if cancer grows faster, it will relapse faster. Hence, decreasing the cancer growth rate will have the strongest effect on slowing cancer relapse. This observation supports the conclusion of [10]. On the other hand, the initial host cell concentration remains significant for affecting cancer relapse. Hence, as an overall strategy, it might be more effective to preserve as many general host cells as possible prior to transplantation.

4.2   Simulating Initial Conditions of Minitransplants

The most natural way to preserve a high host cell population is to reduce the level of chemotherapy. However, reduced chemotherapy also leads to higher leukemia loads before transplantation. Transplants performed under reduced chemotherapy

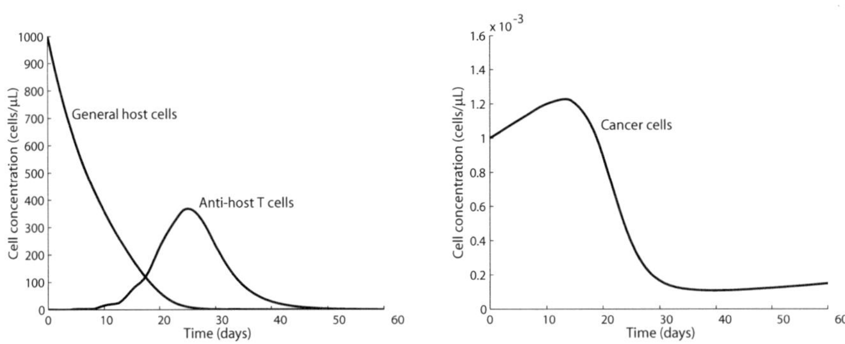

Fig. 8. Time evolution of general host cell, anti-host T cell, and cancer cell populations. Because cells, die at a proportional rate, the cancer cell population may become very low, but it never reaches zero. Hence, the cancer population always eventually relapses. Parameter values are given in the appendix.

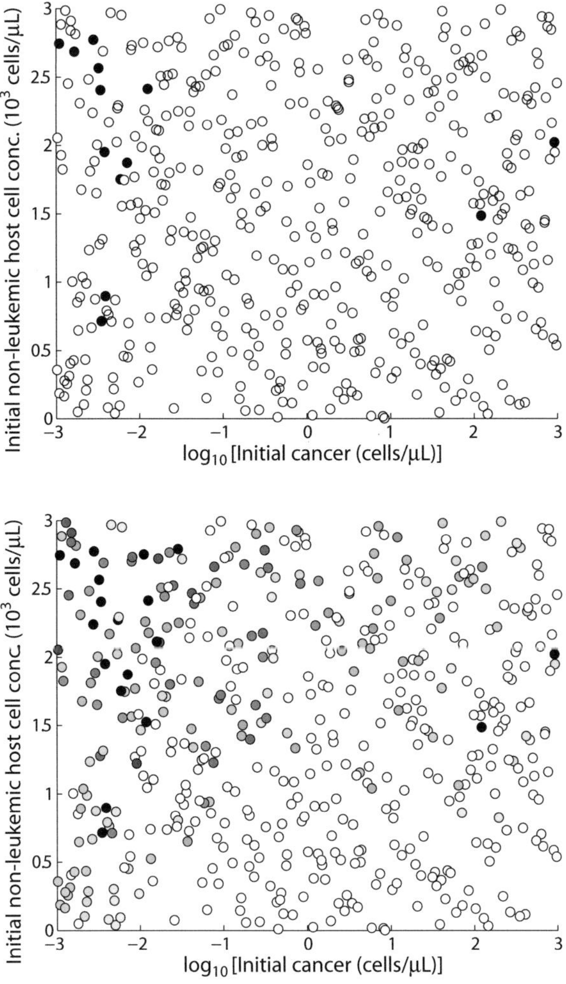

Fig. 9. Scatter plots of transplant outcomes with respect to initial non-leukemic host cell concentrations and cancer cell concentrations. All model parameters are varied using Latin Hypercube Sampling. *Top*: Black indicates cancer elimination, and white indicates cancer relapse. *Bottom*: Grayscale plot where the shade corresponds to the minimum cancer concentration attained after transplant. Black indicates complete elimination. White indicates a minimum cancer concentration of greater than 10^{-2} cells/μL.

are called non-myeloablative, or mini-, transplants. In mini-transplants higher levels of both the general host cell and cancer cell populations are retained.

We apply the model to study the influences of these two populations on the success of the transplant. Since the model is based on a system of deterministic differential equations, the cancer population never attains a 0 value. (See Figure 8 for an example of the time evolution of cancer, T cell, and general host cell populations.) Instead, we say that the cancer population is eliminated if its concentration passes below the concentration of one cell in the body. (We estimate this concentration to be around 1 cell/6L of blood $\sim 10^{-7}$ cells/μL.) We say that a transplant is successful if this threshold is crossed and refer to this barrier as our extinction criterion.

Using the model and applying the extinction criterion, we obtained preliminary results showing how a transplant's success depends on the two parameters: the initial (non-leukemic) host cell concentration and the initial leukemia concentration. Using LHS, we randomly varied these two parameters uniformly between 0 and 3000 cells/μL and logarithmically between 10^{-3} and 10^3 cells/μL, respectively. To obtain a more diverse sample, we simultaneously varied the other model parameters within a range of $\pm 25\%$ of their estimated values.

The results are shown in Figure 9. The x-axes show the initial cancer cell concentrations on a log scale, and the y-axes show the initial (non-leukemic) host cell concentrations. Figure 9(top) simply shows in black and white when cancer cells were eliminated (by the extinction criterion) and when they survived. Figure 9(bottom) shows the same plot in a grayscale, where black indicates a minimum cancer concentration at or below 10^{-7} cells/μL (extinction level) and white indicates a minimum cancer concentration at or above 10^{-2} cells/μL.

As expected, the figures show that it is best to start with almost no leukemia cells and as many non-leukemic host cells as possible. This ideal, however, is not always possible. The figures also show that having very little non-leukemic host cells makes it almost impossible to eliminate cancer, regardless of the initial cancer concentration. In addition, Figure 9(bottom) shows that the anti-leukemia immune response might do better even against a high level of cancer cells, if it has sufficient stimulus from non-leukemic cells. Our goal in this study is to determine ideal conditions for a successful transplant and devise treatment strategies to get as close to this ideal as possible. In some cases, mini-transplants might be more effective than full transplants, especially in cases where full transplantations are risky.

5   Discussion

In this paper, we extended the model of [3] by dividing all target cells into two states: alive and dying. This modification accounts for the delay between the time that a T cell engages a target cell and the time that the target cell actually dies. During this time period, target cells may continue to stimulate other circulating T cells.

Our analysis of the parameter sensitivity indicates that the initial host cell concentration is the most important parameter in influencing the minimum cancer concentration and the cancer relapse time together. A low minimum cancer concentration most likely implies that there is a higher probability that the cancer population was completely eliminated. This analysis supports the conclusion on [3]. Alternatively, a low cancer growth rate hardly affects the minimum cancer concentration, but strongly affects the relapse time. This last result supports the conclusion of the analogous CML model, based on ordinary differential equations, of [10].

We also use Latin hypercube sampling to examine the affects of the initial host cell and cancer cell concentrations on the outcome of a transplant. Our simulations show that it may be advantageous to attempt to decrease a host's cancer load while sparing as many general host cells as possible. Not only will this approach lead to a less toxic treatment before transplantation, but it will also improve the effectiveness of the donor-derived graft-versus-host immune response.

As a future work, we plan to conduct a more thorough analysis of the dynamics of minitransplants. Various strategies to minimize a patient's leukemia load, while preserving a supply of non-leukemic host cells are to optimally lessen the dosage of chemotherapy, administer the drug Gleevec to selectively inhibit cancer proliferation, or to use both methods in cominbination. With modeling, we intend to gain insights into the most effective combination of pre-transplant therapies. We will also consider whether certain post-transplant therapies such as donor lymphocyte infusion or cancer vaccination may enhance the effectivenes of the treatment.

Acknowledgments. The work of Doron Levy was supported in part by the NSF under Career Grant DMS-0133511. The work of Peter S. Kim was supported by the NSF Graduate Research Fellowship Program and the Department of Mathematics at Stanford University. The work of Peter P. Lee was supported by a Research Scholar Award from the American Cancer Society. We would also like to acknowledge the Blood and Marrow Transplantation Division of the Stanford Medical School for providing us with parameters derived from CML patients who underwent allogenic bone marrow transplantation.

References

1. R Antia, C T Bergstrom, S S Pilyugin, S M Kaech, and R Ahmed. Models of CD8+ responses: 1. What is the antigen-independent proliferation program. *J Theor Biol.*, 221(4):585–598, 2003.
2. D L Chao, S Forrest, M P Davenport, and A S Perelson. Stochastic stage-structured modeling of the adaptive immune system. *Proc IEEE Comput Soc Bioinform Conf.*, 2:124–131, 2003.
3. R DeConde, P S Kim, D Levy, and P P Lee. Post-transplantation dynamics of the immune response to chronic myelogenous leukemia. *J Theor Biol.*, 236(1):39–59, 2005.
4. C P Duvall and S Perry. The use of 51-chromium in the study of leukocyte kinetics in chronic myelocytic leukemia. *J Lab Clin Med.*, 71(4):614–628, 1968.
5. A S Fokas, J B Keller, and B D Clarkson. Mathematical model of granulocytopoiesis and chronic myelogenous leukemia. *Cancer Res.*, 51(8):2084–2091, 1991.
6. P Friedl and M Gunzer. Interaction of T cells with APCs: the serial encounter model. *Trends Immunol.*, 22(2):187–191, 2001.

7. Peter P Lee. Unpublished data: generated by the Lee Lab at Stanford Medical School, 2007.
8. T Luzyanina, K Engelborghs, S Ehl, P Klenerman, and G Bocharov. Low level viral persistence after infection with LCMV: a quantitative insight through numerical bifurcation analysis. *Math Biosci.*, 173(1):1–23, 2004.
9. M D McKay, W J Conover, and R J Beckman. A comparison of three methods for selecting values of input variables in the analysis of output from a computer code. *Technometrics*, 21:239–245, 1979.
10. Helen Moore and Natasha K. Li. A mathematical model for chronic myelogenous leukemia (CML) and T cell interaction. *J Theor Biol.*, 225(4):513–523, 2004.
11. C A Schiffer, R Hehlmann, and R Larson. Perspectives on the treatment of chronic phase and advanced phase CML and Philadelphia chromosome positive ALL. *Leukemia*, 17(4):691–699, 2003.
12. P A Stryckmans, L Debusscher, and E Collard. Cell kinetics in chronic granuclotyci leukaemia (CGL). *Clin Haematol.*, 6(1):21–40, 1977.

A   Parameters for Figures

Parameter	Value		
Figure #	7	8	9
ρ	0.0035	0.0035	0.0035
σ	0.0007	0.0007	0.0007
τ	1	1	1
υ	1	1	1
$p_1^{T_H/H}$, $p_1^{T_H/T_D}$, $p_1^{T_D/D}$, $p_1^{T_D/T_C}$, $p_1^{T_D/T_H}$	0.9	0.9	0.9
$p_1^{T_C/C}$, $p_1^{T_H/C}$	0.8	0.8	0.8
p_1^{C/T_C}, p_1^{C/T_H}	0.6	0.6	0.6
$q_1^{T_H/H}$, $q_1^{T_H/T_D}$, $q_1^{T_D/D}$, $q_1^{T_D/T_C}$, $q_1^{T_D/T_H}$	0.5	0.5	0.5
$q_2^{T_H/H}$, $q_2^{T_H/T_D}$, $q_2^{T_D/D}$, $q_2^{T_D/T_C}$, $q_2^{T_D/T_H}$	0.5	0.5	0.5
$q_1^{T_C/C}$, $q_1^{T_H/C}$,	0.25	0.25	0.25
$q_2^{T_C/C}$, $q_2^{T_H/C}$	0.25	0.25	0.25
d_{T_C}, d_{T_H}, d_{T_D}	0.23	0.23	0.23
d_D, d_H	0.1	0.1	0.1
r_C	0.02	0.02	0.02
m_C	2×10^5	2×10^5	2×10^5
n	2 to 6	4	
k	0.001	0.001	0.001
$T_{C,0}$	1	0	0
$T_{H,0}$	1	1	1
D_0	1000	0	0
C_0	0.001	0.001	10^{-3} to 10^3
$T_{D,0}$	10^{-5}	0	0
H_0	1000 to 5000	1000	0 to 3000
S_D	10^{-5}	10^{-5}	10^{-5}
S_H	10^{-7}	10^{-7}	10^{-7}

Modeling of the HIV/AIDS Infection: An Aid for an Early Diagnosis of Patients

Djomangan Adama Ouattara and Claude H. Moog

IRCCyN, UMR-CNRS 6597. École Centrale de Nantes,
1, rue de la Noë - BP 92101 - 44321 Nantes Cedex 03
djomangan-adama.ouattara@irccyn.ec-nantes.fr
claude.moog@irccyn.ec-nantes.fr
http://www.irccyn.ec-nantes.fr

Summary. Mathematical modeling is used for individual patients to help for an early diagnosis of the evolution of the infection. The feasibility of the method is depicted on some patients who start a HAART (Highly Active AntiRetroviral Therapy). It is shown how this mathematical study can be used in the early diagnosis of the immunological failure for HIV patients.

Keywords: HIV, AIDS, identification, identifiability, non-linear systems, therapeutical failures.

1 Introduction

HIV infected patients require a life time therapy since the virus can not be eradicated to date. In real life, therapy interruptions may occur, due to heavy secondary effects. Such therapeutic interruptions may be planned to minimize the secondary effects by using the mathematical tools and models [1, 2]. Mathematical models are designed to analyze and predict the evolution of the disease. The infection dynamics are modeled by mathematical (differential) equations which represent the main kinetics of the infection: the viral load, the healthy and the infected CD4+ T-cells kinetics.

In [3, 4], Perelson *et al.* studied the *in vivo* life span of the virus and of infected T-cells. Ho *et al.* [5] and Wei *et al.* [6] analyze the virus and the CD4+ T-cells turnover in HIV-1 infected patients. When using a model that includes the dynamics of long-lived infected cells, Perelson *et al.* explained the biphasic decay characteristics of the viral load in HIV-1 infected patients [4]. This biphasic decay has been revisited by Arnaout *et al.* [7] who described it as a possible effect of cytotoxic T-lymphocytes. Recently, new contributions by Filter and Xia present a mathematical method for estimating all the parameters of the basic 3-dimensional model of the HIV infection [8]. An application to vaccine readiness was deduced from the estimation of the viral load set point and the settling time for patients from a South Africa cohort.

Herein, we focus on the parameter identification, right after initiation of a therapy to predict the future evolution of the infection and an eventual clinical

I. Queinnec et al. (Eds.): Bio. & Ctrl. Theory: Current Challenges, LNCIS 357, pp. 21–43, 2007.
springerlink.com

failure. It is shown that such a mathematical analysis may help medical doctors for an early clinical diagnosis of HIV infected patients. This result is obtained from standard clinical data, *i.e.* the measurement of the viral load, of the CD4+ T-cells count and/or the CD8+ T-cells. Due to the high cost of data measurements and to physical constraints on the patients, it is not possible to get a large number of data. Thus, the estimation of parameters has to be performed under the constraint of the availability of a minimal number of data. The outline of this paper is as follows.

- In Section 2, an elementary mathematical model of the infection is presented. The basic 3–dimensional model (or 3D model) is described as well as its use in the rest of this work.
- In Section 3, some theoretical backgrounds on identifiability of non-linear systems are presented. We show that the 3D model, under the assumption that the measured CD4 count is the total (infected + uninfected) number of CD4, is identifiable.
- It is presented, in Section 4, the manner how the mathematical analysis of infection should be used for an effective aid for the clinical monitoring of HIV patients.
- Therapeutical failures are defined in Section 5, both from a clinical and a mathematical point of view. Some mathematical characterizations of immunological an virological failures are given.
- One major achievement here is the practical application on real clinical data. This was possible thanks to the recruitment of 6 patients according to a suitable protocol as presented in Section 6. The so-called EDV05 clinical trial has been set up in cooperation with Nantes University Hospital (France).
- The technical aspects of parameter computation are presented in Section 7. A Monte-Carlo method for estimating the parameters of the infection is introduced. This method yields a robust and stable estimation of the parameters from standard clinical data. The implementation of this estimation method in software is presented in Section 8.
- Section 9 is devoted to results obtained from the EDV05 trial. Some parameters of the patients of the trial are presented. Our results show that immunological failures can be early predicted through the mathematical analysis.
- Finally, the last Section will conclude this study.

2   Mathematical Modeling of the Infection

Review the basic modeling of the HIV/AIDS dynamics as described by a 3-dimensional (3D) continuous-time model [9, 3]. This model includes the dynamics of infected CD4+ T-cells, uninfected CD4+ T-cells and virions.

$$\begin{cases} \frac{\mathrm{d}T}{\mathrm{d}t} = s - \delta T - \beta TV + rT\frac{V}{K+V}, \\[2mm] \frac{\mathrm{d}T^*}{\mathrm{d}t} = \beta TV - \mu T^*, \\[2mm] \frac{\mathrm{d}V}{\mathrm{d}t} = kT^* - cV. \end{cases} \tag{1}$$

Model 1: The continuous time 3D model

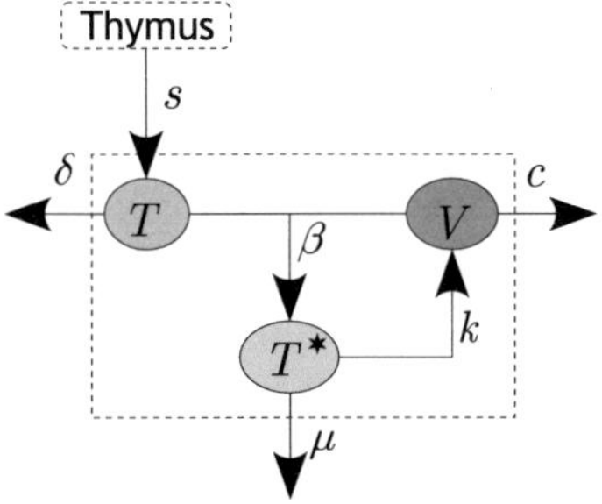

Fig. 1. The 3D model of the HIV/AIDS infection without proliferation of CD4+ T cells

Figure 1 is a schematic representation of the 3D model (without the proliferation term). In the 3D model, T (CD4/mm^3) represents the amount of uninfected CD4+ T-cells, T^* (CD4/mm^3) represents the amount of the infected CD4+ T-cells and V (RNA copies/ml), the free virions. Free virus particles infect uninfected cells at a rate proportional to both T and V (βTV). They are removed from the system at the rate cV. $1/c$ is the natural life span of virions. In model (1) it is assumed that healthy CD4+ T-cells are produced at a constant rate s. This is the simplest way to model the production of CD4+ T-cells. μT^* represents the rate at which infected cells are removed from the system. The term $rT\frac{V}{K+V}$ models the proliferation of CD4 T-cells due to the infection. r is the maximal proliferation rate of the process. K is the half-saturation[1] which is constant in the proliferation process [10, 11].

The 3D model is the simplest way to model the infection process. More sophisticated models exist in the literature. For instance, these models take into account the dynamic of the latently CD4 T-cells, the CD8 T-cells, or viral resistance, etc. The reader can refer to [9, 7, 2, 12] for more details on these models. Only the 3D model (1) will be studied in this paper. Table 1 presents a description of the 8 parameters involved in the 3D model.

---

[1] When $V = K$, then $r\frac{V}{K+V} = \frac{r}{2}$.

Table 1. Description and units of the 3D model parameters. To be conform to the clinical standard we suppose that the CD4 count is in CD4/mm^3 and the viral load is in copies/ml.

Parameters	Definition	Units
s	source of T-cells from the thymus	mm^{-3}/day
δ	death rate of heathy cells	day^{-1}
β	rate of T-cells infection	ml/day
μ	death rate of infected cells	day^{-1}
k	rate of virions production	mm^3/ml/day
c	death rate of virions	day^{-1}
r	maximal proliferation rate	mm^{-3}/day
K	half-saturation constant	ml^{-1}

3 Identifiability Check of the 3D Model

We will consider the 3D model presented by equation (1) with the measured outputs $y_1 = T + T^*$ and $y_2 = V$.

Identifiability theories of non-linear dynamical systems are well presented in [13]. It is also shown that the 3D model (without the proliferation term $rT\frac{V}{K+V}$), is identifiable under the assumption $y_1 = T$ and $y_2 = V$. However, since biological tests of the CD4 T-cells count do not distinguish the infected CD4 T-cells from the healthy CD4 T-cell, this assumption is not realistic. So, the estimation of the parameters of the infection are done here under the assumption that the measured CD4 T-cells count is the total number of CD4 T-cells (infected + non-infected). Therefore, we present here the identifiability of the 3D model (1) under this assumption.

To begin, let us introduce briefly theoretical concepts of the characterization of non-linear systems identifiability, according to [13].

3.1 Characterization of Non-linear Systems Identifiability

Consider the continuous-time and non-linear system

$$\Sigma_\theta \triangleq \begin{cases} \dot{x} = f(x, \theta, u), \\ y = h(x, \theta, u), \end{cases} \tag{2}$$

where $x \in \mathbb{R}^n$, $y \in \mathbb{R}^p$, $u \in \mathbb{R}^m$, $\theta \in \mathbb{R}^q$. Functions f and g are supposed to be meromorphic functions. We will denote by $\mathcal{P}$ an open subset of $\mathbb{R}^q$, by $\mathcal{M}$, an open subset of $\mathbb{R}^n$, and by $\mathcal{C}_\mathcal{U}^k[0, T]$ the set of inputs u that admit k continuous derivatives for $t \in [0, T]$. Similarly, $\mathcal{C}_y^k[0, T]$ will be the set of outputs y that admit k continuous derivatives on $[0, T]$. For simplicity, $\mathcal{C}_\mathcal{U}^1[0, T]$ is denoted $\mathcal{C}_\mathcal{U}[0, T]$.

Identifiability of dynamical systems is about knowing under which conditions the outputs $y(x, \theta_1, u)$ and $y(x, \theta_2, u)$ of the system Σ_θ are distinct for two parameters θ_1 and θ_2 such that $\theta_1 \neq \theta_2$ ([14, 15, 13]).

Definition 1 (distinguishability [15]). *Let $T > 0$, $x_0 \in \mathcal{M} \subset \mathbb{R}^n$. Parameters $\theta_1, \theta_2 \in \mathcal{P} \subset \mathbb{R}^q$ are x_0-indistinguishable on $[0,T]$, if for all admissible input u, all $t \in [0,T]$,*

$$y(t, \theta_1, x_0, u) = y(t, \theta_2, x_0, u).$$

Otherwise, θ_1 and θ_2 are said x_0-distinguishable.

This concept of distinguishability of the parameters of system Σ_θ through the output y implies a local injective map between the parameters and the outputs of Σ_θ. So a generic definition of identifiability can be formulated as follows:

Definition 2 (algebraic identifiability [13]). *System Σ_θ is said algebraically identifiable if there exist $T \geq 0$, an integer $k \geq 0$ and a meromorphic function $\Phi : \mathbb{R}^q \times \mathbb{R}^{(k+1)m} \times \mathbb{R}^{(k+1)p} \to \mathbb{R}^q$ such that*

$$\Phi(\theta, u, \dot{u}, \cdots, u^{(k)}, y, \dot{y}, \cdots, y^{(k)}) = 0 \text{ and } \det\left(\tfrac{\partial \Phi}{\partial \theta}\right) \neq 0, \tag{3}$$

on $[0,T]$ for $(\theta, u, \dot{u}, \cdots, u^{(k)}, y, \dot{y}, \cdots, y^{(k)})$ in an open and dense subset of $\mathcal{P} \times (\mathcal{C}_{\mathcal{U}}[0,T])^{k+1} \times (\mathcal{C}_{\mathcal{Y}}[0,T])^{k+1}$, with (θ, x_0, u) in an open and dense subset of $\mathcal{P} \times \mathcal{M} \times \mathcal{C}_{\mathcal{U}}^k[0,T]$.

If the condition given by equation (3) is true, then there exist a function Φ of y, u, their higher derivatives and θ (and that does not depend on x) such that Φ is locally inversible with respect to θ. So, according to definition (2), system Σ_θ is algebraically identifiable of there exist integers $k_i, s_i \geq 0$ $(i = 1, \cdots, p)$ such that

$$\text{rank} \frac{\partial(y_1^{(k_1)}, \cdots, y_1^{(s_1)}, y_2^{(k_1)}, \cdots, y_2^{(s_2)}, \cdots, y_p^{(k_p)}, \cdots, y_p^{(s_p)})}{\partial \theta} = \dim(\theta), \tag{4}$$

where k_i are the observability indices [16] of the outputs, *i.e.* each output $y_i^{(k_i)}$ can only be written as a function of u, y, their derivatives and θ.

3.2   Application to the Identifiability Check of the 3D Model

It is trivial to show that the 3D model (with $y_1 = T + T^*$ and $y_2 = V$) is observable with observability indices $k_1 = 1$, $k_2 = 2$.

1) Using $T^* = \frac{\dot{y}_2 + c y_2}{k}$, and $T = \frac{1}{\beta y_2}\left(\dot{T}^* + \mu T^*\right) = \frac{1}{\beta y_2}\left(\frac{1}{k}\ddot{y}_2 + \frac{c+\mu}{k}\dot{y}_2 + \frac{\mu c}{k}y_2\right)$,

$$y_1 = \frac{1}{k\beta y_2}\left[\ddot{y}_2 + (c+\mu)\dot{y}_2 + \mu c y_2 + \beta \dot{y}_2 y_2 + \beta c y_2^2\right].$$

So

$$\ddot{y}_2 = k\beta y_1 y_2 - (c+\mu)\dot{y}_2 - \mu c y_2 - \beta \dot{y}_2 y_2 - \beta c y_2^2. \tag{5}$$

$$= \theta_1 \theta_4 y_1 y_2 - (\theta_2 + \theta_3)\dot{y}_2 - \theta_2 \theta_3 y_2 - \theta_4(\dot{y}_2 y_2 - \theta_2 y_2^2), \tag{6}$$

with $\theta_1 = k$, $\theta_2 = c$, $\theta_3 = \mu$, $\theta_4 = \beta$. Parameters $\theta_1, \theta_2, \theta_3, \theta_4$ are algebraically identifiable if

$$\text{rank}\frac{\partial(\ddot{y}_2, y_2^{(3)}, y_2^{(4)}, y_2^{(5)})}{\partial(\theta_1, \theta_2, \theta_3, \theta_4)} = 4, \tag{7}$$

i.e., if

$$\text{rank}\begin{bmatrix} \theta_4 y_1 y_2 & -\dot{y}_2 - \theta_3 y_2 & -\dot{y}_2 - \theta_2 y_2 & M \\ \theta_4(y_1 y_2)^{(1)} & -\ddot{y}_2 - \theta_3 \dot{y}_2 & -\ddot{y}_2 - \theta_2 \dot{y}_2 & \dot{M} \\ \theta_4(y_1 y_2)^{(2)} & -y_2^{(3)} - \theta_3 \ddot{y}_2 & -y_2^{(3)} - \theta_2 \ddot{y}_2 & \ddot{M} \\ \theta_4(y_1 y_2)^{(3)} & -y_2^{(4)} - \theta_3 y_2^{(3)} & -y_2^{(4)} - \theta_2 y_2^{(3)} & M^{(3)} \end{bmatrix} = 4, \tag{8}$$

where

$$M = \theta_1 y_1 y_2 - \dot{y}_2 y_2 - \theta_2 y_2^2. \tag{9}$$

2) Compute

$$\dot{y}_1 = s - \delta T - \mu T^* + rT\frac{y_2}{K + y_2} \tag{10}$$

$$= s - \frac{\delta}{k\beta}\frac{\ddot{y}_2}{y_2} - \frac{\delta(c + \mu)}{k\beta}\frac{\dot{y}_2}{y_2} - \frac{\delta\mu c}{k\beta}\frac{y_2}{y_1} - \frac{\mu}{k}\dot{y}_2$$

$$- \frac{\mu c}{k}y_2 + rT\frac{y_2}{K + y_2}, \tag{11}$$

thus, by substituting $\ddot{y}_2$ in the above equation by its expression (5),

$$\dot{y}_1 = s - \delta y_1 + \frac{\delta - \mu}{k}\dot{y}_2 + \frac{c(\delta - \mu)}{k}y_2 + \frac{r y_2}{k(K + y_2)}(k y_1 - \dot{y}_2 - c y_2). \tag{12}$$

Finally

$$\dot{y}_1 = s - \delta y_1 + \frac{\delta - \mu}{k}\dot{y}_2 + \frac{c(\delta - \mu)}{k}y_2 + \frac{r y_2}{(K + y_2)}A,$$

$$= s - \delta A + \frac{r y_2}{(K + y_2)}A - \frac{\mu}{k}\dot{y}_2 - \frac{c\mu}{k}y_2,$$

$$= \theta_5 - \theta_6 A + \theta_7\frac{y_2}{(\theta_8 + y_2)}A - \frac{\theta_3}{\theta_1}\dot{y}_2 - \frac{\theta_2\theta_3}{\theta_1}y_2, \tag{13}$$

with $\theta_5 = s$, $\theta_6 = \delta$, $\theta_7 = r$, $\theta_8 = K$, $A = y_1 - \frac{1}{\theta_1}\dot{y}_2 - \frac{\theta_2}{\theta_1}y_2$. So, parameters θ_5, θ_6, θ_7, θ_8 are algebraically identifiable if

$$\text{rank}\frac{\partial(\dot{y}_1, y_1^{(2)}, y_1^{(3)}, y_1^{(4)})}{\partial(\theta_5, \theta_6, \theta_7, \theta_8)} = 4, \tag{14}$$

i.e. if,

$$\text{rank}\begin{bmatrix} 1 & -A & B & \theta_7 B_K \\ 0 & -\dot{A} & \dot{B} & \theta_7 \dot{B}_K \\ 0 & -\ddot{A} & \ddot{B} & \theta_7 \ddot{B}_K \\ 0 & -A^{(3)} & B^{(3)} & \theta_7 B_K^{(3)} \end{bmatrix} = 4, \tag{15}$$

where

$$B = \frac{y_2}{K + y_2} A, \tag{16}$$

$$\dot{B} = \frac{K\dot{y}_2}{(K + y_2)^2} A + \frac{y_2}{K + y_2} \dot{A}, \tag{17}$$

$$\ddot{B} = \frac{K\ddot{y}_2(K + y_2) - 2K\dot{y}_2^2}{(K + y_2)^3} A + \frac{2K\dot{y}_2}{(K + y_2)^2} \dot{A} + \frac{y_2}{K + y_2} \ddot{A}, \tag{18}$$

$$B^{(3)} = \left[\frac{K\ddot{y}_2(K + y_2) - 2K\dot{y}_2^2}{(K + y_2)^3} \right]^{(1)} A + 3 \frac{K\ddot{y}_2(K + y_2) - 2K\dot{y}_2^2}{(K + y_2)^3} \dot{A}$$
$$+ \frac{3K\dot{y}_2}{(K + y_2)^2} \ddot{A} + \frac{y_2}{K + y_2} A^{(3)}. \tag{19}$$

$$B_K = \frac{\partial B}{\partial K} = -\frac{y_2}{(K + y_2)^2} A, \tag{20}$$

$$\dot{B}_K = \frac{\partial \dot{B}}{\partial K} = \frac{y_2\dot{y}_2 - K\dot{y}_2}{(K + y_2)^3} A - \frac{y_2}{(K + y_2)^2} \dot{A}, \tag{21}$$

$$\ddot{B}_K = \frac{\partial \ddot{B}}{\partial K} = \frac{(y_2\ddot{y}_2 - 2\dot{y}_2^2 - K\ddot{y}_2)(K + y_2) + 6K\dot{y}_2^2}{(K + y_2)^4} A$$
$$+ 2\frac{y_2\dot{y}_2 - K\dot{y}_2}{(K + y_2)^3} \dot{A} - \frac{y_2}{(K + y_2)^2} \ddot{A}, \tag{22}$$

$$B_K^{(3)} = \frac{\partial B^{(3)}}{\partial K}. \tag{23}$$

To summarize, the 3D model is algebraically identifiable if

$$\mathrm{rank} \frac{\partial(\dot{y}_1, \cdots, y_1^{(4)}, \ddot{y}_2, \cdots, y_2^{(5)})}{\partial(\theta_1, \cdots, \theta_8)} = 8. \tag{24}$$

The total expression of the Jacobian $\Gamma = \frac{\partial(\dot{y}_1, \cdots, y_1^{(4)}, \ddot{y}_2, \cdots, y_2^{(5)})}{\partial(\theta_1, \cdots, \theta_8)}$ is

$$\begin{bmatrix}
P & Q & R & 0 & 1 & -A & B & \theta_7 B_K \\
\dot{P} & \dot{Q} & \dot{R} & 0 & 0 & -\dot{A} & \dot{B} & \theta_7 \dot{B}_K \\
\ddot{P} & \ddot{Q} & \ddot{R} & 0 & 0 & -\ddot{A} & \ddot{B} & \theta_7 \ddot{B}_K \\
P^{(3)} & Q^{(3)} & R^{(3)} & 0 & 0 & -A^{(3)} & B^{(3)} & \theta_7 B_K^{(3)} \\
\theta_4 y_1 y_2 & O & N & M & 0 & 0 & 0 & 0 \\
\theta_4(y_1 y_2)^{(1)} & \dot{O} & \dot{N} & \dot{M} & 0 & 0 & 0 & 0 \\
\theta_4(y_1 y_2)^{(2)} & \ddot{O} & \ddot{N} & \ddot{M} & 0 & 0 & 0 & 0 \\
\theta_4(y_1 y_2)^{(3)} & O^{(3)} & N^{(3)} & M^{(3)} & 0 & 0 & 0 & 0
\end{bmatrix}, \tag{25}$$

with

$$N = -\dot{y}_2 - \theta_2 y_2, \tag{26}$$

$$O = -\dot{y}_2 - \theta_3 y_2, \tag{27}$$

$$P = \left[-\theta_6 + \theta_7 \frac{y_2}{(\theta_8 + y_2)} \right] \left[\frac{1}{\theta_1^2} \dot{y}_2 + \frac{\theta_2}{\theta_1^2} y_2 \right] + \frac{\theta_3}{\theta_1^2} \dot{y}_2 + \frac{\theta_2 \theta_3}{\theta_1^2} y_2, \tag{28}$$

$$Q = \frac{\theta_6}{\theta_1} y_2 - \frac{\theta_7}{\theta_1} \frac{y_2^2}{(\theta_8 + y_2)} - \frac{\theta_3}{\theta_1} y_2, \tag{29}$$

$$R = -\frac{1}{\theta_1} \dot{y}_2 - \frac{\theta_2}{\theta_1} y_2 = \frac{1}{\theta_1} N. \tag{30}$$

The rank condition (24) is satisfied if $rK\dot{y}_2 \neq 0$, *i.e.*

1. the viral load is non-constant.
2. and $r \neq 0$, $K \neq 0$.

If $r = 0$, then parameter K is not identifiable. In this case, the proliferation process does not exists and can be ignored in the model. Then, we can consider the basic 3D model (without proliferation), *i.e.* with only the 6 parameters $s, \delta, \beta, \mu, k, c$. For $K = 0$, remark that δ and r are not distinguishable from the outputs y_1 and y_2. Conditions 1 insures that the condition of *persistent excitation* of the system is satisfied in order to be in the best conditions for the identification of the parameters.

If the rank condition is satisfied, we conclude that 11 measurements (5 of y_1 and 6 of y_2) are necessary to compute the 8 parameters of the 3D model. In fact, the computation of $y_1^{(4)}$ and $y_2^{(5)}$ is possible if we have, at least, 5 samples of y_1 and 6 samples of y_2.

According to the identifiability check of the 3D model, an adapted data sampling can be set up for the mathematical study of the infection. In section 6, a data sampling scheduling design for the estimation of the parameters of the 3D model, and according to this identifiability check, is presented. This trial aimed to study how the mathematical analysis of the infection can improve the clinical monitoring of patients.

4   An Aid for the Clinical Monitoring of HIV Patients

As for the 3D model (1), the more sophisticated mathematical models in [9, 17, 18] of the HIV/AIDS infection involve several parameters that are related to the immunological or to the virological status of the patient. For instance, s is the natural production of CD4 T-cells by the thymus. So, a poor value of s is representative of a badly damaged immune system.

By estimating the parameters of these models, the numerical evaluation of biological phenomena becomes possible: amongst, evaluation of the immune system, of the efficiency of the therapy. Instead of grounding the clinical diagnosis of the patient only on the measurement of CD4+ T-cells count and of the viral load, new additional information can be obtained through a mathematical analysis (see Figure 2). These information are based on specific mathematical *criteria* that give new insights on the infection process. These mathematical criteria should be considered as a complementary tool to help the clinician in the daily monitoring of HIV patients and in making decisions. They should be

1. easily computable,
2. predictive enough of the biological phenomena they attend to describe, and

3. if possible, independent or weakly dependent on the other biological phenomena.

In the following section, it is shown how the immunological failure can be predicted and evaluated. This prediction is based on simple mathematical criteria – derived from the parameters of the models. The case of the virological failure will be presented later.

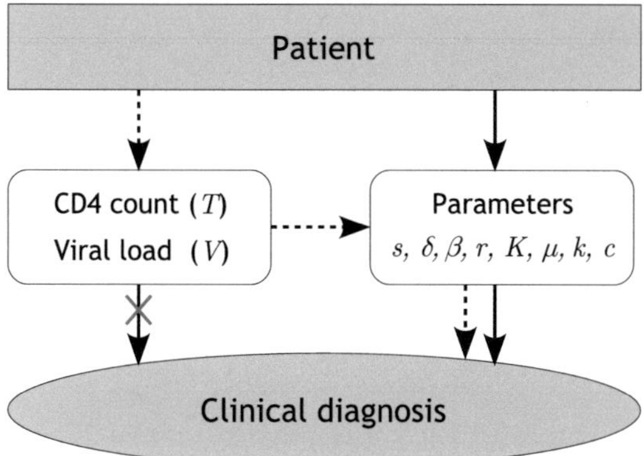

Fig. 2. From the patient to the clinical diagnosis. A mathematical based approach. Instead of deriving the diagnosis of the patient on T and V only, more rich information is used thanks to parameter estimation.

5 Analysis and Prediction of Therapeutical Failures

From a clinical point of view, 4 types of therapeutical failures exist ([19, 20, 11]):

1. **immunological failure:** clinically, the immunological failure is defined when the amount of CD4 T-cells remains below the level of 200 CD4/mm^3 during 6 months of efficient treatment. An efficient treatment is defined as a treatment which is able to decrease the viral load and keep it under the undetectability threshold of 50 copies/ml.

2. **virological failure:** is due to a persistent replication of the viral load under treatment. It usually results from a bad inhibition of HIV replication or viral resistance. So, it can be considered mainly as the consequence of the lack of efficacy of PI (Protease Inhibitors) and/or the lack of efficacy of RTI (Reverse Transcriptase Inhibitors). In real life, virological failures are commonly classified into 3 groups:
   - the *weak virological failure*: $V < 5\,000$ copies/ml. The viral load becomes positive after attaining the undetectability level;
   - the *medium virological failure*: $5\,000$ copies/ml $< V < 30\,000$ copies/ml. Virus replication is more important;
   - the *strong virological failure*: $V > 30\,000$ copies/ml. Virus replication is very strong and the viral load increases dramatically.

3. **biological failure:** is defined when the viral load is higher than 30 000 copies/ml and the CD4 T-cells account is lower than $200/\text{mm}^3$. It can be considered as the superposition of both virological and immunological failures.

4. **clinical failure:** is characterized by the clinical manifestation of opportunistic diseases. This failure is generally related to a biological failure with a low amount of CD4 T-cell and a high level for the viral load. Clinical failure will not be considered here. The effects of opportunistic diseases are not modeled in the 3D model.

5.1   Mathematical Characterization of Immunological Failures

The main parameters that model the immune system status are s (the production of the healthy CD4+ T-cells by the thymus) and δ (the natural death rate of the healthy CD4+ T-cells). With these two parameters, two mathematical criteria can be derived to evaluate the status of the immune system and to make a diagnosis about immunological failures for HIV/AIDS patients. These criteria are presented below.

$\star$ **Criterion** 1: The first criterion is given by the maximal level of CD4+ T-cells that can be reached if a 100% efficient treatment is administrated. This maximal level is obtained when setting $V = 0$ in the first equation of the model (1) and computing its equilibrium point:

$$\Lambda_{\max} = \frac{s}{\delta} \tag{31}$$

Solving $\frac{dT}{dt} = s - \delta T$, one gets $T(t) = [T(0) - \frac{s}{\delta}]e^{-\delta t} + \frac{s}{\delta}$. Then, the CD4 T-cells count will increase exponentially towards $\Lambda_{\max} = \frac{s}{\delta}$.

$\star$ **Criterion** 2: In practise, it is desirable to reach the equilibrium $\Lambda_{\max}$ within a reasonable time. Thus, a more restrictive evaluation of the status of the immune system is provided by the time t_{200} required to reach the critical threshold of 200 $\text{CD4}/\text{mm}^3$.

More generally, denote the time t_Λ as the time required to reach the level Λ of CD4 T-cells count in the case of a 100% efficacy of the therapy. t_Λ is computed as

$$t_\Lambda = \frac{1}{\delta}\left[-\log_e\left(1 - \frac{\delta}{s}\Lambda\right) + \log_e\left(1 - \frac{\delta}{s}T(0)\right)\right] \text{ with } \Lambda < \frac{s}{\delta}. \tag{32}$$

If $\Lambda_{\max} = \frac{s}{\delta} < \Lambda$, then set $t_\Lambda = \infty$.

To compute $\Lambda_{\max}$ and t_{200}, a good estimation of the parameters of the mathematical model is necessary.

5.2   Mathematical Characterization of Virological Failures

Mathematically, the parameters related to the virological failures are k and β. They represent respectively the production rate of new virions and the infectivity of the virus. Thus, they are predictive of the *in vivo* virus replication. In [11], these two parameters were used to study separately the effect of RTI and PI on the infection. To date, currents results on the mathematical characterization of the virological failure did not enable to confirm results obtained in [11]. In fact, as we will see later (in next section), patients of the EDV05 trial do not enable an effective study of this failure.

6   Design of the EDV05 Clinical Trial

The EDV05 trial was initiated in the CHU (University Hospital Centre) of Nantes in February 2005, and 6 patients were included in the first part of the trial. To be in the best condition for the study on the infection, these patients did respect some rigorous inclusion criteria.

6.1   Inclusion Criteria of Patients

The main conditions for each patient (Female or Male) to be included in the trial are as follows:

1. to be infected by the HIV-1 or HIV-2 type virus,
2. to be naive of any treatment at the beginning (at day d_0) of the trial,
3. no HBV or HCV co-infection during the 6 months before the inclusion into the trial,
4. to need starting an antiretroviral treatment at the beginning of the trial.

All the patients start a treatment (not necessarily the same) at the beginning of the trial. Day d_0 is not the same for all the patients. Condition 2 enables avoiding any viral resistance (due to treatments) that are not taken into account in our model. Since, 25% of HIV patients, in France, are co-infected by HCV – 5% for HBV – [19], condition 3 enables avoiding these special cases that are also non-modeled by the 3D model. Table 2 presents the 6 first patients enroled in the trial.

Table 2.   The 6 patients enroled in the EDV05 trial. Day d_0 is the first day of the trial for the patient (the enrollment day). The average age was 39.8 years.

Patient ID	Day d_0	Year of birth (age)
01	14 march 2005	1956 (49 years)
02	14 march 2005	1967 (38 years)
03	22 march 2005	1962 (43 years)
04	22 march 2005	1964 (41 years)
05	04 april 2005	1967 (38 years)
06	02 may 2005	1970 (35 years)

6.2   The Data Sampling Scheduling

Remind that the measured data are the viral load and the total CD4 T-cells count. 11 blood samples are taken during 3 months according to the following

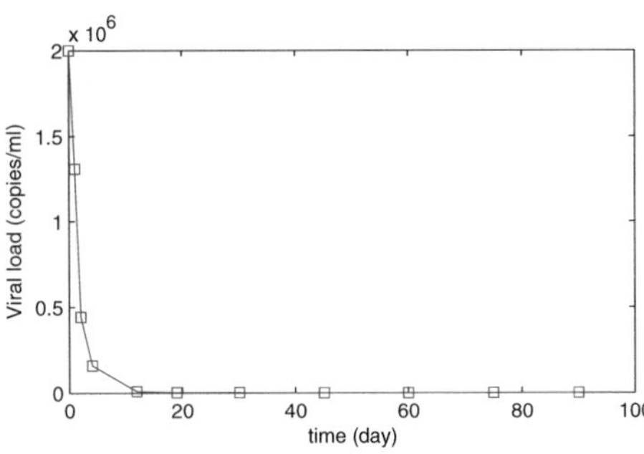

Data	Day of measurement
1	d_0
2	d_1
3	d_2
4	d_4 (or d_6)
5	$d_{12}(\pm 2\,\text{days})$
6	$d_{19}(\pm 2\,\text{days})$
7	$d_{30}(\pm 2\,\text{days})$
8	$d_{45}(\pm 2\,\text{days})$
9	$d_{60}(\pm 2\,\text{days})$
10	$d_{75}(\pm 2\,\text{days})$
11	$d_{90}(\pm 2\,\text{days})$

Fig. 3. Planning of data measurements during the clinical trial. d_0 is the first day of the trial for the patient. d_i is equal to the day $d_0 + i$.

planning described in Figure 3. The day d_i is equal to the day $d_0 + i$. After the initiation of the treatment, at time d_0, the viral load drops exponentially (in the first 3 weeks) before stabilizing below the undetectability level of 50 copies/ml. Remember that all the patients are naive of any treatment *i.e.* without drug resistances. So, the dynamics of the infection are strongly disturbed by the treatment at the beginning of the trial. Thus, data samples are scheduled to be numerous in the first days of the trial, *i.e.* in the transient stage of the infection dynamics, when data contain more information (to satisfy the *persistent excitation* condition). We get 6 samples in this stage (enough to have a first estimation of all the parameters of the 3D model) and the 5 others samples are scheduled on the last days of the trial.

6.3   Viral Load Measurement Method

Since, the measured viral load values depend on the laboratory protocol used to quantify the RNA copies ([21, 22, 23, 24, 25]), it has been decided for each patient to perform all the measurements of the viral loads at the same time with the same protocol. Thus, the 11 blood samples of each patient have been kept frozen (at $-80°$C) until the end of the trial. The measurements were done at the same time using the Roche Taqman 48^{TM} essay, with an undetectability threshold < 50 copies/ml.

7   Identification of the Parameters

Estimation, or identification, of the parameters of the model (1) consists in the computation of the 8 parameters of the 3D model using clinical data.

7.1   The Estimation Procedure: A Monte-Carlo Approach

The identification of the parameters of the infection was studied in [26]. It is presented a simplex based approach to estimate all the parameters of the 3D model. However, because of the lack of data, and the complexity of the problem, the presented algorithm was very sensitive to its initialization. So, manual and progressive calibrations of the algorithm were necessary to have good estimations of the parameters and avoid any mistake. To overcome this problem, we introduce in [27] a new estimation method, based on a Monte-Carlo approach. This method is based on the simplex optimization algorithm of [26] and consists in exploring the parameter space, by random tests, in order to deduce stable and robust (with respect to the initialization of the algorithm) estimations of the 3D model parameters.

Notation:

$\theta \in \mathbb{R}^q$: the parameter vector to estimate.

$\mathcal{P}$: a connected, non-empty compact of $\mathbb{R}^q$.

$\mathbf{J}(\theta)$: an objective (cost) function depending on θ.

$\theta^* \in \mathbb{R}^q$: the real (exact) parameter value of the process we want to identify.

$\hat{\theta} \in \mathbb{R}^q$: the estimated value of θ^*.

Let consider the optimization problem consisting computing

$$\hat{\theta} = \text{Argmin}_\theta \{\mathbf{J}(\theta)\} \tag{33}$$

where

$$\mathbf{J}^2(\theta) = \frac{1}{w_1} \sum_{t=t_0}^{t_N} [y_1(\theta, t) - \hat{y}_1(t)]^2 + \frac{1}{w_2} \sum_{t=t_0}^{t_N} [y_2(\theta, t) - \hat{y}_2(t)]^2 + \mathcal{C}. \tag{34}$$

- $\theta = [s, \delta, \beta, r, K, \mu, k, c]$ for the 3D model (1).
- Time t_0 is the first day of the trial and time t_N is the last.
- $\hat{y}_1(t)$ is the clinical measured CD4 counts at time t and $\hat{y}_2(t)$ is the clinical measured viral load at time t.
- $y_1(\theta, t) = T(\theta, t) + T^*(\theta, t)$ and $y_2(\theta, t) = V(\theta, t)$ are respectively the simulated total CD4 counts and the simulated viral load.
- $w_1 = \dfrac{1}{N+1} \sum_{t_0}^{t_N} [\hat{y}_1(t)]^2$ and $w_2 = \dfrac{1}{N+1} \sum_{t_0}^{t_N} [\hat{y}_2(t)]^2$ are the weights that are used. Note that these weights render $J(\theta)$ independent from the units of $V(t)$ and $T(t) + T^*(t)$.
- $\mathcal{C} = \displaystyle\sum_{i=1}^{p} [M \cdot \max((\theta_i - \text{ub}_i), 0) + M \cdot \max((\text{lb}_i - \theta_i), 0)]$ is a penalty term used to validate constrains on parameters. Each parameter is constrained to be positive and finite. ub_i and lb_i are respectively the upper and the lower bounds of the ith parameter θ_i. M is chosen equal to $1e10$. Note that more sophisticated constraints (for instance: $c > \mu$ or $T(t) > 0 \ \forall t \geq 0$, etc.) can be included in $\mathcal{C}$.

This function is the weighted euclidian distance between the clinical data and the model. Under the condition that $\mathbf{J}$ is a quadratic function, if we choose

$N > 0$ initial conditions (*i.c.*) – according to a uniform law in a domain $\mathcal{P}$ of the parameter space – for the optimization algorithm and compute the N solutions associated to these *i.c.*, therefore, for N large enough, we can prove that the N solutions of the problem are closely distributed in the neighborhood of the search optimum. We can also prove that a robust and stable solution for the optimization problem is the median of the distribution.

Let illustrate this principle by the following example.

Example of a Quadratic Function

Consider $\mathbf{J}(\theta) = 0.01\theta^2 + 0.05\theta + 0.075$, and compute

$$\hat{\theta} = \operatorname{Arg} \min_{\theta}\{\mathbf{J}(\theta)\},$$

with $\theta \in \mathcal{P} = [-30; 30]$. By initializing the simplex algorithm with 5000 initial conditions (*i.c.*) chosen uniformly in the interval $[-30; 30]$, we get the distribution of the computed solutions presented by Figure 4. Denote by $\Theta = [\theta_1, \theta_2, \cdots, \theta_{5000}]$ the vector of all the computed solutions of the 5000 *i.c.* The probability density function of the variable θ is then given by $\mathcal{D}(\theta) = P(\theta - \mathrm{d}\theta \le \theta \le \theta + \mathrm{d}\theta)$. $P(\theta)$ is the probability to have a given solution θ. Figure 4 gives the density function of the solutions θ. The IQR (interquartile range) measures the

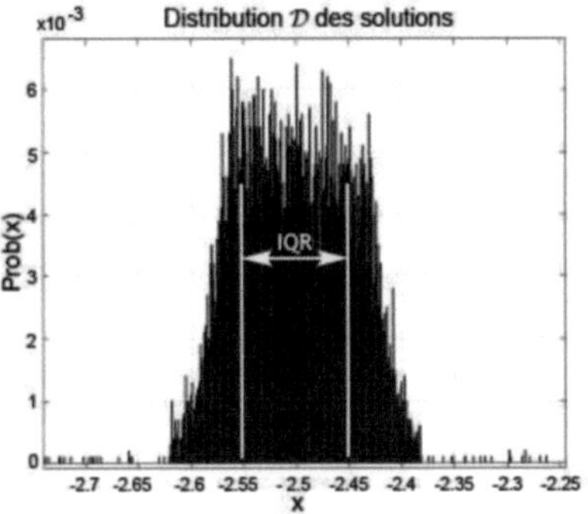

Fig. 4. Density function $\mathcal{D}(\theta)$ of the estimations results with a weak tolerance on x and $\mathbf{J}(\theta)$ equal to 0.25. $\hat{\theta} = \operatorname{median}(\mathcal{D}) = -2.4908$ with IQR = 0.1. IQR is the interquartile range. The real optimum is $\theta^* = -2.5$.

dispersion of the results around the real optimum. It will depend on the chosen precision and the convexity of $\mathbf{J}$. The IQR gives an important information on the confidence on the results. IQR = $Q_3 - Q_1$, where Q_1 (resp. Q_3) is the 1st (resp. the 3rd) quartile of the distribution. Q_2 (the median) is the 2nd quartile.

General Case of Non-quadratic Functions

For the general case which consists in minimizing non-quadratic functions $\mathbf{J}$, if $\mathbf{J}$ is locally smooth in the neighborhood of its minima, then $\mathbf{J}$ can be approximated locally (in the neighborhood of these optima) by quadratic functions. Therefore, this problem can be approximated by the special case of quadratic functions in the neighborhood of the optima. Therefore, this Monte-Carlo approach can be applied, as presented after, to biological systems that are smooth dynamical systems.

Remark 1. If **J** has several local minima in the interval $\mathcal{P}$, then the density function will be characterized by several peaks as in the distribution shown in Figure 4. One peak exists for each local minimum. Then, this approach can be used to identify all the local minima of a function in a given interval. Moreover, if $\theta^* \in \mathcal{P}$ then $\theta^* \in \mathcal{D}(\theta)$. In any cases, $\hat{\theta} = \text{median}(\mathcal{D})$ is a good and stable estimator for θ^*.

Application to the 3D Model

To identify the parameters of the infection, a discrete-time 3D model (35) is used instead of the continuous-time one in the estimation algorithm. This approach enables simplifying the implementation of the process to improve the computation time which is very long with the continuous time model. Note that discrete-time systems, on contrary to continuous-times systems are sequential equations and do require specific solvers (as the Runge-Kutta solver for instance).

The discrete-time 3D model used is the $1^{\text{st}}-$order Taylor approximation (also called Euler approximation) of the 3D continuous-time 3D model and is as follows:

$$\begin{cases} T_{t+1} = s + (1 - \delta)T_t - \beta T_t V_t + rT\frac{V_t}{K+V_t}, \\ \\ T_{t+1}^* = \beta T_t V_t + (1 - \mu)T_t^*, \\ \\ V_{t+1} = kT_t^* + (1 - c)V_t. \end{cases} \tag{35}$$

Model 2: The 1st order discrete-time 3D model

where $T_t = T(t)$, $T_t^* = T^*(t)$, $V_t = V(t)$ for each $t = t_0, t_1, \cdots, t_N$.

According to identifiability theory of discrete-time systems ([28], [29]), we can show (in the same manner than the continuous-time case) that system (35) is identifiable. Moreover, studying relationship between identifiability of continuous-time systems and identifiability of their discretized (by Taylor approximation), we can show that *all* the identifiability properties of the continuous-time 3D model remain true for the $1^{\text{st}}-$order discrete-time one ([29]).

We apply the Monte-Carlo algorithm to an academic case in order to validate the method. For simplicity, we will consider, without lost of generality, the continuous-time 3D model (1), but without proliferation. The academic parameters, ($\theta^* = [s^*, \delta^*, \beta^*, \mu^*, k^*, c^*]$) used here are $s^* = 10$, $\delta^* = 0.01$, $\beta^* = 1e - 7$, $\mu^* = 0.09$, $k^* = 1000$, $c^* = 0.3$. The system is initialized with $T_0 = 1000$ CD4/mm^3, $T_0^* = 50$ CD4/mm^3, and $V_0 = 100$ copies/ml. We generate 200 data points with the continuous-time model and try to recover them with the

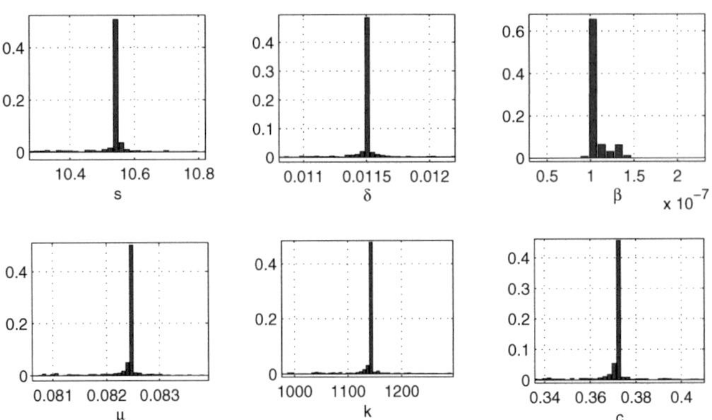

Fig. 5. Distributions of estimated parameters are strongly closed to real parameters $\theta^* = [10, 0.01, 1e-7, 0.09, 1000, 0.3]$ with high confidences (small IQR)

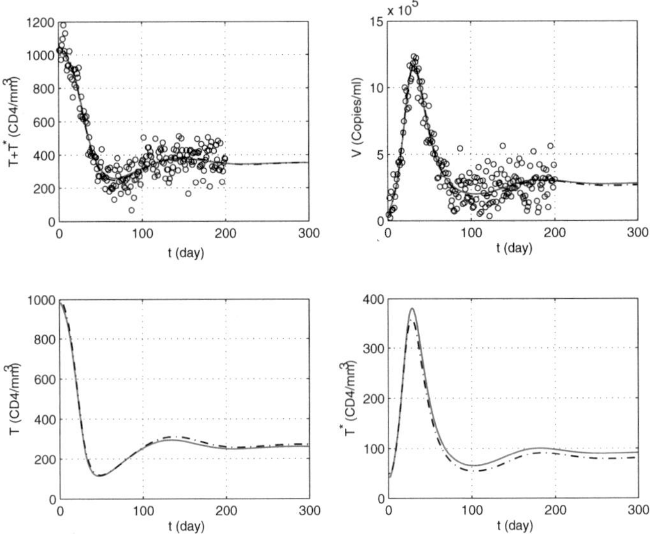

Fig. 6. Simulation of dynamics: Real data without noise are in dark dash-dot lines $(-\cdot)$, data with noise are in blue circles $(\circ)$, and data fittings after estimation of the parameters are in red lines $(-)$

discrete-time one. To test the behavior of the estimation algorithm under data with noise, noises are generated by normally distributed random numbers (with a mean equal to 0) and a standard deviation depending on data amplitudes, and is added to data. Standard deviation used for T is $\sigma_T = 70$. For T^*, this standard deviation is $\sigma_{T^*} = 10$ and $\sigma_V = 1e5$ for V.

The calibration of the algorithm is as follows:

- according to the biological characteristics of the infection, the estimated parameters are constrained to be bounded in the following intervals : $\hat{s} \in [1e-5\,;20]$, $\hat{\delta} \in [2.75e-5\,;0.1]$, $\hat{\beta} \in [1e-20\,;1]$, $\hat{\mu} \in [2.75e-5\,;3]$, $\hat{k} \in [1e-20\,;1e8]$, $\hat{c} \in [2.75e-5\,;10]$. These bounds are use in the penalty term $\mathcal{C}$.
- $N = 1000$ randomizations are performed.
- the domains $\mathcal{P}_\theta$ of each parameters θ, in which the 1000 random initial conditions are generated, are: $\mathcal{P}_s = [-1e-5\,;10]$, $\mathcal{P}_\delta = [1e-3\,;0.03]$, $\mathcal{P}_\beta = [1e-10\,;1e-5]$, $\mathcal{P}_\mu = [0.01\,;0.1]$, $\mathcal{P}_k = [1\,;5000]$, and $\mathcal{P}_c = [0.1\,;3]$.

Figure 5 presents the distribution of the estimated parameters. The fittings of the generated noised data are drawn in Figure 6. Presented results show that the estimation method is able to restitute with a strong accuracy the parameters of 3D model under noised data. The reader can also refer to [26], where the robustness of the simplex optimization algorithm under noised data was studied and shown. The numerical values of the estimated parameters and their IQR are in Table 3.

Table 3. Values of parameters estimates. $\text{CI}_{50\%} = [Q_1\,;Q_2]$ is the 50% confidence interval.

Param.	Estimates $\hat{\theta}$	IQR	$\text{CI}_{50\%}$
$\hat{s}$	10.55	0.93	$[9.62;10.55]$
$\hat{\delta}$	0.011	0.002	$[0.009;0.012]$
$\hat{\beta}$	$1.03e{-}7$	$2.38e{-}8$	$[1.03e{-}7;1.27e{-}7]$
$\hat{\mu}$	0.082	0.005	$[0.077;0.082]$
$\hat{k}$	1140.84	410.68	$[730.30;1141.00]$
$\hat{c}$	0.37	0.06	$[0.31;0.37]$

Remark 2. The method was also tested with additive gaussian white noise on data (with the `awgn()` function of `Matlab`®). It is also able to provide good restitutions of the parameters for signal-to-noise ratios $SNR \geq 10$.

8 A Software for the Parameters Computation

The above methodology is now included in a software prototype (described in Figure 7) available to medical doctors. This mathematical tool should enable them to perform easy computations of the model parameters for each patient, and get advantage of this additional information. It consists in a Web interface [30] using a kernel – implemented directly with `C/C++` – for parameter computations. It allows the management of a large number of patients. Each patient data and the estimation results can be saved on the Web server for further analysis.

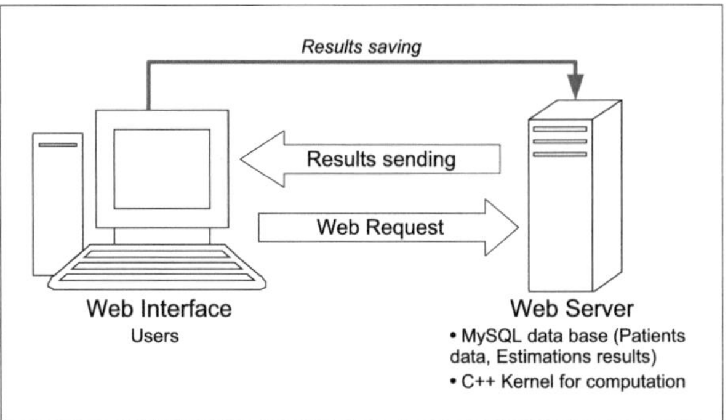

Fig. 7. Schematic description of the software functioning. Available at `http://www.irccyn.ec-nantes.fr/hiv`

9   Results on the EDV05 Clinical Trial

The estimation algorithm has been applied to the 6 first patients included in the trial. According to the estimation results, 1 patient (patient 03) has been diagnosed with an immunological failure. The others patients have good immune systems. Tables 4 and 5 present the estimations results of s and δ for all the patients of the trial. Parameters are progressively estimated with $6, 7, \cdots, 11$ samples to follow the evolution of the parameters and evaluate the accuracy of the prediction of the algorithm.

$\star$ *Case* 1 (*patients with good immune systems*):
Patients 01, 02, 04, 05 and 06 have good immune systems. The estimate of $\Lambda_{\max} = \frac{s}{\delta}$ is above the critical level of 200 $CD4/mm^3$ for the 5 patients. After 6 blood samples, the average value of $\Lambda_{\max}$ [min ; max] is estimated equal to 621.65 [372.73 ; 1084.00]. After the 11 blood samples (90 days), it is estimated equal to 471.14 [301.52 ; 579.09]. Referring to the (temporal) mean-based variability $\rho = \frac{\text{std}}{\text{mean}} \times 100\%$ for parameters s and δ, for each patient, one conclude that the estimation method provides a good prediction of the immune system status after 6 blood samples. The average value of ρ [min ; max] for s is 11.2% [9.2% ; 14.4%]. This value is equal to 10.5% [16.1% ; 5.6%] for δ. No patients, in this group, were clinically diagnosed (by clinicians) with an immunological failure during and at the end of the trial. Note that it is not necessary here to estimate t_{200} since the amount of CD4 is always over the level of 200 $CD4/mm^3$ for these patients.

Figures 8 and 9 present parameters distributions and data fittings for patient 01. Note that the fitting of the CD4 count is less accurate than the fitting of the viral load. In fact, on contrary to the viral load, the dynamic of the CD4 count is very sensitive to physiological states of the patients (stress, tire, etc.) and daily infections by others pathogenic agents. As a result, its dynamic can reveal high

Table 4. $\Lambda_{\max}$ shows a good immune system status for these patients. `mean` is the average value of all parameters estimates (computed from 6, 7, ..., and 11 samples. `std` is the standard deviation. ρ is the mean-based variability of the parameters estimates.

▷ Patient: 01

Samples	6	7	8	9	10	11	mean±std (ρ)
s	6.46	4.61	4.33	5.29	5.40	5.62	5.28±0.76 (14.4%)
δ	1.07e-2	1.24e-2	1.39e-2	1.18e-2	9.49e-3	1.15e-2	1.16e-2±1.50e-3 (12.9%)
$\Lambda_{\max}$	603.74	371.77	311.51	448.31	569.02	488.70	465.51±112.39 (24.1%)

▷ Patient: 02

Samples	6	7	8	9	10	11	mean±std (ρ)
s	8.52	7.76	7.66	7.18	6.83	6.37	7.39±0.76 (10.3%)
δ	7.86e-3	1.01e-2	9.93e-3	1.00e-2	9.88e-3	1.10e-2	9.61e-3±1.11e-3 (11.5%)
$\Lambda_{\max}$	1084.00	768.32	771.40	718.00	691.30	579.09	782.82±165.40 (20.7%)

▷ Patient: 04

Samples	6	7	8	9	10	11	mean±std (ρ)
s	5.35	4.59	4.83	5.56	5.36	3.98	4.95±0.60 (12.1%)
δ	1.16e-2	1.30e-2	1.15e-2	1.21e-2	1.25e-2	1.32e-2	1.23e-2±7.08e-4 (06.6%)
$\Lambda_{\max}$	461.12	353.08	420.00	459.50	428.80	301.52	404.02±63.75 (15.8%)

▷ Patient: 05

Samples	6	7	8	9	10	11	mean±std (ρ)
s	5.72	5.54	4.12	4.15	4.46	4.37	7.43±0.71 (09.6%)
δ	9.75e-3	1.02e-2	1.18e-2	1.07e-2	1.10e-2	1.06e-2	1.07e-2±7.01e-4 (05.6%)
$\Lambda_{\max}$	586.67	543.14	349.15	387.85	405.45	412.26	447.42±94.61 (21.1%)

▷ Patient: 06

Samples	6	7	8	9	10	11	mean±std (ρ)
s	4.51	4.49	5.27	5.48	4.53	5.19	4.91±0.45 (09.2%)
δ	1.21e-2	1.08e-2	8.53e-3	7.91e-3	1.05e-2	9.04e-3	9.81e-3±1.58e-3 (16.1%)
$\Lambda_{\max}$	372.73	415.74	6.17.82	692.79	431.43	574.12	517.44±128.62 (24.9%)

Table 5. After 6 samples (in 21 days), the mathematical criterion $\Lambda_{\max}$ (= $146.48/\text{mm}^3 < 200/\text{mm}^3$) predicts an immunological failure for patient 03. This diagnosis is mathematically confirmed at 7 samples (in 30 days) by $\Lambda_{\max} = 74.38/\text{mm}^3 \ll 200/\text{mm}^3$.

▷ Patient: 03

Samples	6	7	8	9	10	11	mean±std (ρ)
s	2.14	1.22	1.19	0.47	1.06	0.54	1.103±0.60 (54.4%)
δ	1.46e-2	1.64e-2	1.53e-2	1.64e-2	1.59e-2	1.53e-2	1.56e-2±7.12e-4 (4.7%)
$\Lambda_{\max}$	146.58	74.39	77.78	28.66	66.67	35.29	71.56±42.08 (58.8%)

frequencies dynamics (non-predictive of the HIV infection) if the sample step is small enough. This result suggests that the sample step for the CD4 count should be larger in future trials.

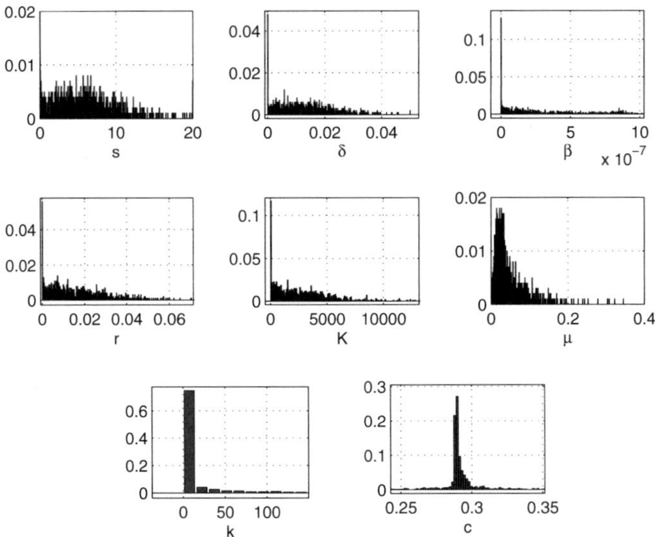

Fig. 8. Parameters distribution for patient 01. Estimation with 11 samples. 1000 randomizations are performed. Results denote a high confidence in the estimate of parameter c ($c = 0.29$) with $\mathrm{IQR_c} = 4.8e-4$. However, spread distributions of β or K suggest a low confidence in their estimates.

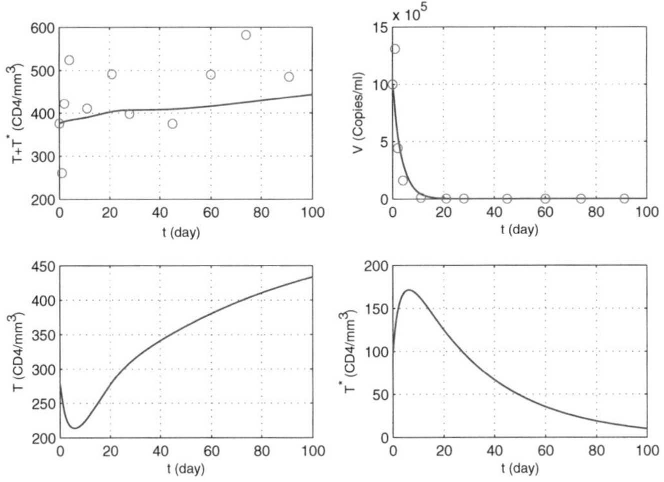

Fig. 9. Data fittings for patient 01

$\star$ *Case 2 (patient in immunological failure)*:
Immunological failure was diagnosed for patient 03. Table 5 presents the estimates of s and δ for this patient. According to the value of $\Lambda_{\max}$

$(= 146.58/\text{mm}^3 < 200/\text{mm}^3)$, an immunological failure is diagnosed, for this patient, after only 21 days of treatment (6 blood samples). After 1 month (7 blood samples), the level of $\Lambda_{\max}$ $(= 74/\text{mm}^3 \ll 200/\text{mm}^3)$ shows a severe immune deficiency for this patient. Note that the delay required here to make the mathematical diagnosis of immunological failure for this patient is very short. Only 21 days instead of 6 months as it is for standard clinical monitorings. This immunological failure was clinically confirmed by clinicians, during the trial, after 3 months of monitoring.

10 Conclusion

The mathematical analysis of the HIV infection (through the basic 3D model), for an effective aid for the clinical monitoring of patients, has been studied in this paper. We briefly present theoretical backgrounds on identifiability of non-linear systems and we show that the 3D model is identifiable under the realistic condition that the count of CD4 T-cells is the total amount of CD4 T-cells (infected + uninfected CD4 T-cells). The conditions for having good estimations of the 3D model parameters were also deduced. The application of this model to 6 real patients cases of the EDV05 clinical trial is also presented. It was possible thanks to a *had hoc* Monte-Carlo estimation method (based on the simplex algorithm) that we introduced. This method allows having stable and robust computations of the 3D model parameters. As a consequence, its implementation in a user-friendly software becomes possible. The goal of this tool is to allow clinicians to perform easy computations of the parameters of the infection to improve the therapeutical care of patients.

The results obtained from the EDV05 trial show that the immunological failure can be mathematically characterized and predicted. The case of the virological failure was also studied but current results do not enable us concluding the feasibility it mathematical prediction.

These results open new perspectives in this domain. For instance, the application of the identified models for the optimal control of the infection kinetics becomes possible. It will involve control theories of non-linear systems: the reader can refer to [31, 32, 33, 2, 17, 34] for details on application of control theories to some models of the HIV/AIDS infection.

References

1. A. M. Jeffrey. *A control theoretic approach to HIV/AIDS drug dosage desing and timing the initiation of therapy.* PhD thesis, University of Preotria, South Africa, July 2006.
2. R. Zurakowski and R. A. Teel. A model predictive control based scheduling method for HIV therapy. *Journal of Theoretical Biology*, pages 368–382, July 2006.
3. A. S. Perelson and P. W. Nelson. Mathematical analysis of HIV-1 dynamics in vivo. *SIAM Review*, 41(1):3–44, 1999.
4. A. S. Perelson et al. Decay characteristics of HIV-1 infected compartment during combination therapy. *Nature*, 387:188–191, 1997.

5. David D. Ho et al. Rapid turnover of plasma virion and CD4 lymphocytes in HIV-1 infection. *Nature*, 373:123–126, January 1995.

6. Xiping Wei et al. Viral dynamics in human immunodeficiency virus type 1 infection. *Nature*, 373:117–122, January 1995.

7. R. Arnaout, M. A. Nowak, and D. Wodarz. HIV-1 dynamics revisited: Biphasic decay by cytotoxic T lymphocyte killing. *Proc. Royal Society*, 267:1347–1354, 2000.

8. R. A. Filter, X. Xia, and I. K. Gray. Dynamic HIV/AIDS parameter estimation with application to a vaccine readiness study in southern Africa. *IEEE Transactions on Biomedical Engineering*, 52(5):284–291, May 2005.

9. M. A. Nowak and R. M. May. *Virus dynamics: Mathematical principles of immunology and virology*. Oxford University Press, 2002.

10. D. Kirschner and G. F. Webb. Understanding drug resistance for monotherapy treatment of HIV infection. *Bulletin of Mathematical Biology*, 59(4):763–785, 1997.

11. D. A. Ouattara. Mathematical analysis of the HIV-1 infection: Parameter estimation, therapies effectiveness, and therapeutical failures. In *27th Annual International Conference of the IEEE Engineering in Medecine and Biology Society (EMBC'05)*, Shanghai, China, September 2005.

12. D. Wodarz et al. A new therory of cytotoxic T-lymphocyte memory: Implication for the HIV treatement. *Phil. Trans. R. Soc. Lond.*, 355:329–343, 2000.

13. X. Xia and C. H. Moog. Identifiability of nonlinear systems with application to HIV/AIDS models. *IEEE Transactions on Automatic Control*, 48(2):330–336, February 2003.

14. Eric Walter. *Identifiability of Parametric Models*. Pergamon Press, London, U.K., 1987. Edited, Updated and Expanded papers of the 7th IFAC/IFCOR Symposium on Identification and System Parameter Estimation, July 1985 in York, England.

15. E. T. Tunali and T.-J. Tarn. New results for identifiability of nonlinear systems. *IEEE Transactions on Automatic Control*, AC-32(2):146–154, Febr. 1987.

16. G. Conte, C. H. Moog, and A. M. Perdon. *Nonlinear Control Systems*. Springer-Verlag, London, U.K., 1999.

17. J. Kim, W. H. Kim, H. B. Chung, and C. C. Chung. Constant drug dose leading long-term non-progressor for HIV-infected patients with RTI and PI. In *44th IEEE Conference on Decision and Control, and the European Control Conference 2005*, Seville, Spain, December 2005.

18. D. Wodarz and M. A. Nowak. CD8 memory, immuonodominance, and antigenic escape. *Eur. J. Immunol.*, 30:2704–2712, 2000.

19. JF. Delfraissy. *Prise en charge des personnes infectées par le VIH : Recommandations du groupe d'experts*. Médecine Science, Paris, flammarion edition, 2004. Available at `http://www.ladocumentationfrancaise.fr/brp/notices/044000467.shtml`.

20. U.S. Dept. Health and Human Services. Guidelines for the use of antiretroviral agents in HIV-1-infected adults and adolescents, May 2006. Available at `http://www.aidsinfo.nih.gov/guidelines`.

21. I. T. Prud'homme et al. Amplicor HIV Monitor, NASBA HIV-1 RNA QT and Quantiplex HIV RNA version 2.0 viral load assays: a Canadian evaluation. *Journal of Clinical Virology*, 11:189–202, 1998.

22. A. Berger et al. Comparative evaluation of the COBAS Amplicor HIV-1 Monitor^{TM} ultrasensitive test, the new COBAS AmpliPrep/COBAS Amplicor HIV-1 Monitor^{TM} and the versant HIV RNA 3.0 assays for quantitation of HIV-1 RNA in plasma samples. *Journal of Clinical Virology*, 33:43–51, 2005.

23. R. Galli, L. Merrick, M. Friesenhahn, and R. Ziermann. Comprehensive comparison of the Versant® HIV-1 RNA 3.0 (bDNA) and COBAS Amplicor HIV-1 Monitor® 1.5 assays on 1000 clinical specimens. *Journal of Clinical Virology*, 34:245–252, 2005.

24. K. Israel-Ballard et al. Taqman rt-pcr and Versant® HIV-1 RNA 3.0 (bDNA) assay quantification of HIV-1 RNA viral load in breast milk. *Journal of Clinical Virology*, 34:253–256, 2005.

25. Fiches Techniques de la Firme Roche. Available at `http://www.roche-diagnostics.fr`.

26. R. A. Filter and X. Xia. A penalty function to HIV/AIDS model parameter estimation. In *13th IFAC Symposium on System Identification*, Rotterdam, 2003.

27. D. A. Ouattara, F. Bugnon, F. Raffi, and C. H. Moog. Parameter identification of an HIV/AIDS model. In *13th International Symposium on HIV and Emerging Infectious Diseases*, Toulon, France, September 2004.

28. S. Nõmm. *Realization and Identifiability of Discret-time Nonlinear Systems*. PhD thesis, Tallinn University of Technology (Estonia), Ecole Centrale de Nantes (France), 2004. ISBN: 9985-59-440-1 / ISSN: 1406-4723.

29. D. A. Ouattara. *Modélisation de l'infection par le VIH, identification et aide au diagnostic*. PhD thesis, Ecole Centrale de Nantes & Université de Nantes, Nantes, France, September 2006.

30. IRCCyN Web software for the computation HIV infection parameters. Available at `http://www.irccyn.ec-nantes.fr/hiv`.

31. D. Kirschner, S. Lenhart, and S. Serbin. Optimal control of the chemotherapy of HIV. *J. Math. Biol*, 35:775–792, 1997.

32. M. A. Jeffrey, X. Xia, and I. K. Graig. When to initiate HIV therapy : A control theoretic approach. *IEEE Transactions on Biomedical Engineering*, 50(11): 1213–1220, 2003.

33. F. Biafore and C. E. D'Attellis. Exact linearisation and control of an HIV-1 predator-prey model. In *27th Annual International Conference of the IEEE Engineering in Medecine and Biology Society*, Shanghai, China, September 2005.

34. D. A. Ouattara and C. H. Moog. Identification, linéarisation et commande optimale du modèle 3D de l'infection VIH-1. In *Conférence International Francophone d'Autotmatique, CIFA 2006*, Bordeaux, France, 2006.

A Multiscale Model for the Selection Control of Ovulatory Follicles

Nki Echenim, Frédérique Clément, and Michel Sorine

INRIA Rocquencourt, BP 105, 78153 Le Chesnay Cedex, France

Summary. The biological meaning of follicular development is to free fertilizable oocytes at the time of ovulation. The selection of ovulatory follicles in mammal ovaries is an FSH dependent selection process. In this paper, we design a multiscale model of follicular development, where selection arises from the feedback between the ovaries and the pituitary gland and appeals to control theory concepts. Each ovarian follicle is characterized by a 2D density function giving an age and maturity-structured description of its cell population. The control intervenes in the velocity and loss terms of the conservation law ruling the changes in the density. This leads to some new control problems.

1 Introduction

The development of ovarian follicles is a crucial process for reproduction in mammals, as the biological meaning of folliculogenesis is to free fertilizable oocyte(s) at the time of ovulation. A better understanding of follicular development is both a clinical and zootechnical challenge; it is required to improve the control of anovulatory infertility in women, as well as ovulation rate and ovarian cycle chronology in domestic species.

Within all the developing follicles, very few actually reach the ovulatory size; most of them undergo a degeneration process, known as atresia [1]. The ovulation rate (number of ovulatory follicles) results from an FSH (Follicle Stimulating Hormone)-dependent follicle selection process. FSH acts on the cells surrounding the oocyte, the granulosa cells, and controls both their commitment towards either proliferation, differentiation or apoptosis and their ability to secrete hormones such as estradiol. The whole estradiol output from the ovaries is responsible for exerting a negative feedback on FSH release. Following the subsequent fall in plasmatic FSH levels, most of the follicles undergo atresia and only the ovulatory ones survive in the FSH-poor environment.

The previous mathematical models dealing with follicular development can be splitted into two main approaches. The first one takes into account the mechanisms underlying follicular development on the cellular and molecular scales,

I. Queinnec et al. (Eds.): Bio. & Ctrl. Theory: Current Challenges, LNCIS 357, pp. 45–52, 2007.
springerlink.com   © Springer-Verlag Berlin Heidelberg 2007

considering separately ovulatory and atretic trajectories. It aims at characterizing and understanding FSH-induced changes in granulosa cells [2, 3] and FSH signal transduction [4]. The other approach is concerned with the process of follicle selection from the viewpoint of follicular population dynamics [5]. A phenomenological and macroscopic function separates ovulatory follicular trajectories from atretic ones. Only the best-fitted follicles can survive.

In this paper, we aim at merging the molecular and cellular mechanistic description introduced by the former approach with the competition process dealt with in the latter. To build such a model, we use both multiscale modeling and control theory concepts. For each follicle, the cell population dynamics is ruled by a first order conservation law with variable coefficients which describes the changes in age and maturity of the granulosa cell density. A multiscale control term representing FSH signal intervenes both in the velocity and loss terms of the conservation law. We distinguish between a local control, specific to each follicle (micro scale), and a global control that results from the ovarian feedback (macro scale).

In Section 2, the conservation laws describing the model for follicle selection are presented. Section 3 presents physiological situations that arise from the numerical simulations of the equations. The final section discusses control problems associated to the model.

2   Controlled Conservation Laws for the Follicle Selection Model

A follicle is characterized by its granulosa cell population, made up of proliferating cells, running along the cell cycle and undergoing mitosis (whereby a mother cell gives birth to two daughter cells), and differentiated cells, having left the cell cycle. Cells may also engage a dying program, known as apoptosis. Cells age and maturity levels enable to distinguish between proliferating, differentiated or apoptotic cells.

According to biological knowledge, we introduce 3 main cellular phases:

- Phase 1 corresponds to cells going along the cell cycle but not yet committed to mitosis. A local control term u_f acts on their maturation and aging velocities. The cells stay in phase 1 until their age (modulo the cell cycle length) reaches the maximum phase age, a_1, when they enter phase 2, or until their maturity reaches a threshold value, γ_s, when they enter phase 3.
- Phase 2 corresponds to the cells committed to complete the whole cell cycle, whatever the control signal may be. They undergo mitosis once their age (modulo the cell cycle length) reaches the maximum phase age, a_2, when they enter phase 1 once again.
- Phase 3 corresponds to differentiated cells, which have exit the cell cycle in an irreversible way, after their maturity has reached the threshold γ_s. The local control u_f only acts on their maturation velocity.

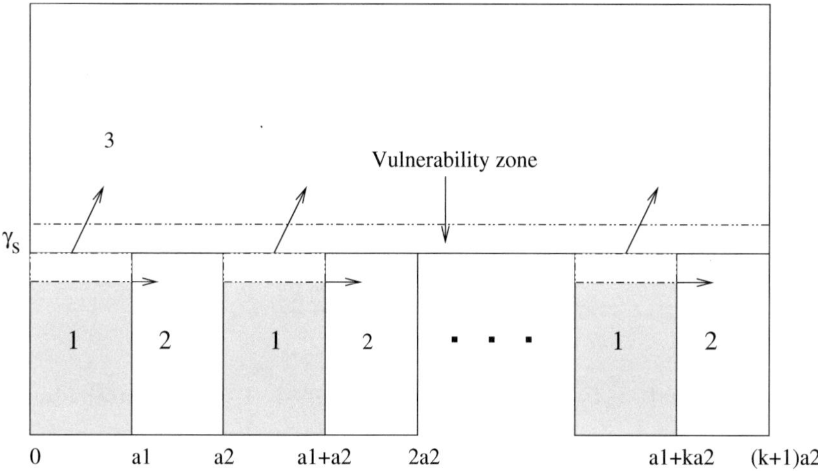

Fig. 1. Cellular phases on the age-maturity plan. *The cell cycle consists of the cyclic phase 1-phase 2-phase 1 pathway. When it enters phase 1 from phase 2, a mother cell gives birth to two daughter cells. Cell differentiation corresponds to the one-way flow from phase 1 into phase 3. a_2 is the cell cycle length. Cells can undergo apoptosis in the vulnerability zone.*

In phase 1 and phase 3, cells can undergo apoptosis. Such a phenomenon occurs for a specific interval of maturity values, if the global resource in FSH, U, is poor, compared to a reference value [1]. Cellular apoptosis finally leads a follicle to atresia. The phases and vulnerability zone are illustrated Figure 1.

For a given follicle, f, the cell density function, $\phi_f(a, \gamma, t)$, evolves with different dynamics according to the cellular phase. The generic form of the conservation law for ϕ_f is:

$$\forall a \in [0, +\infty), \quad \forall \gamma \in [0, +\infty)$$
$$\frac{\partial \phi_f}{\partial t} + \frac{\partial g_f(a, \gamma, u_f)\phi_f}{\partial a} + \frac{\partial h_f(a, \gamma, u_f)\phi_f}{\partial \gamma} = -\lambda(a, \gamma, U)\phi_f$$

The variable a denotes the cell age, γ the cell maturity and t denotes time. Both the global and the local control terms, U and u_f, act on the velocity and loss terms of the conservation law. The g_f and h_f functions are respectively the aging and maturation velocities, and λ is the loss term. They differ from one phase to another, according to:

$$\text{In phase 1:} \quad \forall (a[a_2], \gamma) \in [0, a_1] \times [0, \gamma_s]$$
$$g_f(u_f) = \beta'(b_{af}u_f + g_{a1})$$
$$h_f(\gamma, u_f) = -\beta\gamma^2 + \beta(c_1\gamma + c_2)(1 - \exp(-u_f/\overline{u}))$$
$$\lambda(\gamma, U) = \Omega(\gamma)\frac{|U - U_{max}|_-}{U_{max}}$$

In phase 2: $\forall (a[a_2], \gamma) \in [a_1, a_2] \times [0, \gamma_s]$

$$g_f(u_f) = 1$$
$$h_f(\gamma, u_f) = 0$$
$$\lambda(\gamma, U) = 0$$

In phase 3: $\forall (a, \gamma) \in [0, \infty) \times [\gamma_s, \infty)$

$$g_f(u_f) = 1$$
$$h_f(\gamma, u_f) = -\beta\gamma^2 + \beta(c_1\gamma + c_2)(1 - \exp(-u_f/\overline{u}))$$
$$\lambda(\gamma, U) = \Omega(\gamma)\frac{|U - U_{max}|_-}{U_{max}}$$

Where $|x|_- = \max(-x, 0)$, $\Omega(\gamma) = K \exp\left(-\left(\frac{\gamma - \gamma_s}{\gamma}\right)^2\right)$, and all parameters are real positive numbers.

The initial conditions for each follicle are given by: $\phi_f(a, \gamma, 0) = \Gamma_f(a, \gamma)$.

The transfer conditions between each cellular phase are flux continuity conditions defined by:

$$g_f(u_f)\phi_f(a_1^-, \gamma, t) = \phi_f(a_1^+, \gamma, t),$$
$$2\phi_f(a_2^-, \gamma, t) = g_f(u_f)\phi_f(0, \gamma, t),$$
$$\phi_f(a, \gamma_s^-, t) = \phi_f(a, \gamma_s^+, t).$$

The feedback exerted by the ovaries on the secretion of the hormonal control FSH yields a close-loop system (cf. Figure 2). Define the observation operator:

$$M(\phi_f)(t) = \int \gamma \int \phi_f(a, \gamma, t)dad\gamma$$

This moment of ϕ_f corresponds to the global maturity in a follicle. Applying the same operator to $\phi = \sum_f \phi_f$, gives the global maturity in the ovary $M(\phi)$.

The global control term U can be interpreted as the plasmatic FSH levels and it is a decreasing function of the feedback exerted by $M(\phi)$ [6]. Its dynamics can be described classically as:

$$\frac{dU[M(\phi)(t)]}{dt} = S[M(\phi)] - kU[M(\phi)] + U_0(t)$$

$S[M(\phi)]$ is a decreasing sigmoid function of $M(\phi)$ representing FSH release. $kU[M(\phi)]$ is the clearance of FSH and $U_0(t)$ represents a potential exogenous entry in FSH.

The local control term u_f is a proportion of U, depending on follicular maturity [7].

$$u_f[M(\phi_f), M(\phi)] = b_f[M(\phi_f)]U[M(\phi)]$$
$$b_f \leq 1$$

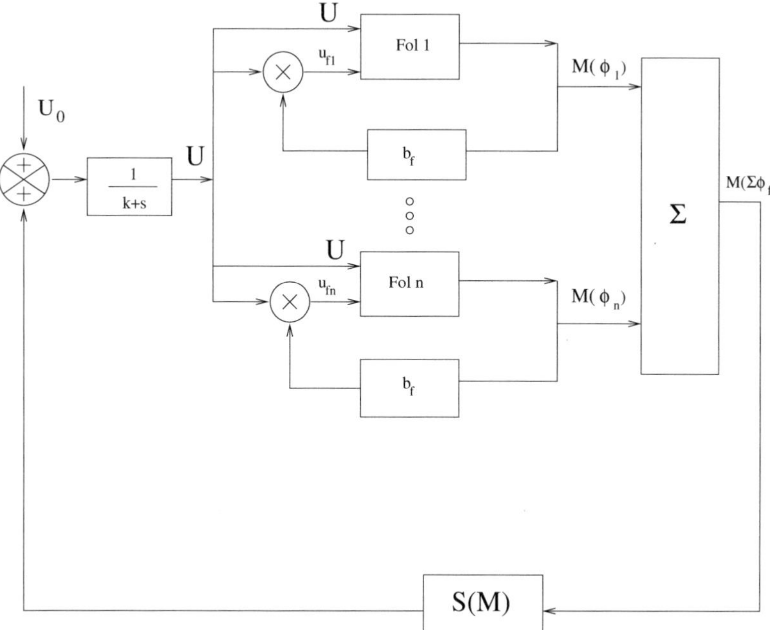

Fig. 2. The ovarian system in close-loop. *The global feedback control U is modulated on the follicle scale (Fol i), by the follicular maturity, $M(\phi_i)$, to define the local control u_{fi}.*

Ovulation is triggered as soon as the ovarian maturity reaches a threshold M_s so that ovulation time is defined by:

$$T \; / \; M(\phi)(T) = M_s$$

The follicles are then sorted; the ovulatory follicles are those whose maturity has overcome a threshold denoted M_{fs} [8]. $N = Card\{f/M(\phi_f)(T) \geq M_{fs}\}$ is thus the ovulation rate.

3   Numerical Simulations

Even if the framework of the model applies for most mammalian species, the numerical simulation is dedicated to the ovine species. This species is indeed particularly interesting regarding the follicle selection process, as the ovulation rate in the ewe may vary from 0 to more than 6 follicles. The numerical calibration called as much as possible for data available in the literature. When no data was available, an heuristic identification was undergone, so that the simulation outputs in terms of cell number fitted desired outputs.

We simulated a "competition" process between two follicles with different aging and maturation velocities. The top panel of Figure 3 illustrates the simulation

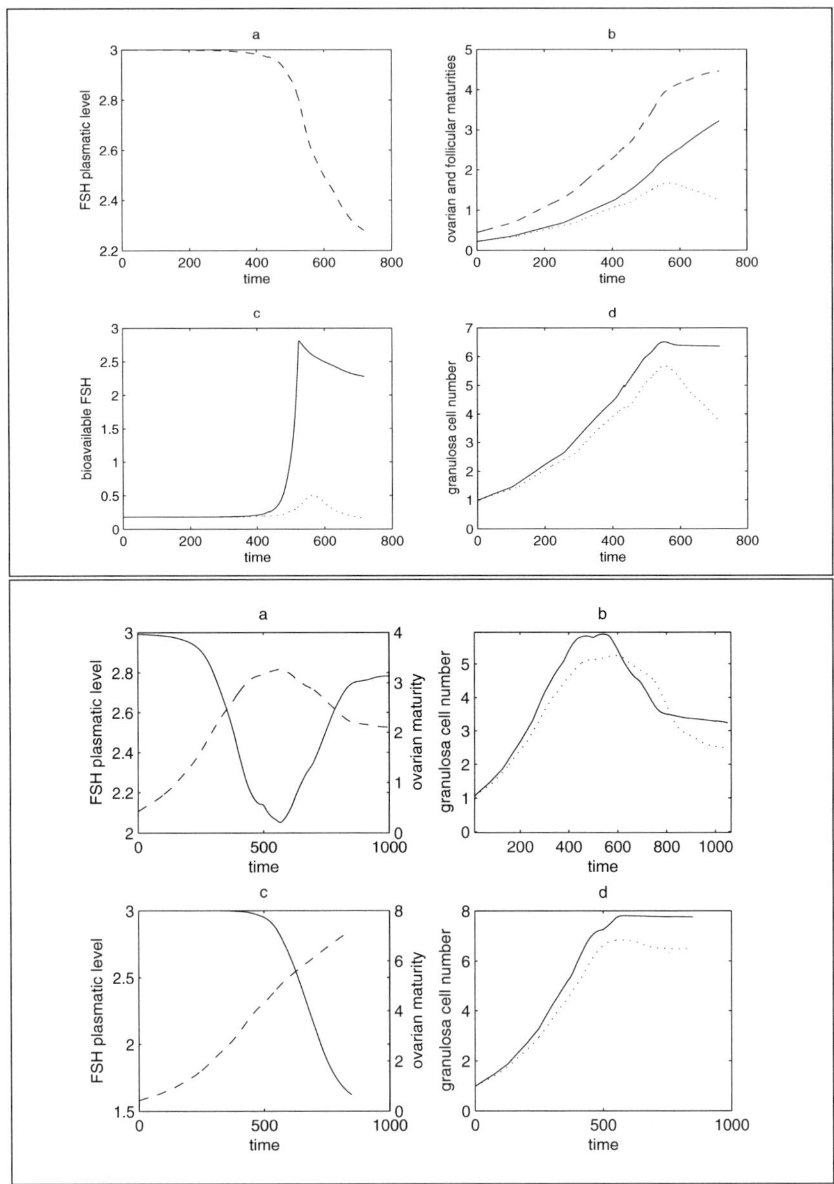

Fig. 3. Different selection cases. Dashed lines correspond to ovarian scale outputs, solid lines correspond to follicle 1, dotted lines to follicle 2. Top panel: Selection process within a cohort of two follicles, follicle 1 follows an ovulatory trajectory, follicle 2 an atretic one. Top panel on the top: early fall in plasmatic FSH, both follicles are anovulatory. Top panel on the bottom: late fall in plasmatic FSH, both follicles are ovulatory.

outputs, on the ovarian (global maturity, global control) as well as follicular (cell number, follicular maturity, local control) scales. One of the follicles follows an ovulatory trajectory, while the other follows an atretic trajectory. The results highlight the effect of the local control, which confines the cells of the atretic follicle in the area of vulnerability towards apoptosis, whereas it lets the cells from the other follicle escape from this area.

We also studied the effect of changing the sensitivity of FSH release to ovarian feedback. In case of a high sensitivity to ovarian feedback, we notice a premature fall in the global control U. None of the follicles resist to this fall, and the ovarian maturity $M(\phi)$ fails to reach the threshold to trigger ovulation, which results in an anovulatory situation (see Figure 3, bottom panel, top). In case of a low sensitivity to ovarian feedback, we notice a premature late fall in the global control U. Both follicles ovulate, as their maturity $M(\phi_f)$ raises enough to reach the follicular threshold for ovulation before FSH levels fall (see Figure 3, bottom panel, bottom).

We have also investigated the effect of adding exogenous FSH. We first simulated the selection process within a cohort of five follicles with different initial conditions and velocity parameters. The resulting ovulation rate was 2. In a next simulation using same initial conditions and parameter values, the fall in FSH was compensated by an exogenous entry U_0. The resulting ovulation rate became 3, as one of the former atretic follicles had been rescued by the exogenous entry.

4 Discussion and Perspectives

We designed a multiscale model for the FSH-controlled process of ovulatory follicle selection. A multiscale control is introduced, the local control operates on the micro scale, in a follicle-specific manner, while the global control issues from a macro scale information, defined on the ovary level. The two-way interaction between the micro and macro scales calls for the moments of the density function. Our approach thus merges the mechanistic viewpoint focusing on FSH effects on granulosa cells, with the population dynamics viewpoint focusing on the ovarian-mediated interactions between follicles. A more precise description of the model can be found in [9].

The model behavior, as assessed by numerical simulations, is consistent with physiological knowledge, regarding the effect of the local control modulation in determining a follicle's fate (ovulatory or atretic) and the effect of the sensitivity of FSH release to ovarian feedback in determining the ovulation output (mono-, poly- or anovulation). The trial simulation with exogenous FSH administration suggests that a finely tuned FSH control may be used to reach an *a priori* prescribed ovulation rate.

Both ovulation triggering and follicular ovulation depend on the reaching of a target, which defines a reachability problem under current research.

References

1. G.S. Greenwald and S.K. Roy. Follicular development and its control. In E. Knobil and J.D. Neill, editors, *The Physiology of Reproduction*, pages 629–724. Raven Press, New York, 1994.
2. F. Clément, M.-A. Gruet, P. Monget, M. Terqui, E. Jolivet, and D. Monniaux. Growth kinetics of the granulosa cell population in ovarian follicles: an approach by mathematical modelling. *Cell Prolif.*, 30:255–270, 1997.
3. F. Clément. Optimal control of the cell dynamics in the granulosa of ovulatory follicles. *Math. Biosci.*, 152:123–142, 1998.
4. F. Clément, D. Monniaux, J. Stark, K. Hardy, J-C Thalabard, S. Franks, and D. Claude. Mathematical model of fsh-induced camp production in ovarian follicles. *Am. J. Physiol. (Endocrinol. Metab.)*, 281:E35–E53, 2001.
5. H.M. Lacker and E. Akin. How do the ovaries count? *Math. Biosci.*, 90:305–332, 1988.
6. G.E. Mann, B.K. Campbell, A.S. McNeilly, and D.T. Baird. The role of inhibin and oestradiol in the control of fsh secretion in the sheep. *J. Endocrinol.*, 133:381–391, 1992.
7. K.P. McNatty, M. Gibb, D. Dobson, D.C. Thurley, and J.K. Findlay. Changes in the concentration of gonadotropic and steroidal hormones in the antral fluid of ovarian follicle throughout the oestrous cycle of the sheep. *Aust. J. Biol. Sci.*, 34:67–80, 1981.
8. R.J. Scaramuzzi, N.R. Adams, D.T. Baird, B.K. Campbell, J.A. Downing, J.K. Findlay, K.M. Henderson, G.B. Martin, K.P. McNatty, A.S. McNeilly, and C.G. Tsonis. A model for follicle selection and the determination of ovulation rate in the ewe. *Reprod. Fertil. Dev.*, 5:459–478, 1993.
9. N. Echenim, D. Monniaux, M. Sorine, and F. Clément. Multi-scale modeling of the follicle selection process in the ovary. *Math. Biosci.*, 198:57–79, 2005.

Robotics Insights for the Modeling of Visually Guided Hand Movements in Primates

Philippe Souères[1], Christophe Jouffrais[2], Simona Celebrini[3], and Yves Trotter[3]

[1] LAAS-CNRS,
   Université de Toulouse, 7 Avenue du Colonel Roche,
   31077 Toulouse Cedex 4, France
   `soueres@laas.fr`
[2] IRIT,
   Université Toulouse 3, Université Toulouse 1, INPT, CNRS,
   Université Paul Sabatier, 31062 Toulouse Cedex 9, France
   `jouffrai@irit.fr`
[3] CerCo,
   Université Toulouse 3, CNRS,
   Faculté de Medecine de Rangueil, 31062 Toulouse Cedex 9, France
   `simona.celebrini@cerco.ups-tlse.fr, yves.trotter@cerco.ups-tlse.fr`

Summary. In this chapter we focus on the modeling of cortical activity related to planning and control of visually guided reaching hand movements in primates. We bring a new light to this problem by considering the visual-servoing framework in robotics. A review of representative theories and models describing the neuronal processes related to 3D representation of space, motor control and visuomotor integration in Neuroscience is first presented. The kinematics and dynamics of manipulators and the basics of visual-servoing techniques in robotics are then recalled. In particular, for the control of a robotic arm with a deported camera, we underline the fact that the task-Jacobian is dependent on all the joints of the kinematic chain linking the camera to the end-effector. This point suggests that the motor activity during a visually guided movement of the hand cannot be completely encoded within a body-centered reference frame as claimed by numerous models in Neuroscience. Finally, we present an experimental result showing the existence of gaze-related signals in the monkey premotor cortex during visually guided reaching. This result corroborates the idea that the eye position with respect to the head and the head position with respect to the body, which belong to the kinematic chain linking the eye to the hand, must also be coded by neurons in premotor and motor cortex.

1 Introduction

Several complementary approaches such as Neurophysiology, Psychophysics or Clinical studies are used in Neurosciences to characterize the activity of different cortical areas involved in the representation of space and the control of movements. Based on the analysis of experimental results, neuroscientists try

I. Queinnec et al. (Eds.): Bio. & Ctrl. Theory: Current Challenges, LNCIS 357, pp. 53–75, 2007.

to elaborate models to describe the functioning of the *Central Nervous System* (CNS), and characterize the nature of data processing performed in each area of the brain. For this modeling task, the point of view of roboticians may be of great interest. Indeed, the problem of perception and motor control is central in robotics. To conceive robotic systems able to navigate autonomously, engineers need to develop repeatable and robust algorithms to deal with geometry, mechanics and that allow to cope with many constraints such as real-time data processing, noisy measurements, etc. As they are used to take into account these physical constraints for the design of artificial systems, roboticians can be of some help for interpreting experimental results in Neuroscience in view of constructing models.

In this chapter we focus on the modeling of cortical activity related to planning and control of visually guided reaching hand movements in primates. To control such movements the CNS needs to deal with the sensory information that gives the spatial position of the target and must coordinate the activation of different muscles involved in arm control. This task requires different kinds of data processing and coordinates transformations that allow to integrate multisensory data for space representation and define sensorimotor links.

In a first trend of opinion, numerous authors proposed that the planning and control of voluntary movements was based on the coding of the hand movement direction with respect to a body-centered reference frame [25], [26], [15]. This way of thinking was corroborated by the interpretation of reaching experiments in monkeys and inspired by the geometric models of manipulator that link the extrinsic representation of hand position in shoulder-centered reference frame and the intrinsic articular joint parameters of the arm. More recent models propose that the reference input for control is the difference vector between the hand and the target expressed in eye-centered reference frame [13], [18], [46]. This latter idea seems to be consistent with current models that suggest a multisensory spatial representation in eye-centered coordinates for reaching [42]. However, though many authors agree that the difference vector between the hand and the target is encoded with respect to the eye, most part part of them still claim that the motor control of visually guided arm movements is independent from eye-position [46].

Our objective is to bring a new light to this question by considering the recent visual-servoing framework in robotics. We propose to identify the hand-eye difference vector, expressed in eye-centered reference frame, with a vision-based task function. On this basis, the reaching movement can be described as a visual servoing task with deported camera. Underlining the fact that the task-Jacobian is a function of each joint of the eye-to-hand kinematic chain, mathematical arguments can be derived to show that the motor control not only depends on the angular joint parameters of the arm but also on the position of the head and the gaze direction.

The chapter is organized as follows: A review of representative theories and models describing the neuronal processes related to 3-dimensional (3D) representation of space, motor control and visuomotor integration in Neuroscience is first presented in sections 2. In section 3 we recall the models of manipulator kinematics and dynamics and give an introduction to the visual-servoing framework. On this basis, a discussion is proposed to consider possible analogies with the control of visually guided arm movements in primates. In particular we propose geometric arguments to justify the existence of cortical neurons sensitive to eye-position in cortical premotor and motor areas. Finally, an experimental result showing the existence of gaze-related signals in the monkey premotor cortex during visually guided reaching is presented in section 4.

2   The Primate Visually Guided Hand Movement

Neuroscientists interested in the human ability to reach visual targets need to understand the brain visual processes related to 3D reconstruction of the environment and, more specifically, target localization in this 3D reconstruction. They also need to understand the neuronal processes underlying planning and execution of hand movements as well as the biomechanical constraints that the brain has to overcome. Our aim is not to thoroughly describe the outcome of all these fields of research. We present here selected results across these different fields that are necessary to understand how neuroscientists think this very frequent action is achieved.

2.1   Multisensory Integration for 3D Representation of the Scene

In human and non human primates, the cortical representation of the surrounding world is essentially visual, as revealed by the large amount of cortex dedicated to vision. Therefore, our internal representation of space, necessary to execute the majority of actions in our everyday life, mainly relies on the visual system. But to be really efficient in every situation, our brain has to take into account other available information like our position in space and its changes in time, oculomotor information etc., to allow a unified and stable perception of the environment. How this unified and stable percept is processed by the brain remains a major question in Neurosciences.

The image of an object located in the visual field falls on both retinae and stimulates the photoreceptors. Visual information about that object is then transmitted from the retinal ganglion cells through the optic nerve to different sub-cortical relays, the main of them being the lateral geniculate nucleus that in turn projects its afferents to the primary visual cortex (area V1). Different attributes of the object, like its shape, depth, color or motion,... will be analyzed by different cortical areas located in one of the two main visual pathways. 1) The ventral pathway, classically called the "what pathway", is mainly dedicated to the analysis of the shape and color of objects to allow their recognition. It starts in the first visual cortical area, area V1, and finishes in the temporal cortex after several analysis stages. 2) The dorsal pathway, called the "where pathway", is dedicated to the analysis of motion and localization of objects in space. It starts in area V1 and ends in the parietal cortex after several steps of visual processing in the different visual areas specialized for the "where coding". The receptive fields of visual neurons (i.e the portion of visual space viewed by a single neuron) are really small in area V1, less than 1 degree, and can cover the entire visual field in temporal or parietal cortex. Thus, the neuronal processing performed in the successive areas in both pathways allows the integration of the visual information over an increasingly larger portion of the visual space, leading to a more global representation of the visual scene. It also allows the analysis of more and more complex properties of the visual scene from simple attributes like bars in V1 to faces in the temporal cortex and complex motion in parietal cortex.

To produce any simple motor act such as grasping an object viewed in the immediate environment, we need to localize objects in the visual space, whatever the position of the eyes in the orbit and of the head on the trunk. Consequently, depending on the reference frame that is used, our brain needs to combine information about the position of the images of the objects on the retinae with information about the position of the eyes and of the head. The location and the nature of this neural processing is still poorly understood. Furthermore, retinae are not homogeneous, only the center of the retina called the fovea, is able to detect fine details. As a result, when we are visually exploring our environment, our eyes have to move continuously in their orbit so that the foveae can be oriented towards the different regions of interest of the scene. The eyes are usually moving together with the head and the primate ocular system has to stabilize the gaze via two mechanisms, the vestibulo-ocular (VOR) and optokinetic (OKR) reflexes. The VOR relies on vestibular inputs to compensate for head movement relative to the visual environment. The OKR relies on motion detectors in the retina to compensate for image movement. The consequence of both mechanisms is the image immobility (stability) on the retinas, for a duration long enough to allow signal transduction. The saccades quickly move the eyes between image stabilizations when these reflexes reach their limits.

This dynamic aspect of the localization problem implies that successive "snapshots" must be memorized and combined in order to build a complete representation of the visual scene. This also means that any mechanism involved in this localization processing has to be reactive enough to update new spatial position several times per second.

Clinical studies in humans have described some difficulties in locating visual targets in space following lesions of the posterior parietal cortex [3]. Electrophysiological studies performed in behaving monkeys showed that this cortical area integrates visual and oculomotor signals; light sensitive cells respond with a certain neural gain depending on where the monkey is looking at.

For many years the parietal cortex was believed as being the only cortical site for processing space localization, following the pioneering work of Andersen and col. [1], [2]. Since then, eye position signals have been observed in many cortical areas. In the ventral pathway, effects of eye position have been demonstrated in area V4 [11] and in the inferotemporal cortex [40]. But most studies have focused on the dorsal pathway more involved in localizing objects in space. These works have shown that neurons modulate their visual activity as a function of eye position in several areas in parietal cortex [19], [24], [27], in medio- temporal areas [39] and in V3A [23]. Recent evidences showed that even primary visual cortex, area V1, classically thought as purely visual, participates as well to the neural localization processing by integrating retinal information with extraretinal information about direction of gaze [52].

Though our surrounding space is not flat, almost all the above studies have investigated a localization processing limited to a 2D space, because of the visual stimuli and the protocols used. Since the eyes are horizontally separated, the image of an object viewed in depth is projected on the retinae on slightly different locations, and the brain is able to measure this angular difference, called horizontal disparity, to reconstruct the position in depth relative to (behind, in or in front of) the fixation point. Many visual cortical cells, present in several cortical visual areas, are specialized in detecting and coding horizontal disparity and are classically considered as the neural substrate of depth perception. Studies performed in the primary visual cortex have extended the previous findings on neural spatial localization mechanisms to the third dimension by testing specifically disparity coding and varying the location of visual fixation in depth [50], [51]. For a large majority of V1 cells, horizontal disparity coding was shown to be dependent on the vergence angle in area V1 of awake behaving monkeys in such a way that this 3D property could be coded at a given distance of fixation but absent at others. Changing the distance of fixation, and thus the vergence angle, in absence of visual stimulation also affects the level of spontaneous activity of about half of cells, especially at short distances of fixation. Thus the effect of eye position on visual activity must result from the integration of visual and nonvisual cues. Among these nonvisual cues, the proprioceptive and motor signals coming from the oculomotor system are undoubtedly involved in this cortical processing.

The main conclusion of these studies performed in V1 [50], [51], [52], is that V1 neurons encode the position of objects in depth around the fixation point only for certain position of that fixation point in the 3D space. Bringing together the results on the effects of the viewing distance and those of the gaze direction in area V1, the authors come up with the proposal that cortical properties such as orientation and retinal disparity selectivities, that define shapes and volumes of objects, are optimally expressed in a limited range of 3D gaze directions so that information about stimuli in the V1 area is conveyed by cell populations only when the object is present within restricted volumes of space. These modules should be regarded as 3D fields and as being a part of the neural substrate that is involved in sensory-motor transformations for 3D space localization.

The neural modulation observed in the different visual areas could be the basis for encoding, at the population level, the position of objects in multiple coordinates, such as in an eye-, head- (see above) and even body-centered representation [12]. However, clear explicit evidence of such encoding at the single cell level, leading to the expression of a pure head- or body-reference frames, has not been demonstrated yet.

Although visual information seems to play a major role, spatial localization mechanisms must integrate not only visual cues, but also auditory, somesthesic, and vestibular cues. For example, it is necessary to be able to localize a sound in our visual surrounding space. An efficient way to achieve this ability is to have all sensory modalities coded in a common reference frame in order to allow them to participate coherently to the neural representation of space. A study, performed in calcarine V1 of an awake behaving monkey confirms this hypothesis by showing a decrease of latency of neuronal response when auditory and visual stimuli are coherently combined in the visual space [53]. Furthermore, another result obtained in parietal cortex reported that auditory information can be represented in eye centered coordinates [48].

Taken together, these studies suggest that visual information and, perhaps, information from other sensory modalities, are available and coded together with eye positions at all stages of the visual processing. These results imply that all visual areas are to some extent involved in the neural 3D space localization processing. Information about eye position and head position (in parietal cortex) are implicitly represented at the population level but this does not rule out the possibility that a common reference frame centered on the eyes might be used to facilitate communication between visual and non visual cortical areas.

The distributed representation in the different areas composing the hierarchical visual pathways has the advantage to allow information about stimuli of different spatial scales and coding of a large range of visual attributes to be integrated in a similar way with non visual signals. However, some evidences suggest that the parietal cortex should play a more central role in 3D representation of space by computing a more global and integrated representation of the visual scenes [4].

2.2 Motor Command

A complete description of the motor system is beyond the scope of this section. We will make a simplified depiction of the system anatomy, i.e. the muscles and the parts of the nervous system involved in a voluntary arm movement (main cortical areas, subcortical nuclei, and spinal cord) with the connections between them. We will then put an emphasis on the main models that describe the coding of arm movements at the spinal and cortical levels. To move the arm, the CNS must generate force slowly with springlike actuators (muscles) that act against a skeleton. It must also analyze inputs from sensory transducers that provide feedback. In addition, the force produced by the muscle not only depends on the command but also on the configuration (muscle length) and the recent history of the limb (muscle fatigue).

There are many parts of the nervous system that are involved in the motor command (spinal cord, cerebellum and other brainstem nuclei, thalamus, basal ganglia, and many areas of the cerebral cortex. See Figure 1). Spinal cord has four divisions - cervical, thoracic, lumbar, and sacral - that contain, on the one side, motor neurons connected to the muscles, and, on the other side, sensory neurons receiving signals from the receptors in the skin, joints and muscles. Motor neurons connected to the muscular apparatus of the hindlimb are in the cervical division of the spinal cord. Spinal cord is linked to the brain (brainstem and forebrain) through sensory and motor pathways. There are many motor pathways descending down from the brain to the spinal cord. The main descending pathway is the pyramidal system projecting directly from the motor cortical areas (M1, PM, SMA and CMA; see Figure 1) onto the spinal neurons. This pathway allows the control of voluntary arm movement. The other descending pathways link different subcortical structures (cerebellum, and basal ganglia for instance) to the spinal cord neurons and are more important in the automatic control of posture, coordination, etc. These cortical and subcortical areas are linked to each other through multiple connections. To describe the physiology of the motor system for reaching, it is important to understand the lowest level of the control that results from activation of neurons in the spinal cord. The stretch reflex is a remarkable example of such a basic control loop which plays an essential part in the control of posture. Beyond this reflex, numerous modulus that constitute the basic elements for the control of limbs - such as central pattern generators (GPCs) that allow a scheduled activation of muscles for walking - are located in the spinal cord. Experiments done by Bizzi et al [7] on spinalized animals have shown that microactivations of the spinal cord induce time varying force fields with a unique equilibrium point that allow to position the end-effector of limbs. To control the paw motion, they showed that the muscle activation produces a gradual shift of this equilibrium point from the initial to the final end-effector position, providing an efficient way to reject external perturbations. As proposed in [37], these force fields could constitute computational primitives used by the central nervous system (CNS) for generating a grammar of motor behavior. How do the cortical motor areas that project onto spinal motor neurons select and combine such primitives to execute a voluntary motion of the hand? To solve this problem, the CNS needs to relate the extrinsic configuration of the target and the end-effector, with the intrinsic parameters of the arm, i.e. its angular and inertia parameters. For roboticians these problems are referred to as kinematics and dynamics (see section 2). As stated in section 1.2, a multisensory representation of space, which allows to code the target position in eye-centered coordinates, is available in PPC. On the other hand, numerous authors propose that the intrinsic parameters of the arm are coded in the motor cortex.

Most neurophysiologists agree that PPC and the frontal motor areas play an important role in visually guided arm movements (including processes such as localization of visual target, visuomotor integration and arm movement planning) (see [30], [54], for reviews). Currently, there is still a debate on the nature of variables controlled in the frontal motor areas, called the "muscles" versus "movement" debate. In the "muscles" side of the debate, there are arguments to show that individual neurons of the primary motor area control single muscles, whereas, in the "movement" side people claim that the control mainly concerns groups of muscles. This discussion is closely related to the control architecture and the location of potential modulus devoted to the computation of kinematics and dynamics. Related to this issue is the determination of the reference

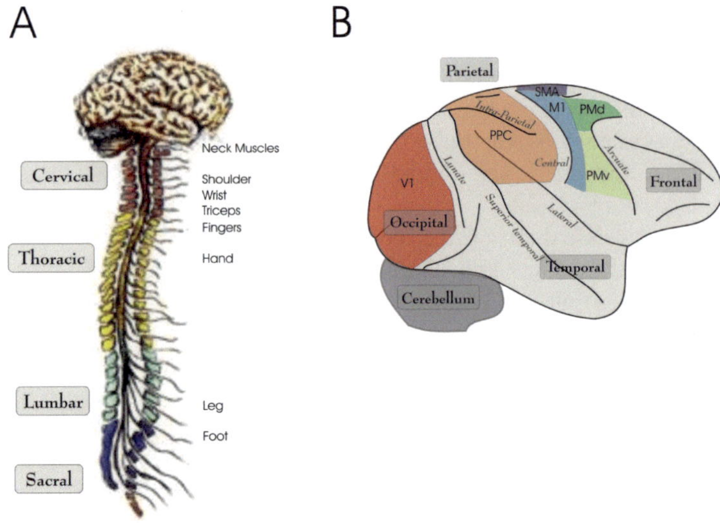

Fig. 1. A- Lateral view of the human brain and spinal cord. The four sections of the spinal cord are indicated on the left. The nerves that connect the spinal cord to the upper limb are in the cervical section. B- Lateral view of a macaque monkey brain showing four lobes of the cerebral cortex and the cerebellum. The main areas involved in visually guided reaching movements are mentioned (Visual, Parietal, and Motor areas). Many areas that are mentioned in the text are not shown because they are hidden: V2 and V3 lie in the Lunate Sulcus; MT and PO in the Superior Temporal Sulcus; the basal ganglia and the thalamus are under the cerebral cortex. Abbreviations: V1, Primary Visual cortex; PPC, Posterior Parietal Cortex; PMv and PMd, Premotor ventral and dorsal; SMA, Supplementary Motor Area; M1, Primary Motor. Main sulci are indicated in italics.

frame used to encode locations. Coordinate frames may be intrinsic with encoded variables such as joint angles, joint angular velocity, or torques generated by the muscles. On the other side, they may be extrinsic, in a Cartesian frame of reference, based on a particular origin in space (fixation point, fovea, head, shoulder, hand, target, etc.). The models of robotics manipulators and the associated control theory have strongly influenced the elaboration of computational models in Neuroscience. As the control of the end-effector is usually defined with respect to the manipulator basis, the shoulder is often considered as the natural reference frame for motor control [14], [26]. Though these authors propose an extrinsic coding of end-effector position, these results can be interpreted in terms of intrinsic coordinates [33], [44]. On this basis, many authors agree that intrinsic parameters describing the configuration of joints between the hand and the shoulder are somehow encoded in the motor cortex. Contrary to robotics arms, these intrinsic parameters are not given by the direct measure of articular joints but deduced from the length of muscles. Besides this intrinsic representation of the arm configuration, the position of the hand and the target are known to be encoded in extrinsic coordinates. Different reference frames have been proposed to this end. Numerous authors have brought evidences showing that, during the construction of a motor plan and the execution of a visually guided reaching movement, the CNS encodes a difference vector between the hand and the target in eye-centered coordinates [13], [18], [42]. Considering the role of vision for 3D representation of space in primates (see section 1.2), the coding of this difference vector in eye-centered reference frame seems rather natural. At this stage, an important question arises. How does the CNS control the intrinsic coordinates of the arm, which encode the configuration of the successive links with respect to the shoulder, by considering the difference vector expressed in eye-centered coordinates. This question, which is referred to as visuomotor transformation, will be considered in the next section.

2.3 Visuomotor Integration

Many studies in the last 20 years tried to link the eye-centered visual representation of the target with the body-centered reference frame used to execute arm movements. The first prominent results were obtained by Andersen and Mountcastle [1]. They recorded single neurons in the posterior parietal area 7a that were coding simultaneously the position of the target on the retina and the eye position in the orbit. Andersen and colleagues then proposed a computational model showing that these two signals were sufficient to code for the head-related position of the target [2], [55]. By analogy, they made the hypothesis that such a mechanism would provide the body-centered position of the target by addition of the signal corresponding to head position on the trunk (Figure 2). However, very few neurons encode target location in a true head-centered reference frame. In such a reference frame, the receptive field of a single neuron should encode the same location in space independently of movements and/or position of the eyes. In fact, the neuronal processes underlying the putative transformations from

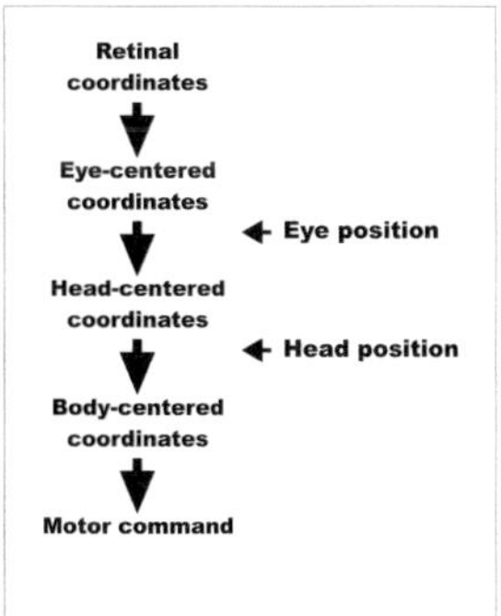

Fig. 2. Schema showing how body-centered coordinates of a visual target can be computed from its retinal coordinates and extra-retinal signals. The body-centered localization would be an ideal representation for planning an arm movement towards the target.

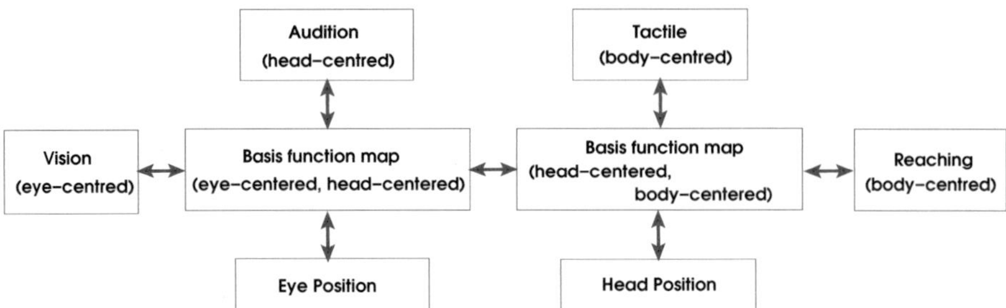

Fig. 3. Schematic representation of a basis function network for reaching towards visual, auditory and tactile targets. The first basis function map encodes auditory and tactile targets in eye- and head-centered coordinates. This map communicates with a second map which encodes targets in head- and body-centered coordinates.

retina to head and body centered reference frames appear to be distributed over large populations of neurons. The visual and eye position signals interact to form "planar gain fields" in which the amplitude of the visual response varies linearly with eye position [2]. Brotchie and colleagues [12] showed that the head position signals necessary to compute body-centered coordinates of the target also exist in the parietal cortex at the level of population of cells. They conclude that this distributed representation could be the final stage for coding locations in space, or could be used as an intermediate step in the construction of body-centered receptive fields. Altogether, these results tended to prove that the model of sequential coordinates transformation is computationally plausible, based on the existence of required signals related to retinal localization of the target, eye position in the orbit and head position on the trunk. Furthermore, the anatomic pathway from the visual cortex to the motor cortex goes through parietal areas; and lesions of the parietal cortex of humans and monkeys induce visuomotor disorders like optic ataxia (inability to reach an object perceived in space [34]).

More recently, Pouget et al showed that reaching movements could be coded in eye-centered reference frame [42]. Their data indicates that the position of reaching targets is represented in eye-centered coordinates regardless of the sensory modality used to guide the movement (audition or vision). In addition, they conclude that, these representations are updated after each eye or head movement. As this recoding of auditory target localization in an eye-centered reference frame is counter intuitive in the scheme of the model shown in Figure 2, they proposed a computational model that reconciles multisensory integration and sensory guided arm movements [41]. This model relies on the combination of basis functions and attractor dynamics. Figure 3 illustrates their basis function network that is designed to reach targets regardless of the input sensory modality. Interestingly, the network can predict the position of a target based on its position in another sensory modality, and also predict the consequences of a reaching movement in all sensory modalities. However, for Pouget et al. [42], the body-centered frame of reference is essential because it is used

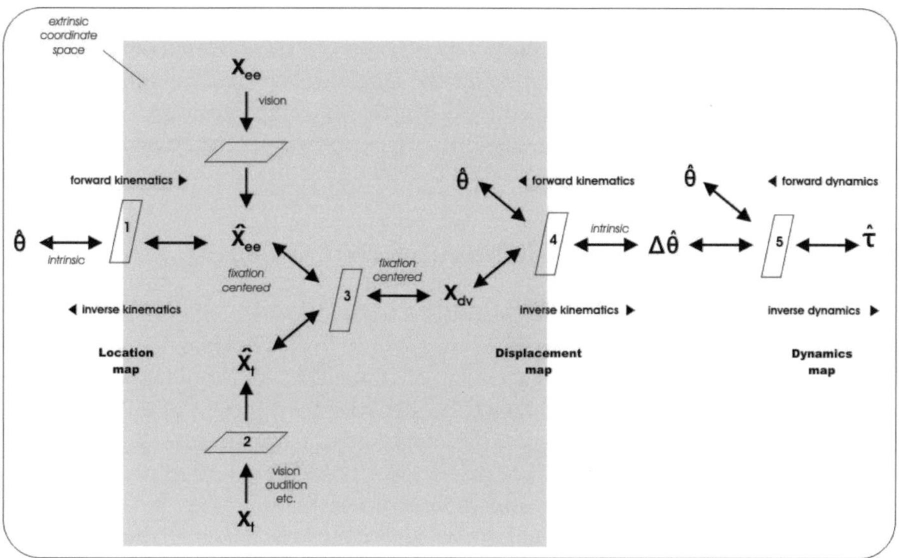

Fig. 4. Schematic of a series of computations for planning a reaching movement. Joint sensors provide a measure of arm configuration θ and a camera provides an estimate of hand and target locations, x_{ee} and x_t, in camera-centered coordinates. A network with bidirectional connections (1) aligns joint- and camera-based estimates of end effector location. From the estimates of target and end-effector locations, another network (3) computes a difference vector x_{dv} between target and end effector. An additional network (4) aligns this difference vector with a joint-rotation vector $\Delta\theta$. This final transformation depends on the arm's configuration θ. In some cases, you might want to move to a target that you cannot see, but that relates to a visual target. The network (2) serves this purpose. (From Shadmehr and Wise [46], MIT Press, with permission).

in the reaching module (right side of the network). R. Shadmehr and S. Wise [46] proposed an overall model inspired from robotics showing the different computations required for planning a reaching movement (Figure 4). The first step consists in computing a difference vector between estimated target location from visual sensors, and estimated end-effector (the hand) location from proprioceptive sensors. Once this difference vector has been computed, the system can work out the joint-coordinate representation that correspond to the actual movement of the arm. The final step consists in alignment between the joint rotations and force commands. As in the Pouget et al. model presented earlier, the authors added a network (network 2 in figure 4) allowing a multimodal representation of the target (for example, a bee may be a visual and auditory target for a reaching movement). The conclusion of these recent works is that movement planning is made in a fixation-centered (Shadmehr and Wise) or an eye-centered (Andersen and colleagues) reference frame. In an attempt to anatomically localize the networks involved in visuomotor integration, it is now accepted that all these computations are made in the dense parieto-frontal network [54]. There is also an agreement to say that the kinematics parameters are most encoded in the parietal areas whereas dynamics is most encoded in the frontal areas close to the central sulcus. In addition, it is very interesting to note that all these authors are strongly influenced by the robotics of simple two-joints effectors, considering that the motor command (delta theta) of the primate arm only addresses the elbow and shoulder joints.

3   Robot Manipulators Modeling and Control

In this section we first recall some theoretical results about modeling and control of robotic arms. In a first subsection we introduce the notions of operational and joint coordinates and the statement of kinematic and dynamic equations of manipulators which were developed by roboticians during the 80's and had important impact in the neuroscientific community interested in the coding of voluntary arm movements. These models have contributed to give rise to the concept of intrinsic and extrinsic spaces in Neuroscience. In a second subsection we present the more recent visual servoing techniques which allow to design closed-loop controllers with visual data as input. In view of these results, we discuss in a third subsection the possible analogies with visually guided reaching movement in primates. In particular, we propose a model to explain the existence of neurons in motor and premotor areas whose activity is modulated by eye position during visually guided hand movements.

3.1   Classical Models of Manipulators

Depending on the nature of the task to be performed, different mechanical structures have been proposed by roboticians for the design of manipulators. We focus here on the important class of open-loop kinematic chains, which include a sequence of rigid links connected by articulated joints, between the basis and the end-effector, and roughly follow the structure of primates arms. A general framework has been developed during the 80's to model the geometry, the kinematics and the dynamics of such serial linkages which have proved to be highly coupled and nonlinear [20], [47], [5]. Figure 5 represents a manipulator arm modeled as a serial linkage of rigid bodies including prismatic and revolute joints. Two sets of variables are usually introduced to model the kinematics of such a manipulator: the *operational coordinates* $X = (x_n, y_n, z_n, \phi, \theta, \psi)$ describe both the position of the end-effector O_n and its orientation with respect to a frame attached to the basis of the arm (ϕ, θ and ψ stand for a parametrization of orientation such as the Euler angles), whereas the *joint coordinates* $Q = (q_1, q_2, \ldots, q_n)$ represent the measure of the successive joints. By considering the serial transformation between the basis and the end-effector, one can write the kinematic equation of the arm under the form $X = f(Q)$. This relation expresses the configuration of the end-effector as a function of the links involved in the kinematic chain. Now, to find the joint displacement that is necessary to move the end-effector to a specified configuration, one needs to solve the converse problem, called *inverse kinematics problem*. However, as for dexterity reasons the number of joints,

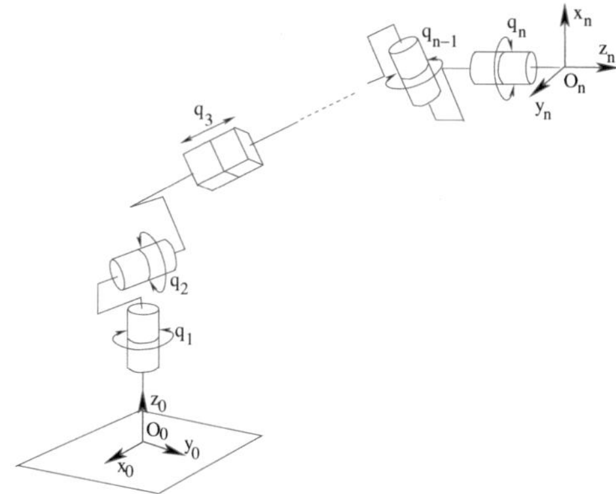

Fig. 5. Example of robotic manipulator with n links

n, is usually greater than six, the problem is not directly invertible and optimization techniques are required to characterize a solution among the possible joint configurations [38]. The geometric relationship between the situation of the end-effector and the joint coordinates is not sufficient to model instantaneous variations of the structure. To this end it is necessary to compute the derivative of the kinematic equation to obtain a differential relationship between the joint displacement $\dot{Q}(t) = dQ/dt$, and the end-effector velocity $\nu(t) = dX/dt$. This relation is expressed through the definition of the robot Jacobian $J(Q)$ as follows:

$$\nu = J(Q)\dot{Q} \tag{1}$$

As for the geometric problem, the converse problem called *inverse instantaneous kinematics* needs to be solved to determine the variation of joint coordinates, which allows to drive the end-effector along a prescribed trajectory. It is important at this stage to remark that the Jacobian $J(Q)$ is function of the n joint coordinates, which need to be reevaluated at each point. This computation, which is time-consuming, states a difficult problem for real-time control.

Once we know from the kinematic study how to move the manipulator joints in order to drive the end-effector towards a target with a prescribed velocity, we must consider the manipulator *dynamics* to design an adequate control. The dynamics expresses the relationship between the arm configuration, the mass and inertia of different links, and the joint torques exerted by the actuators. This relation usually described by a differential equation is called *equation of motion*, and usually deduced from the Newton-Euler equations of mechanics or from the Lagrangian formulation. It expresses the actuator torques Γ as a function of the joint coordinates and their first and second order derivatives as follows :

$$\Gamma = M(Q)\ddot{Q} + B(Q, \dot{Q}), \tag{2}$$

In this equation, $M(Q)$ is the so-called inertia matrix and $B(Q, \dot{Q})$ gathers the effects of gravity, Coriolis and centrifugal forces and frictions at the joint level. On this basis, the closed loop control can be designed according to the diagram represented in Figure 6. In this scheme, the desired configuration of the end-effector comes as the input and the inverse kinematics is used to express the corresponding reference value of joint coordinates. The control torque Γ is then defined as a function of the gap ΔQ between the current value of joint coordinates, provided by the measure of sensors, and the desired one. This closed-loop control model

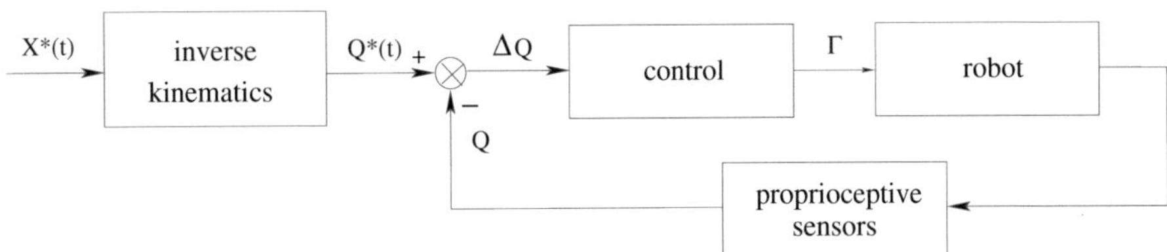

Fig. 6. Classical diagram for the control of manipulators

has been considered with great interest by neuroscientists. Indeed, the dynamic link between articular parameters and spatial position of the end-effector offers a mathematical model for human arm control. The notions of operational and generalized coordinates have contributed to the definition of two complementary spaces in Neuroscience: an extrinsic space given by the endpoint of movement and an intrinsic space given by the joint angle and muscle activations needed to achieve the movement endpoint [36]. Several models of control inspired by this scheme have been proposed to describe the motor activity during a reaching task [29]; the authors identifying the basis of the arm with the human shoulder and the end-effector with the hand.

As long as the position of the target with respect of the robot basis $X^\star$ is perfectly known, proprioceptive sensors are sufficient to determine the current value of articular coordinates $Q(t)$ to implement the control scheme of Figure 5. In practice, the configuration of the target is not *a priori* known and the use of exteroceptive measurement is required. Artificial vision offers an efficient way to cope with this question. The basics of visual servoing techniques, which allow to combine vision and control in robotics are recalled in the next paragraph.

3.2 The Visual Servoing Scheme

Thanks to the high frequency of the CCD cameras which equip the robots today, it is possible to design closed-loop controls in which the error to regulate is directly expressed as a function of visual data. This approach, called *visual servoing*, is based on the definition of the so-called *interaction matrix* which defines a differential link between a vector of visual data and the relative position of the target with respect to the camera [21]. The *task function* formalism, which was introduced in [43], allows to define such vector functions of sensory data within a rigorous mathematical framework. Task functions can be viewed

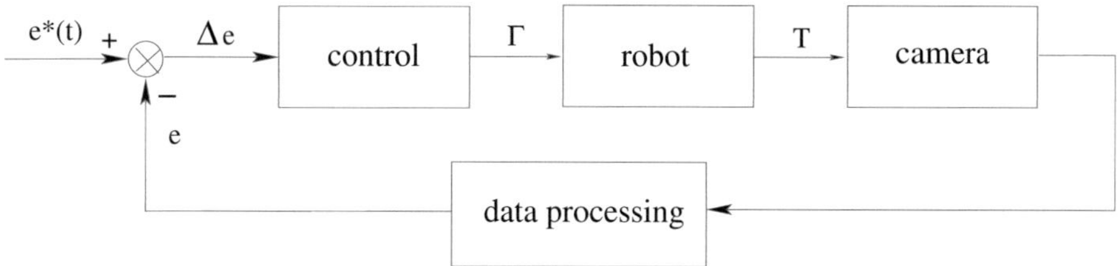

Fig. 7. The visual servoing scheme

as a generalization of operational coordinates. On this basis, a new control diagram can be defined, in which the reference input is no more the position of the end-effector but a vector function $e^\star$ of visual data, which corresponds to image features at the expected position of the camera. In this control diagram, which is represented in Figure 7, the current value of the task function e is deduced from an extraction algorithm in the camera image plane. According to this diagram, as the control is directly defined as a closed-loop function of visual data, the robustness with respect to measure uncertainty and modeling errors is naturally increased. Different visual-servoing techniques have been proposed which consider either bidimensional features in the camera image plane (image-based visual servoing) [16], or an evaluation of the target configuration with respect to the camera (position-based visual servoing) [35]. Depending on applications, the camera may be deported or attached to the end-effector of the robot. In both cases, it is possible to express the differential link between the vector of visual data e and the vector of angular coordinates Q by a relation of the form:

$$\dot{e} = \varphi(Q)\dot{Q} \tag{3}$$

This equation can be viewed as a generalization of equation (1). In this relation, $\varphi(Q)$ is defined by the product of three matrices as follows:

$$\varphi(Q) = CLJ(Q) \tag{4}$$

where C is a combination matrix, which allows to consider more than n visual signals, L is the interaction matrix which depends on the nature of visual features, and $J(Q)$ is the robot Jacobian introduced by equation (1). As a result $\varphi(Q)$ is clearly dependent on the successive joints $q_1, \ldots, q_n$ of the kinematic chain.

3.3 Discussion

Considering this last development, it is now interesting to go back to the problem of modeling visually guided arm movements in primates. To illustrate our purpose, let us look at the drawing in Figure 8, which represents a human being executing a reaching hand movement towards a visual target. According to the development presented in §2.1, the 3D position of the target is initially encoded with respect to the eye-centered reference frame F_E. If we compare this task with the one of driving the end-effector of a robotic arm towards a visual target, there are basically two ways of modeling the problem.

- In the first one, we may suppose that a localization process allows to reconstruct the position of the target with respect to a body-centered frame F_S, attached to the shoulder. Such a body-centered representation can be obtained by combining the visual position of the target in eye-centered coordinates with the proprioceptive measurement of eye position in the orbit (i.e. with respect to the head) and of head position on the trunk (i.e related to the body). On this basis, it is possible to implement a control scheme for the arm, similar to the one represented in Figure 6. In this case, the reference input X is defined by the difference vector between the position of the hand and the position of the target, expressed in the reference frame F_S. This first control approach seems to correspond rather well with a part of models in Neuroscience which suggest that motor cortical neurons encode hand movements in a body-centered reference frame [15], [25], [26]. However, from an automatic control point of view, the transformation of visual data into a "blind" body-centered reference frame, which is at the basis of this control scheme, does not appear very robust. Indeed, the need to use proprioceptive signals to express the target position with respect to the basis of the arm introduces inaccuracy in the control loop. It is a well-known result in robotics that sensor-based control schemes, for which the error to regulate is directly expressed in terms of sensory data, are more robust with respect to modeling and measurement errors than methods involving a step of state reconstruction. Furthermore, though early ideas about the common reference frame for target and hand representation in the CNS focused on body-centered reference frame, more recent research however suggest that both data are encoded in eye-centered frame [13], [18] [41]. All these elements suggest a second control scheme.

- The second way of modeling the reaching task, is to consider that the reference input is still the difference vector between the position of the hand and the target, but now, expressed in eye-centered coordinates. This idea, which is today shared by numerous authors is illustrated by the control diagram proposed by Shadmehr and Wise [46], which is reported in figure 4 (left part). To compute the location of the hand with respect to the eye-centered frame F_E, the CNS could use visual and proprioceptive information about the arm, as illustrated by the network 1 in Figure 4. If the image of the hand falls on the retina, its location can be directly determined with respect to this frame. However, if the hand is not in the field of view, the proprioceptive information of the arm but also the neck and eye are necessary to determine its position with respect to F_E. Note that proprioceptive measurements where also necessary, in the first control scheme, to compute the position of the hand with respect to the shoulder frame F_S.

On this basis, the reaching movement can be viewed as a visual servoing task with deported camera and represented by the diagram of figure 7. By analogy, the task function e can be defined as the hand-target difference vector expressed in frame F_E. In that case, the vision-based control of the hand is related to the variation of the complete kinematic chain from the eye to the hand. This kinematic chain is represented in Figure 8 as a sequence of dotted-lines ellipsoids linking the eye frame F_E to the hand frame F_H. According to the formalism introduced in subsection 3.2, a vision-based controller for this system would be expressed as a function of all the joints of this chain. Indeed, the matrix $\varphi(Q)$ that appear in equation (3), is dependent on the whole set of joint coordinates Q, at each time.

By analogy with this model, if we agree with the idea that the difference vector is encoded in eye-centered coordinate, we could expect that the motor neurons involved in the control of the visually guided hand movement be sensitive to the variation of any joints of the kinematic chain that links the eye to the hand. In particular their activity must be dependent on the direction of gaze and the orientation of the head with respect to the trunk. However, although the existence of neurones encoding the joint parameters of the arm has been showed in the primary motor cortex M1 and in the premotor areas PMv and PMd [44], [33],[49], [45], the influence of eye position or head position on the activity of motor neurons has only been marginally reported [8], [31], [17]. More surprisingly, considering the fact that the difference vector is encoded in eye-centered coordinates, some authors claim that the activity of motor neurons is invariant with respect to variation of eye position. This reasoning, which still reflects the idea that premotor and motor cortical neurons encode the reaching motion in a shoulder-centered reference frame, is illustrated by the control diagram reported in Figure 4. Indeed, in this figure, the computation of the displacement map represented by the network 4 - that correspond to equation (2) - only involves the joint parameters $\hat{\theta}$ of the arm. In other terms, according to this figure, the knowledge of the estimation of the arm configuration $\hat{\theta}$ and the expression of the difference vector in eye-centered coordinates are sufficient for the CNS to compute the variation of joint parameters of the arm, $\Delta\hat{\theta}$, that allow to move the hand towards the target. Our model points out the insufficiency of this representation. According to our reasoning, this computation would actually be possible if the head and the eye remained fixed with respect to the trunk and the head respectively. Otherwise, as the difference vector x_{dv} is expressed in eye centered coordinates, any variations of the eye or head position would modify the expression of the Jacobian matrix $J(Q)$ and therefore the control of all the joint parameters of the chain.

We believe that this analogy with vision-based control schemes provides a different light to explain the existence of neurons, in premotor and motor areas, whose activity is modulated by gaze direction, as reported by several authors. One of these results is presented in the next section. It concerns experimental results in monkeys showing the activity modulation of movement related neurons, in the dorsal premotor cortex, depending on eye position.

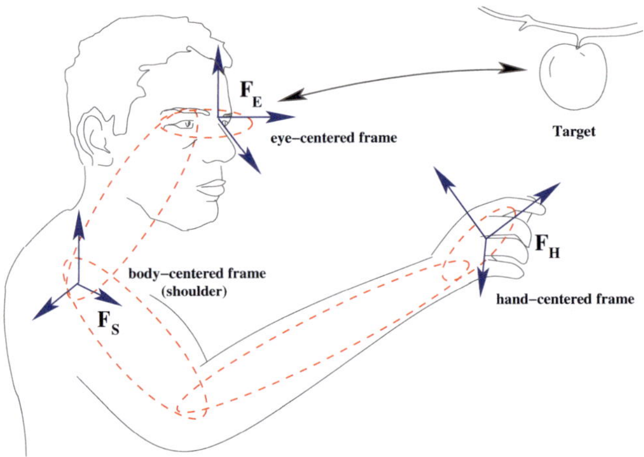

Fig. 8. Complete eye-hand kinematic chain involved in a visually guided reaching movement and different reference frames

4   Gaze-Related Signals in the Primate Premotor Cortex

In the last 15 years, much attention has been devoted to the role of sensori-motor areas in coordinate transformations. The aim of most of these studies was to show that, in agreement with a sequential transformation from eye- to body-centered coordinates, premotor and motor cortical neurons encode hand movement in a shoulder-centered reference frame [25], [26], [15]. Noteworthily, eye movements were not monitored in these studies. Graziano and colleagues specifically searched for eye independent responses in the premotor cortex [28]. They concluded that premotor neurons "are in arm-centered, not retinocentric, coordinates". However, in the legend of figure 3 (p.1056), they noted that "the magnitude of the response varied with eye position". Similarly, Fogassi and colleagues, whose aim was to find somato-centered neurons in the premotor cortex, reported eye-dependent modulation of the single cells responses (see [22], figure 8 p. 149). At the same period, Boussaoud and colleagues [8] [9], [10] showed that the vast majority of both PMv and PMd neurons combine at least two directional parameters: arm movement direction and gaze direction in space. In their subsequent study [32], they trained monkeys to make center-out arm movements towards eight peripherals targets while fixating a point on the video monitor (Figure 9). As they systematically varied the position of the fixation point, they observed the gaze-related modulation of single neurons activity during the preparation and the execution of identical arm movements. Two examples of PMd neurons are presented in Figure 10. The graph in A shows that the preparation-related activity is maximum when gaze angle is 45° and when the upcoming movement direction is 291°. If gaze is shifted towards 225°, the directional preference of the cell is slightly shifted (337°) and less significant. The graph in B shows another example of PMd cell whose preferred direction is drastically changing with gaze direction. For this neuron, the preferred direction jumps from 194° to 316° with opposite gaze angles. At the population level, all these studies agree that more than 90% of the PMd neurons exhibit a preferred direction for arm movement. They also agree that movement-related (including selection, preparation and execution) discharge of more than 70% of the cells is modulated by gaze signals. In continuity with this work, Baker and colleagues [6] showed that gaze-related modulation of cortical motor activity also occurs in humans. Altogether these results show that eye-position is encoded by premotor dorsal neurons.

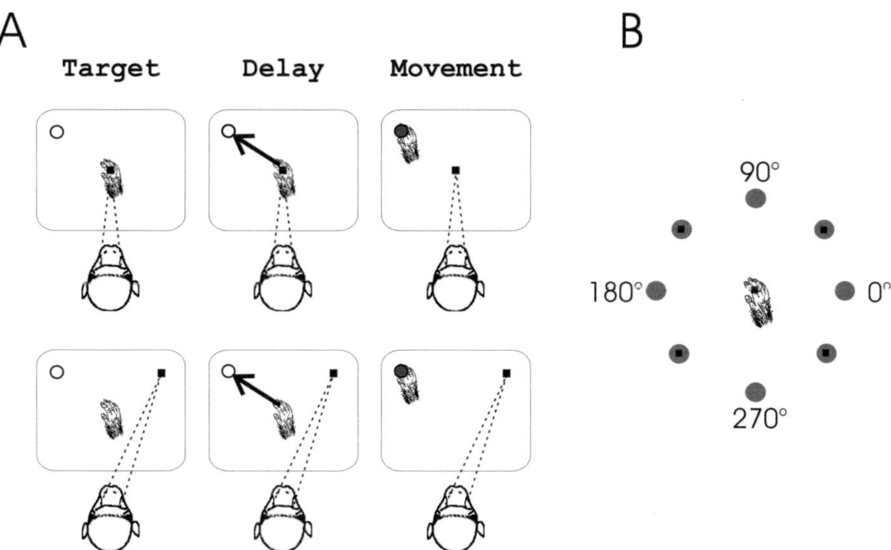

Fig. 9. Schematic representation of the behavioral task. The monkey is facing a video monitor with a touch sensitive screen. While the monkey is looking at the fixation point, with its hand resting at the center of the screen, a target appears. The monkey has to wait for the target dimming to make a pointing movement. The two lines illustrate two different trials in which gaze direction differs. The three columns correspond to the target presentation, the delay preceding arm movement (the monkey is preparing the movement but it is waiting for a go signal) and arm movement execution (the monkey makes a pointing movement towards the target without moving the eyes). In these two different trials, the monkey prepares the same arm movement but with two different gaze directions. B. Positions of the eight targets and the five fixation points that were used in the protocol. For each trial, the selection of one target and one fixation spot was pseudo-randomized. This design allowed Jouffrais et al. [32] to record PMd cells involved in arm movement planning and execution with systematic movement directions and gaze positions.

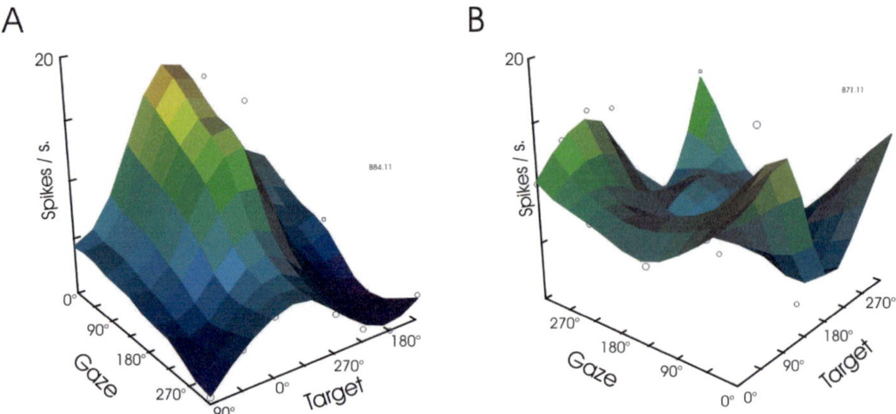

Fig. 10. Effect of gaze angle on the preparation-related activity of two PMd cells. Mean discharge rate of the cells during the delay period is represented as a function of target and gaze angles. Eight target positions and five fixation points were recorded. A. Example of a PMd cell whose discharge is maximal for a target located at 291°. This marked preference for 291° target is dramatically reduced and partially shifted (337°) when gaze is oriented towards 270°. B. Example of a PMd cell whose preferred arm movement direction is inverted (from 316° to 194°) when gaze angle shifts from 45° to 225°. Preferred directions were computed for each gaze direction with two methods giving highly similar results: 1- maximum value of a sinusoid fitted on the discharge rates as a function of target position; 2- center of mass of the 8 discharge rates. Statistical significance was assessed with a bootstrapping test. See Jouffrais et al. [32] for details.

5   Conclusion

This chapter illustrates the interest of exchanges between neuroscientists and roboticians in the search for modeling the neuronal activity related to the perception of space and the control of motion. Using the visual-servoing formalism used in robotics, we brought interesting elements for modeling visually guided hand movements in primates. Considering that the Jacobian of a vision-based task devoted to the control of a robotic arm depends on each articular joint, we proposed a model that accounts for the existence of neurons selective to gaze direction in the premotor area PMd. The same reasoning suggests that cortical motor neurons involved in arm movement control encode not only eye position signals but also head position signals (head with respect to the trunk). To our knowledge, such experimental results have not been reported so far. An interesting consequence of this result is that the eye-centered frame appears to be a common referential for observation and control. Indeed, this frame is used for coding the difference vector between the hand and the target and constitutes the basis of the kinematic chain involved in the control of the arm. Does the eye-centered frame also constitute the origin of the kinematic chains that make up the body and hindlimbs? This still open question is of great interest for the control of humanoid robots. Indeed, to control the complex structure of this kind of robot, it is necessary to stabilize particular reference frames which are not easy to determine a priori.

References

1. Andersen R A, Mountcastle V B, (1983) The influence of the angle of gaze upon the excitability of the light-sensitive neurons of the posterior parietal cortex, J. Neurosci. 3:532-548.
2. Andersen R, Essick G K, Siegel R M (1985) Encoding of spatial location by posterior parietal neurons, Science 230:456–458.
3. Andersen R A, Gnadt J W (1989) Posterior parietal cortex, Rev. Oculomot. Res. 3:315-335.
4. Andersen R A (1997). "Multimodal integration for the representation of space in the posterior parietal cortex." Philos Trans R Soc Lond B Biol Sci 352(1360): 1421-8.
5. Asada H, Slotine J J, (1986) Robot Analysis and control, New York, Wiley.
6. Baker J T, Donoghue J P, Sanes J N, (1999) Gaze direction modulates finger movement activation patterns in human cerebral cortex, J. Neurosci., 19: 10044-10052.
7. Bizzi E, Mussa-Ivaldi F, Gitzer (1991) Computations underlying the execution of movemenyt: a biological perspective. Science 253: 287-291.
8. Boussaoud D, Barth T M, Wise S P (1993) Effects of gaze on apparent visual responses of frontal cortex neurons, Exp. Brain Res., 93: 423-434.
9. Boussaoud D (1995) Primate premotor cortex: Modulation of preparatory neuronal activity by gaze angle, J. Neurophysiol., 73:886-890.
10. Boussaoud D, Jouffrais C, Bremmer F (1998) Eye position effects on the neuronal activity of dorsal premotor cortex in the macaque monkey, J. Neurophysiol., 80 1132-1150.
11. Bremmer F (2000) Eye position effects in macaque area V4. Neuroreport 11(6): 1277-83.
12. Brotchie P R, Andersen R A, Snyder L H, Goodman S J (1995) Head position signals used by parietal neurons to encode locations of visual stimuli, Nature, 375:232-235.
13. Bueno C A, Jarvis M R, Batista A P, Andersen R A, (2002) Direct visuomotor transformations for reaching. Nature 416, 632-636.
14. Caminiti R, Johnson P B, Urbano A (1990) Making arm movements within different parts of space: dynamic aspects in the primate motor cortex. Journal of Neuroscience 10, 2039-2058
15. Caminiti R, Johnson P B, Galli C, Ferraina S, Burnod Y (1991) Making arm movements within different parts of space: the premotor and motor cortical representation of a coordinate system for reaching to visual targets, J. Neurosci., 11:1182-1197.
16. Chaumette F (1990) La relation vision-commande: theorie et application des tches robotiques, PhD Thesis, University of Rennes France, IRISA.
17. Cisek P, Kalaska J F (2002) Modest gaze-related discharge modulation in monkey dorsal premotor cortex during a reaching task performed with free fixation. J. Neurophysiol. 88: 1064-1072.

18. Cohen Y E, Andersen R A (2002) A common reference frame for movement plans in the posterior parietal cortex. Nat Rev Neurosci. 3(7):553-562.

19. Colby C L, Duhamel J R, Goldberg M E (1993), Ventral intraparietal area of the macaque: Anatomic location and visual response properties. J. Neurophysiol. 69:902-914.

20. Craig J J (1989) Introduction to robotics: Mechanics and control, 2nd Edition. Reading MA: Addison Wesley.

21. Espiau B, Chaumette F, Rives P (1992) A new approach to visual servoing in robotics, IEEE Trans. On Robotics and Automation, 8(6):313-326.

22. Fogassi L, Gallese V, Fadiga L, Luppino G, Matelli M, Rizzolatti G (1996) Coding of peripersonal space in inferior premotor cortex (area F4), J. Neurophysiol., 76:141-157.

23. Galetti C, Battaglini P P (1989) Gaze-dependent visual neurons in area V3A of monkey prestiate cortex. J. Neurosci. 9:1112-1125.

24. Galetti C, Battaglini P P, Fattori P (1995) Eye position influence on the parieto-occipital area PO (V6) of the macaque monkey. Eur. J. Neurosci. 7:2486-2501.

25. Georgopoulos A P, Kalaska J F, Caminiti R, Massey J T (1982) On the relations between the direction of two-dimensional arm movements and cell discharge in primate motor cortex, J. Neurosci., 2:1527-1537.

26. Georgopoulos A P, Schwartz A B, Kettner R E (1986). Neuronal population coding of movement direction. Science 233, 1416-1419.

27. Gnadt J W, Mays L E (1995) Neurons in monkey parietal area LIP are tuned for eye-movement parameters in three-dimensional space. J. Neurophysiol. 73:280-297.

28. Graziano M S, Yap G S, Gross C G (1994) Coding of visual space by premotor neurons., Science, 266:1054-1057.

29. Hollerbach J M, Flash T (1982) Dynamic interactions between limb segments during planar arm movements, Biol. Cybernet. 44:67-77.

30. Johnson P B, Ferraina S, Bianchi L, Caminiti R (1996) Cortical networks for visual reaching: physiological and anatomical organization of frontal and parietal lobe arm regions. Cerebral Cortex 6:102-119

31. Jouffrais C, Boussaoud D (1999) Neuronal activity related to eye-hand coordination in the primate premotor cortex. Exp Brain Res 128: 205-209.

32. Jouffrais C, Rouiller E M, Boussaoud D (2000) The dorsal premotor cortex : gaze signals and directional coding. Abstract of the Society for Neuroscience 26(1), 180. New Orleans, USA.

33. Kakei S, Hoffman D S, Strick P L (1999) Muscle and movement representation in the primary motor cortex. Science 285: 2136-2139

34. Karnath H O, M. T. Perenin (2005) Cortical control of visually guided reaching: Evidence from patients with optic ataxia, Cereb. Cortex, 15: 1561-1569.

35. Martinet P, GalliceJ, Khadraoui D (1996) Vision based control law usin 3D visual features, World Automatic Congres, WAC'96, Robotics and Manufacturing Systems, Montpellier France, Vol. 3:497-502.

36. Moran D W, Schwartz A B (1999) Motor cortical representation of speed and direction during reaching, J. Neurophysiol. 82:2676-2692.

37. Mussa-Ivaldi F A, Bizzi E (2000) Motor learning through the combination of primitives," Phil. Tran. R. Soc. Lond. B 355:1755-1769.

38. Nakamura Y (1991) Advanced Robotics: Redundancy and Optimization, Reading MA, Addison Wesley.

39. Newsome W T, Wurtz R H, Komatsu H (1988) Relation of cortical areas MT and MST to pursuit eye movements. II. Differentiation of retinal from extraretinal inputs. J. Neurophysiol. 60:604-644.

40. Nowicka A, Ringo J L (2000) Eye position-sensitive units in hippocampal formation and in inferotemporal cortex of the macaque monkey." Europ. J. Of Neurosc. 12: 751-759.

41. Pouget A, Deneve S, Duhamel J R (2002) A computational perspective on the neural basis of multisensory spatial representations, Nat. Rev. Neurosci., 3: 741-747.

42. Pouget A, Ducom J C, Torri J, Bavelier D, (2002) Multisensory spatial representations in eye-centered coordinates for reaching. Cognition 83:1-11.

43. Samson C, Le Borgne M, Espiau B (1991) Robot Control, The task function approach, Clarendon Press, Oxford.

44. Scott S H, Kalaska J F (1997) Reaching movements with similar hand paths but different arm orientations, I, Activity of individual cells in motor cortex. J Neurophysiol 77, 826-852.

45. Shen L, Alexander G E (1997) Preferential representation of instructed target location versus limb trajectory in dorsal premotor area. J. Neurophysiol. 77: 1195-1212.

46. Shadmehr R, Wise S P (2005) The Computational Neurobiology of Reaching and Pointing : A Foundation for Motor Learning, The MIT Press.

47. Spong M W, Vidyasagar M (1989) Robot dynamics and control, New York, Wiley.

48. Stricanne B, Andersen R A, Mazzoni P (1996) Eye-centered, head-centered, and intermediate coding of remembered sound locations in area LIP. J Neurophysiol. 76(3):2071-2076.

49. Schwartz A B, Moran D W, Reina G A (2004) Differential representation of perception and action in frontal motor cortex. Science 303, 380-383.

50. Trotter Y, Celebrini S, Stricanne B, Thorpe S, Imbert M (1992) Modulation of neural stereoscopic processing in primate area V1 by the viewing distance, Science 257:1279-1281.

51. Trotter Y, Celebrini S, Stricanne B, Thorpe S, Imbert M (1996) Neural processing of stereopsis as a function of viewing distance in primate visual cortical area V1. J. Neurophysiol. 76:2872-2885.

52. Trotter Y, Celebrini S (1999) Gaze direction controls response gain in primary visual-cortex neurons, Nature 398:239-242.

53. Wang Y, Celebrini S, Trotter Y, Barone P (2005) Multisensory integration in the behaving monkey: behavioral analysis and electrophysiological evidence in the primary visual cortex. Soc. Neurosc. Washington.

54. Wise S P, Boussaoud D, Johnson P B, Caminiti R (1997). Premotor and parietal cortex: corticocortical connectivity and combinatorial computations. Annual Review of Neuroscience 20, 25-42.

55. Zipser D, Andersen R A (1988) A back-propagation programmed network that simulates response properties of a subset of posterior parietal neurons, Nature, 331:679-684.

Part II

Models for System Analysis

Monotone and Near-Monotone Systems

Eduardo D. Sontag

Rutgers University, New Brunswick, NJ, USA

Summary. This paper provides an expository introduction to monotone and near-monotone biochemical network structures. Monotone systems respond in a predictable fashion to perturbations, and have very robust dynamical characteristics. This makes them reliable components of more complex networks, and suggests that natural biological systems may have evolved to be, if not monotone, at least close to monotone. In addition, interconnections of monotone systems may be fruitfully analyzed using tools from control theory.

1 Introduction

In cells, biochemical networks consisting of proteins, RNA, DNA, metabolites, and other species, are responsible for control and signaling in development, regulation, and metabolism, by processing environmental signals, sequencing internal events such as gene expression, and producing appropriate cellular responses. The field of systems molecular biology is largely concerned with the study of such networks. Often, as in control theory, biochemical networks are viewed as interconnections of simpler subsystems. This paper discusses recent work which makes use of *topology* (graph structure) as well as *sign* information regarding subsystems and their interconnection structure in order to infer properties of the complete system.

It is broadly appreciated that behavior is critically dependent on network topology as well as on the signs (activating or inhibiting) of the underlying feedforward and feedback interconnections [1, 2, 3, 4, 5, 6, 7, 8, 9, 10, 11]. For example, Figures 1(a-c) show the three possible types of feedback loops that involve two interacting chemicals,. A mutual activation configuration is shown in Figure 1(a): a positive change in A results in a positive change in B, and vice-versa. Configurations like these are associated to signal amplification and production of switch-like biochemical responses. A mutual inhibition configuration is shown in Figure 1(b): a positive change in A results in repression of B, and repression of B in turn enhances A. Such configurations allow systems to exhibit multiple discrete, alternative stable steady-states, thus providing a mechanism for memory. Both (a) and (b) are examples of positive-feedback systems [12, 13, 14, 15, 16, 17, 18, 19]. On the other hand, activation-inhibition

I. Queinnec et al. (Eds.): Bio. & Ctrl. Theory: Current Challenges, LNCIS 357, pp. 79–122, 2007.

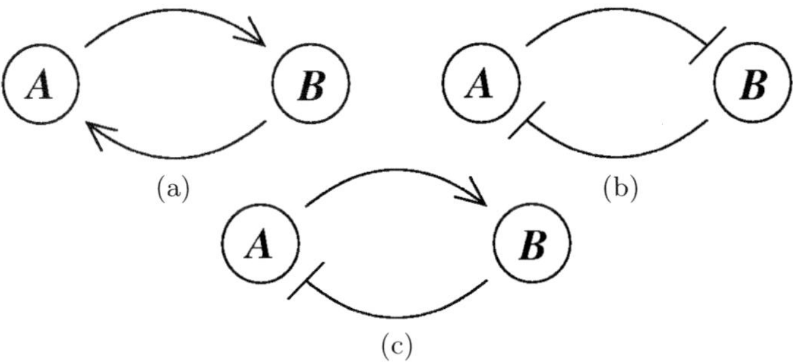

Fig. 1. (a) Mutual activation. (b) Mutual inhibition. (c) Activation-inhibition.

configurations like in Figure 1(c) are necessary for the generation of periodic behaviors such as circadian rhythms or cell cycle oscillations, by themselves or in combination with multi-stable positive-feedback subsystems, as well as for adaptation, disturbance rejection, and tight regulation (homeostasis) of physiological variables [20, 21, 22, 23, 11, 7, 24, 25, 9, 26, 27, 28]. Compared to positive-feedback systems, negative-feedback systems are not "consistent," in a sense to be made precise below but roughly meaning that different paths between any two nodes should reinforce, rather than contradict, each other. For (c), a positive change in A will be resisted by the system through the feedback loop. Consistency, or lack thereof, also plays a role in the behavior of graphs without feedback; for example [10, 29, 30] deal with the different signal processing capabilities of consistent ("coherent") compared to inconsistent feedforward motifs.

A key role in the work to be discussed here will be played by consistent systems and subsystems. We will discuss the following points:

- Interesting and nontrivial conclusions can be drawn from (signed) network structure alone. This structure is associated to purely stoichiometric information about the system and ignores fluxes. Consistency, or close to consistency, is an important property in this regard.
- Interpreted as dynamical systems, consistent networks define *monotone systems*, which have highly predictable and ordered behavior.
- It is often useful to analyze larger systems by viewing them as interconnections of a small number of monotone subsystems. This allows one to obtain precise bifurcation diagrams without appeal to explicit knowledge of fluxes or of kinetic constants and other parameters, using merely "input/output characteristics" (steady-state responses or DC gains). The procedure may be viewed as a "model reduction" approach in which monotone subsystems are viewed as essentially one-dimensional objects.
- The possibility of performing a decomposition into a small number of monotone components is closely tied to the question of how "near" a system is to being monotone.
- We argue that systems that are "near monotone" are biologically more desirable than systems that are far from being monotone.

- There are indications that biological networks may be much closer to being monotone than random networks that have the same numbers of vertices and of positive and negative edges.

The Need for Robust Structures and Robust Analysis Tools

In contrast to many areas of applied mathematics and engineering, the study of dynamics in cell biology must take into account the often huge degree of uncertainty inherent in models of cellular biochemical networks, which arises from environmental fluctuations or from variability among cells of the same type. From a mathematical analysis perspective, this uncertainty translates into the difficulty of measuring the relevant model parameters such as kinetic constants or cooperativity indices, and hence the impossibility of obtaining a precise model.

This means that it is important to develop tools that are "robust" in the sense of being able to lead to useful conclusions from information regarding the *qualitative* features of the network, and if possible not the precise values of parameters or even the forms of reactions. This goal is hard to attain, since dynamical behavior may be subject to phase transitions (bifurcations) which critically depend on parameter values. Nevertheless, and perhaps surprisingly, there have been many successes in finding rich classes of chemical network structures for which such robust analysis is indeed possible. One approach is that of graph-theoretic ideas associated to *complex balancing and deficiency theory*, pioneered by Clarke [31], Horn and Jackson [32, 33], and Feinberg [34, 35, 36]. Another approach, pioneered by Hirsch and Smith [37, 38], relies upon the theory of *monotone systems*, and has a similar goal of drawing conclusions about dynamical behavior based only upon structure. This direction has been enriched substantially by the introduction of *monotone systems with inputs and outputs*: as standard in control theory [24], one extends the notion of monotone system so as to incorporate input and output channels [39]. Once inputs and outputs

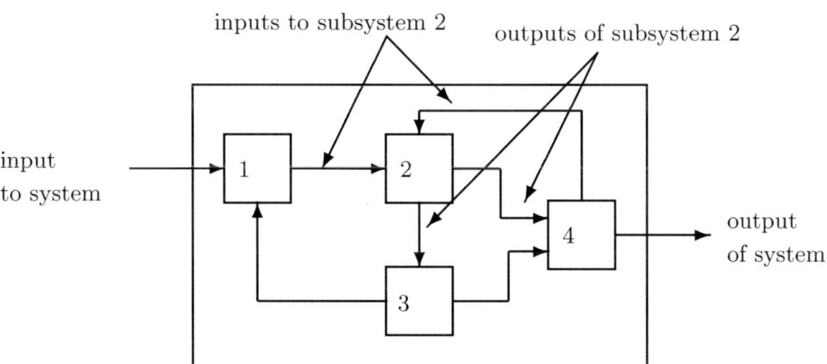

Fig. 2. A system composed of four subsystems

are introduced, one can study interconnections of systems (Figure 2), and ask what special properties hold if the subsystems are monotone [39, 40, 19].

2 Consistent Graphs, Monotone Systems, and Near-Monotonicity

We now introduce the basic notions of monotonicity and consistency. The present section deals exclusively with graph-theoretic information, which is derived from stoichiometric constraints. Complementary to this analysis, bifurcation phenomena can be sometimes analyzed using a combination of these graphical techniques together with information on steady-state gains; that subject is discussed in Section 3. The discussion is informal; Section 3 has more rigorous mathematical statements, presented in the more general context of systems with external inputs and outputs.

The systems considered here are described by the evolution of states, which are time-dependent vectors $x(t) = (x_1(t), \ldots, x_n(t))$ whose components x_i represent concentrations of chemical species such as proteins, mRNA, or metabolites. In autonomous differential equation ("continuous-time") models, one specifies the rate of change of each variable, at any given time, as a function of the concentrations of all the variables at that time:

$$\frac{dx_1}{dt}(t) = f_1(x_1(t), x_2(t), \ldots, x_n(t))$$

$$\frac{dx_2}{dt}(t) = f_2(x_1(t), x_2(t), \ldots, x_n(t))$$

$$\vdots$$

$$\frac{dx_n}{dt}(t) = f_n(x_1(t), x_2(t), \ldots, x_n(t)),$$

or just $dx/dt = f(x)$, where f is the vector function with components f_i. We assume that the coordinates x_i of the state of the system can be arbitrary non-negative numbers. (Constraints among variables can be imposed as well, but several aspects of the theory are more subtle in that case.) Often, one starts from a differential equation system written in the following form:

$$\frac{dx}{dt}(t) = \Gamma R(x),$$

where $R(x)$ is a q-dimensional *vector of reactions* and Γ is an $n \times q$ matrix, called the *stoichiometry matrix*, and either one studies this system directly, or one studies a smaller set of differential equations $dx/dt = f(x)$ obtained by eliminating variables through the use of conserved stoichiometric quantities.

We will mostly discuss differential equation models, but will also make remarks concerning difference equation ("discrete time") models. The dynamics of these are described by rules that specify the state at some future time $t = t_{k+1}$ as a function of the state of the system at the present time t_k. Thus, the ith coordinate evolves according to an update rule:

$$x_i(t_{k+1}) = f_i(x_1(t_k), x_2(t_k), \ldots, x_n(t_k))$$

instead of being described by a differential equation. Usually, $t_k = k\Delta$, where Δ is a uniform inter-sample time. One may associate a difference equation to any given differential equation, through the rule that the vector $x(t_{k+1})$ should equal the solution of the differential equation when starting at state $x(t_k)$. However, not every difference equation arises from a differential equation in this manner. Difference equations may be more natural when studying processes in which measurements are made at discrete times, or they might provide a macroscopic model of an underlying stochastic process taking place at a faster time scale.

One may also study more complicated descriptions of dynamics that those given by ordinary differential and difference equations; many of the results that we discuss here have close analogs that apply to more general classes of (deterministic) dynamical systems, including reaction-diffusion partial differential equations, which are used for space-dependent problems with slow diffusion and no mixing, delay-differential systems, which help model delays due to transport and other cellular phenomena in which concentrations of one species only affect others after a time interval, and integro-differential equations [37, 38, 41, 42]. In a different direction, one may consider systems with external inputs and outputs [39].

The Graph Associated to a System

There are at least two types of graphs that can be naturally associated to a given biochemical network. One type, sometimes called the *species-reaction graph*, is a bipartite graph with nodes for reactions (fluxes) and species, which leads to useful analysis techniques based on Petri net theory and graph theory [43, 44, 45, 46, 47, 48, 49, 50]. We will not discuss species-reaction graphs here. A second type of graph, which we will discuss, is the *species graph G*. It has n nodes (or "vertices"), which we denote by $v_1, \ldots, v_n$, one node for each species. No edge is drawn from node v_j to node v_i if the partial derivative $\frac{\partial f_i}{\partial x_j}(x)$ vanishes identically, meaning that there is no direct effect of the jth species upon the ith species. If this derivative is not identically zero, then there are three possibilities: (1) it is ≥ 0 for all x, (2) it is ≤ 0 for all x, or (3) it changes sign depending on the particular entries of the concentration vector x. In the first case (activation), we draw an edge labeled $+$, $+1$, or just an arrow $\rightarrow$. In the second case (repression or inhibition), we draw an edge labeled $-$, -1, or use the symbol $\dashv$. In the third case, when the sign is ambiguous, we draw both an activating and an inhibiting edge from node v_j to node v_i.

For continuous-time systems, no self-edges (edges from a node v_i to itself) are included in the graph G, whatever the sign of the diagonal entry $\partial f_i / \partial x_i$ of the Jacobian. For discrete-time systems, on the other hand, self-edges are included (we later discuss the reason for these different definitions for differential and difference equations).

When working with graphs, it is more convenient (though not strictly necessary) to consider only graphs G that have no multiple edges from one node to another (third case above). One may always assume that G has this property,

Fig. 3. Replacing direct inconsistent effects by adding a node

by means of the following trick: whenever there are two edges, we replace one of them by an indirect link involving a new node; see Fig. 3. Introducing such additional nodes if required, we will suppose from now on that no multiple edges exist.

Although adding new edges as explained above is a purely formal construction with graphs, it may be explained biologically as follows. Often, ambiguous signs in Jacobians reflect heterogeneous mechanisms. For example, take the case where protein A enhances the transcription rate of gene B if present at high concentrations, but represses B if its concentration is lower than some threshold. Further study of the chemical mechanism might well reveal the existence of, for example, a homodimer that is responsible for this ambiguous effect. Mathematically, the rate of transcription of B might be given algebraically by the formula $k_2a^2 - k_1a$, where a denotes the concentration of A. Introducing a new species C to represent the homodimer, we may rewrite this rate as $k_2c - k_1a$, where c is the concentration of C, plus an new equation like $dc/dt = k_3a^2 - k_4c$ representing the formation of the dimer and its degradation. This is exactly the situation in Fig. 3.

Spin Assignments and Consistency

A *spin assignment* Σ for the graph G is an assignment, to each node v_i, of a number σ_i equal to "$+1$" or "-1" (a "spin," to borrow from statistical mechanics terminology). In graphical depictions, we draw up-arrows or down-arrows to indicate spins. If there is an edge from node v_j to node v_i, with label $J_{ij} \in \{\pm 1\}$, we say that this edge is *consistent with the spin assignment* Σ provided that:

$$J_{ij}\sigma_i\sigma_j = 1$$

which is the same as saying that $J_{ij} = \sigma_i\sigma_j$, or that $\sigma_i = J_{ij}\sigma_j$. An equivalent formalism is that in which edges are labeled by "0" or "1," instead of 1 and -1 respectively, and edge labels J_{ij} belong to the set $\{0, 1\}$, in which case consistency is the property that $J_{ij} \oplus \sigma_i \oplus \sigma_j = 0$ (sum modulo two).

We will say that Σ *is a consistent spin assignment for the graph G* (or simply that G is consistent) if every edge of G is consistent with Σ. In other words, for any pair of vertices v_i and v_j, if there is a positive edge from node v_j to node v_i, then v_j and v_i must have the same spin, and if there is a negative edge connecting v_j to v_i, then v_j and v_i must have opposite spins. (If there is no edge from v_j to v_i, this requirement imposes no restriction on their spins.)

In order to decide whether a graph admits any consistent spin assignment, it is not necessary to actually test all the possible 2^n spin assignments. It is very easy to prove that there is a consistent assignment if and only if *every undirected loop in the graph G has a net positive sign*, that is to say, an even number, possibly zero, of negative arrows. Equivalently, any two (undirected) paths between two nodes must have the same net sign. By undirected loops or paths, we mean that one is allowed to transverse an edge either forward or backward. A proof of this condition is as follows. If a consistent assignment exists, then, for any undirected loop $v_{i_1}, \ldots, v_{i_k} = v_{i_1}$ starting from and ending at the node v_{i_1}, inductively one has that:

$$\sigma_{i_1} \;=\; Q_{i_1, i_{k-1}} \, Q_{i_{k-1}, i_{k-2}} \, \cdots, \, Q_{i_2, i_1} \, \sigma_{i_1}$$

where $Q_{ij} = J_{ij}$ if we are transversing the edge from v_j to v_i, or $Q_{ij} = J_{ji}$ if we are transversing backward the edge from v_j to v_i. This implies (divide by σ_{i_1}) that the product of the edge signs is positive. Conversely, if any two paths between nodes have the same parity, and the graph is connected, pick node v_1 and label it "+" and then assign to every other node v_i the parity of a path connecting v_1 and v_i. (If the graph is not connected, do this construction on each component separately.)

This positive-loop property, in turn, can be checked with a fast dynamic programming-like algorithm. For connected graphs, there can be at most two consistent assignments, each of which is the reverse (flip every spin) of the other.

Monotone Systems

A dynamical system is said to be *monotone* if there exists at least one consistent spin assignment for its associated graph G. Monotone systems [51, 52, 37] were introduced by Hirsch, and constitute a class of dynamical systems for which a rich theory exists. (To be precise, we have only defined the subclass of systems that are *monotone with respect to some orthant order*. The notion of monotonicity can be defined with respect to more general orders).

Consistent Response to Perturbations

Monotonicity reflects the fact that a system responds consistently to perturbations on its components. Let us now discuss this property in informal terms. We view the nodes of the graph shown in Figure 4(a) as corresponding to variables in the system, which quantify the concentrations of chemical species such as activated receptors, proteins, transcription factors, and so forth. Suppose that a perturbation, for example due to the external activation of a receptor represented by node 1, instantaneously increases the value of the concentration of this species. We represent this increase by an up-arrow inserted into that node, as in Figure 4(b). The effect on the other nodes is then completely predictable from the graph. The species associated to node 2 will decrease, because of the inhibiting character of the connection from 1 to 2, and the species associated to node 3 will increase (activating effect). Where monotonicity plays a role is in insuring that the concentration of the species corresponding to node 4 will also increase. It increases both because it is activated by 3, which has increased, and because it is inhibited by 2, so that less of 2 implies a smaller inhibition effect. Algebraically, the following expression involving partial derivatives:

$$\frac{\partial f_4}{\partial x_3}\frac{\partial f_3}{\partial x_1} + \frac{\partial f_4}{\partial x_2}\frac{\partial f_2}{\partial x_1}$$

(where f_i gives the rate of change of the ith species, in the differential equation model) is guaranteed to be positive, since it is a sum of positive terms: $(+)(+)+ (-)(-)$. Intuitively, the expression measures the sensitivity of the rate of change dx_4/dt of the concentration of 4 with respect to perturbations in 1, with the two terms giving the contributions for each of the two alternative paths from node 1 to node 4. *This unambiguous global effect holds true regardless of the actual values of parameters such as kinetic constants, and even the algebraic forms of reactions, and depends only on the signs of the entries of the Jacobian of f.* Observe that the arrows shown in Figure 4(b) provide a consistent spin-assignment for the graph, so the system is monotone.

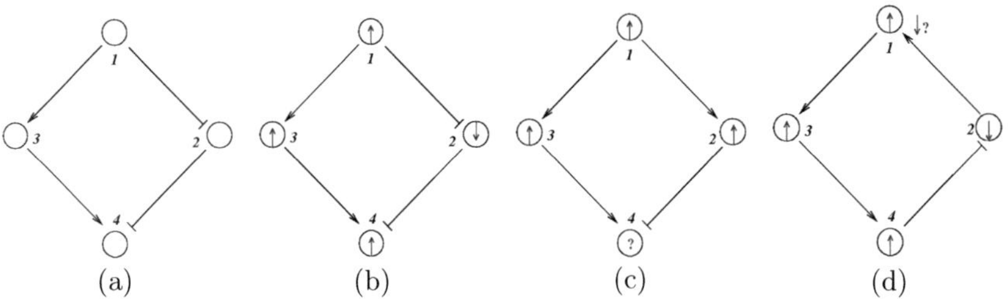

Fig. 4. (a) and (b) graph and consistent assignment, (c) and (d) no possible consistent assignments

In contrast, consider next the graph in Figure 4(c), where the edge from 1 to 2 is now positive. There are two paths from node 1 to node 4, one of which (through 3) is positive and the other of which (through 2) is negative. Equivalently, the undirected loop $1, 3, 4, 2, 1$ ("undirected" because the last two edges are transversed backward) has a net negative parity. Therefore, the loop test for consistency fails, so that there is no possible consistent spin-assignment for this graph, and therefore the corresponding dynamical system is not monotone. Reflecting this fact, the net effect of an increase in node 1 is ambiguous. *It is impossible to conclude from the graphical information alone* whether node 4 will be repressed (because of the path through 2) or activated (because of the path through 3). There is no way to resolve this ambiguity unless equations and precise parameter values are assigned to the arrows.

To take a concrete example, suppose that the equations for the system are as follows:

$$\frac{dx_1}{dt} = 0 \qquad \frac{dx_2}{dt} = x_1 \qquad \frac{dx_3}{dt} = x_1 \qquad \frac{dx_4}{dt} = x_4(k_3 x_3 - k_2 x_2),$$

where the reaction constants k_2 and k_3 are two positive numbers. The initial conditions are taken to be $x_1(0) = x_4(0) = 1$, and $x_2(0) = x_3(0) = 0$, and we ask how the solution $x_4(t)$ will change when the initial value $x_1(0)$ is perturbed. With $x_1(0) = 1$, the solution is $x_4(t) = \exp \alpha t^2/2$, where $\alpha = k_3 - k_2$. On the other hand, if $x_1(0)$ is perturbed to a larger value, let us say $x_1(0) = 2$, then $x_4(t) = \exp \alpha t^2$. This new value of $x_4(t)$ is larger than the original unperturbed value $\exp \alpha t^2/2$ provided that $\alpha > 0$, but it is smaller than it if, instead, $\alpha < 0$. In other words, the sign of the sensitivity of x_4 to a perturbation on x_1 cannot be predicted from knowledge of the graph alone, but it depends on whether $k_2 < k_3$ or $k_2 > k_3$. Compare this with the monotone case, as in Figure 4(a). A concrete example is obtained if we modify the x_2 equation to $dx_2/dt = 1/(1 + x_1)$. Now the solutions are $x_4(t) = \exp \beta_1 t^2$ and $x_4(t) = \exp \beta_2 t^2$ respectively, with $\beta_1 = k_3/2 - k_2/4$ and $\beta_2 = k_3 - k_2/6$, so we are guaranteed that x_4 is larger in the perturbed case, a conclusion that holds true no matter what are the numerical values of the (positive) constants k_i.

The uncertainty associated to a graph like the one in Figure 4(c) might be undesirable in natural systems. Cells of the same type differ in concentrations of ATP, enzymes, and other chemicals, and this affects the values of model parameters, so two cells of the same type may well react differently to the same "stimulus" (increase in concentration of chemical 1). While such epigenetic diversity is sometimes desirable, it makes behavior less predictable and robust. From an evolutionary viewpoint, a "change in wiring" such as replacing the negative edge from 1 to 2 by a positive one (or, instead, perhaps introducing an additional inconsistent edge) could lead to unpredictable effects, and so the fitness of such a mutation may be harder to evaluate. In a monotone system, in contrast, a stimulus applied to a component is propagated in an unambiguous manner throughout the circuit, promoting a predictably consistent increase or consistent decrease in the concentrations of all other components.

Similarly, consistency also applies to feedback loops. For example, consider the graph shown in Figure 4(d). The negative feedback given by the inconsistent path $1, 3, 4, 2, 1$ means that the instantaneous effect of an up-perturbation of node 1 feeds back into a negative effect on node 1, while a down-perturbation feeds back as a positive effect. In other words, the feedback loop acts against the perturbation.

Of course, negative feedback as well as inconsistent feedforward circuits are important components of biomolecular networks, playing a major role in homeostasis and in signal detection. The point being made here is that inconsistent networks may require a more delicate tuning in order to perform their functions.

In rigorous mathematical terms, this predictability property can be formulated as *Kamke's Theorem*. Suppose that $\Sigma = \{\sigma_i, i = 1, \ldots, n\}$ is a consistent spin assignment for the system graph G. Let $x(t)$ be any solution of $dx/dt = f(x)$. We wish to study how the solution $z(t)$ arising from a perturbed initial condition $z(0) = x(0) + \Delta$ compares to the solution $x(t)$. Specifically, suppose that a positive perturbation is performed at time $t = 0$ on the ith coordinate, for some index $i \in \{1, \ldots, n\}$: $z_i(0) > x_i(0)$ and $z_j(0) = x_j(0)$ for all $j \neq i$. For concreteness, let us assume that the perturbed node i has been labeled by $\sigma_i = +1$. Then, Kamke's Theorem says the following: for each node that has the same parity (i.e., each index j such that $\sigma_j = +1$), and for every future time t, $z_j(t) \geq x_j(t)$. Similarly, for each node with opposite parity ($\sigma_j = -1$), and for

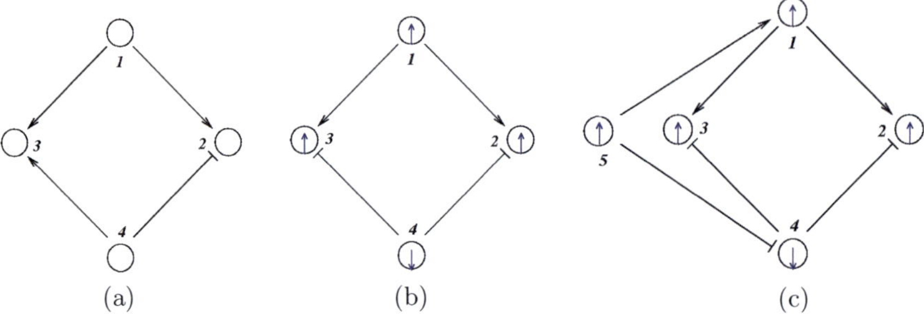

Fig. 5. (a) inconsistent, (b) consistent, (c) adding node to consistent network

every time t, $z_j(t) \le x_j(t)$. (Moreover, one or more of these inequalities must be strict.) This is the precise sense in which an up-perturbation of the species represented by node v_i unambiguously propagates into up- or down-behavior of all the other species. See [37] for a proof, and see [39] for generalizations to systems with external input and output channels.

For difference equations (discrete time systems), once that that self-loops have been included in the graph G and the definition of consistency, Kamke's theorem also holds; in this case the proof is easy, by induction on time steps.

Consistent graphs can be embedded into larger consistent ones, but inconsistent ones cannot. For example, consider the graph shown in Figure 5(a). This graph admits no consistent spin assignment since the undirected loop $1, 3, 4, 2, 1$ has a net negative parity. Thus, there cannot be any consistent graph that includes this graph as a subgraph. Compare this with the graph shown in Figure 5(b). Consistency of this graph may well represent consistency of a larger graph which involves a yet-undiscovered species, such as node 5 in Figure 5(c). Alternatively, and from an "incremental design" viewpoint, this graph being consistent makes it possible to consistently add node 5 in the future.

Removing the Smallest Number of Edges so as to Achieve Consistency

Let us call the *consistency deficit (CD)* of a graph G the smallest possible number of edges that should be removed from G in order that there remains a consistent graph, and, correspondingly, a monotone system.

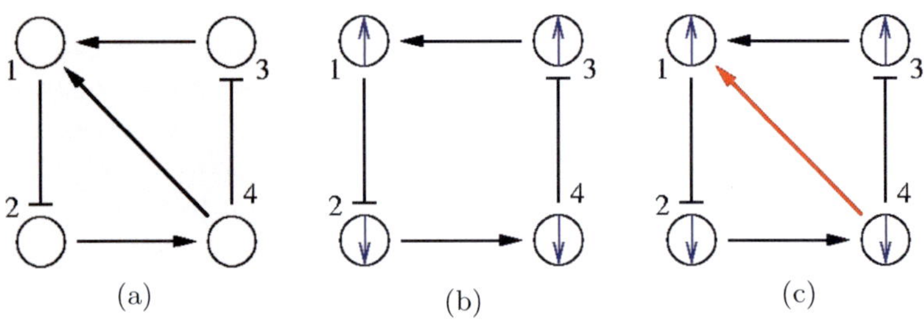

Fig. 6. (a) inconsistent graph, (b) consistent subgraph, (c) one inconsistent edge

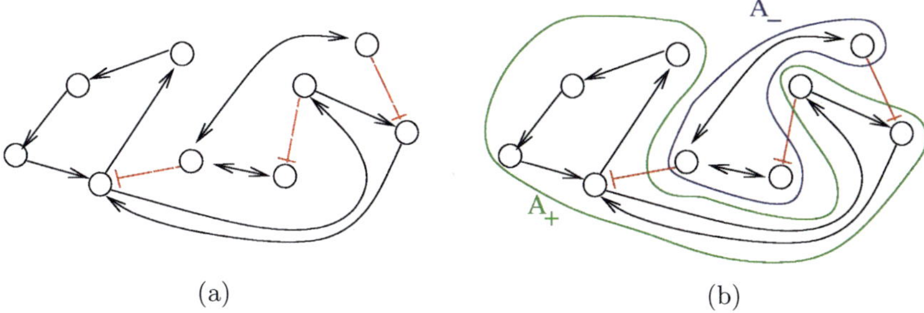

(a)   (b)

Fig. 7. (a) Consistent graph; (b) partition into A_1 and A_{-1}

As an example, take the graph shown in Figure 6(a). For this graph, it suf-fices to remove just one edge, the diagonal positive one, so the CD is 1. (In this example, the solution is unique, in that no other single other edge would suf-fice, but for other graphs there are typically several alternative ways to achieve consistency with a minimal number of deletions.)

After deleting the diagonal, a consistent spin assignment Σ is: $\sigma_1 = \sigma_3 = 1$ and $\sigma_2 = \sigma_4 = -1$, see Figure 6(a). (Another assigment is the one with all spins reversed: $\sigma_1 = \sigma_2 = -1$ and $\sigma_3 = \sigma_4 = 1$.) If we now bring back the deleted edge, we see that in the original graph only the one edge from node 1 to node 4 is inconsistent for the spin assignment Σ (Figure 6(c)).

This example illustrates a general fact: minimizing the number of edges that must be removed so that there remains a consistent graph is equivalent to finding a spin assignment Σ for which the number of inconsistent edges (those for which $J_{ij}\sigma_i\sigma_j = -1$) is minimized.

Yet another rephrasing is as follows. For any spin assignment Σ, let A_1 be the subset of nodes labeled $+1$, and let A_{-1} be the subset of nodes labeled -1. The set of all nodes is partitioned into A_1 and A_{-1}. (In Figure 6(c), we have $A_1 = \{1,3\}$ and $A_{-1} = \{2,4\}$.) Conversely, any partition of the set of nodes into two subsets can be thought of as a spin assignment. With this interpretation, a consistent spin assigment is the same as a partition of the node set into two subsets A_1 and A_{-1} in such a manner that all edges between elements of A_1 be positive, all edges between elements of A_{-1} be positive, and all edges between a node in A_1 and a node in A_{-1} be negative, see Fig. 7. More generally, computing the CD amounts to finding a partition so that $n_1 + n_{-1} + p$ is minimized, where n_1 is the number of negative edges between nodes in A_1, n_{-1} is the number of negative edges between nodes in A_{-1}, and p is the number of positive edges between nodes in A_1 and A_{-1}.

A very special case is when the graph has all of its edges labeled negative, that is, $J_{ij} = -1$ for all i,j. Stated in the language of partitions, the CD problem amounts to searching for a partition such that $n_1 + n_{-1}$ is minimized (as there are no positive edges, $p = 0$). Moreover, since there are no positive edges, $n_1 + n_{-1}$ is actually the total number of edges between any two nodes in A_1 or in A_{-1}. Thus, $N - (n_1 + n_{-1})$ is the number of remaining edges, that is, the number of edges between nodes in A_1 and A_{-1}. Therefore, minimizing $n_1 + n_{-1}$ is the same as maximizing $N - (n_1 + n_{-1})$. This is precisely the standard "MAX-CUT" problem in computer science.

As a matter of fact, not only is MAX-CUT a particular case, but, conversely, it is possible to reduce the CD problem to MAX-CUT by means of the following trick. For each edge labeled $+1$, say from v_i to v_j, delete the edge but insert a new node w_{ij}, and two negative edges, one from v_i to w_{ij} and one from w_{ij} to v_j:

$$v_i \rightarrow v_j \qquad \rightsquigarrow \qquad v_i \dashv w_{ij} \dashv v_j \,.$$

The enlarged graph has only negative edges, and it is easy to see that the minimal number of edges that have to be removed in order to achieve consistency is the same as the number of edges that would have had to be removed in the original graph. Unfortunately, the MAX-CUT problem is NP-hard. However, the paper [53] gave an approximation polynomial-time algorithm for the CD problem, guaranteed to solve the problem to within 87.9% of the optimum value, as an adaptation of the semi-definite programming relaxation approach to MAX-CUT based on Goemans and Williamson's work [54]. (Is not enough to simply apply the MAX-CUT algorithm to the enlarged graph obtained by the above trick, because the approximation bound is degraded by the additional edges, so the construction takes some care.)

Relation to Ising Spin-Glass Models

Another interpretation of CD uses the language of statistical mechanics. An *Ising spin-glass model* is defined by a graph G together with an "interaction energy" J_{ij} associated to each edge (in our conventions, J_{ij} is associated to the edge from v_j to v_i). In binary models, $J_{ij} \in \{1, -1\}$, as we have here. A spin-assignment Σ is also called a (magnetic) "spin configuration." A "non-frustrated" spin-glass model is one for which there is a spin configuration for which every edge is consistent [55, 56, 57]. This is the same as a consistent assignment for the graph G in our terminology. Moreover, a spin configuration that maximizes the number of consistent edges is one for which the "free energy" (with no exterior magnetic field):

$$H(\Sigma) \;=\; -\sum_{ij} J_{ij}\sigma_i\sigma_j$$

is minimized. This is because, if Σ results in $C(\Sigma)$ consistent edges, then $H(\Sigma) = -C(\Sigma) + I(\Sigma) = T - 2C(\Sigma)$, where $I(\Sigma)$ is the number of non-consistent edges for the assignment Σ and $T = C + I$ is the total number of edges; thus, minimizing $H(\Sigma)$ is the same as maximizing $C(\Sigma)$. A minimizing Σ is called a "ground state". (A special case is that in which $J_{ij} = -1$ for all edges, the "anti-ferromagnetic case". This is the same as the MAX-CUT problem.)

Near-Monotone Systems May Be "Practically" Monotone

Obviously, there is no reason for large biochemical networks to be consistent, and they are not. However, when the number of inconsistencies in a biological interaction graph is small, it may well be the case that the network is in fact consistent in a practical sense. For example, a gene regulatory network represents all *potential* effects among genes. These effects are often mediated by proteins which themselves need to be activated in order to perform their function, and this activation will, in turn, be contingent on the "environmental" context: extracellular ligands, additional genes being expressed which may depend on cell type or developmental stage, and so forth. Thus, depending on the context, different subgraphs of the original graph describe the system, and these graphs may be individually consistent even if the entire graph, the union of all these subgraphs, is not. As an illustration, take the system in Figure 4(c). Suppose that under environmental conditions A, the edge from 1 to 2 is not present, and under non-overlapping conditions B, the edge from 1 to 3 is not be present. Then, under either conditions, A or B, the graph is consistent, even though, formally speaking, the entire network is not consistent.

The closer to consistent, the more likely that this phenomenon may occur.

Some Evidence Suggesting Near-Monotonicity of Natural Networks

Since consistency in biological networks may be desirable, one might conjecture that natural biological networks tend to be consistent. As a way to test this hypothesis, the CD algorithm from [53] was run on the yeast *Saccharomyces cerevisiae* gene regulatory network from [10], downloaded from [58]. (The authors of [10] used the YPD database [59]. Nodes represent genes, and edges are directed from transcription factors, or protein complexes of transcription factors, into the genes regulated by them.) This network has 690 nodes and 1082 edges, of which 221 are negative and 861 are positive (we labeled the one "neutral" edge as positive; the conclusions do not change substantially if we label it negative instead, or if we delete this one edge). The algorithm in [53] provides a CD of 43. In other words, *deleting a mere 4% of edges makes the network consistent.* Also remarkable is the following fact. The original graph has 11 components: a large one of size 664, one of size 5, three of size 3, and six of size 2. All of these components *remain connected* after edge deletion. The deleted edges are all from the largest component, and they are incident on a total of 65 nodes in this component.

To better appreciate if a small CD might happen by chance, the algorithm was also run on random graphs having 690 nodes and 1082 edges (chosen uniformly), of which 221 edges (chosen uniformly) are negative. It was found that, for such random graphs, about 12.6% (136.6 ± 5) of edges have to be removed in order to achieve consistency. (To analyze the scaling of this estimate, we generated random graphs with N nodes and $1.57N$ edges of which $0.32N$ are negative. We found that for $N > 10$, approximately $N/5$ nodes must be removed, thus confirming the result for $N = 690$.) Thus, the CD of the biological network is roughly 15 standard deviations away from the mean for random graphs. Both topology (i.e., the underlying graph) and actual signs of edges contribute to this near-consistency of the yeast network. To justify this assertion, the following numerical experiment was performed. We randomly changed the signs of 50 positive and 50 negative edges, thus obtaining a network that has the same number of positive and negative edges, and the same underlying graph, as the original yeast network, but with 100 edges, picked randomly, having different signs. Now, one needs 8.2% (88.3 ± 7.1) deletions, an amount in-between that obtained for the original yeast network and the one obtained for random graphs. Changing more signs, 100 positives and 100 negatives, leads to a less consistent network, with 115.4 ± 4.0 required deletions, or about 10.7% of the original edges, although still not as many as for a random network.

Decomposing Systems into Monotone Components

Another motivation for the study of near-monotone systems is from decomposition-based methods for the analysis of systems that are interconnections of monotone subsystems. One may "pull out" inconsistent connections among monotone components, in such a manner that the original system can then be viewed as a "negative feedback" loop around an otherwise consistent system (Figure 8). In this interpretation, the number of interconnections among monotone components corresponds to the number of variables being fed-back.

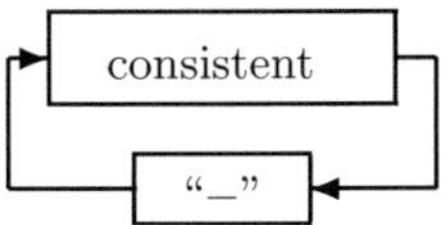

Fig. 8. Pulling-out inconsistent connections

For example, let us take the graph shown in Figure 6(a). The procedure of dropping the diagonal edge and seeing it instead as an external feedback loop can be modeled as follows. The original differential equation $dx_1/dt = f_1(x_1, x_2, x_3, x_4)$ is replaced by the equation $dx_1/dt = f_1(x_1, x_2, x_3, u)$, where the symbol u, which represents an external input signal, is inserted instead of the state variable x_4. The consistent system in Figure 8 includes the remaining four edges, and the "negative" feedback (negative in the sense that it is inconsistent with the rest of the system) is the connection from x_4, seen as an "output" variable, back into the input channel represented by u. The closed-loop system obtained by using this feedback is the original system, now viewed as a negative feedback around the consistent system in Figure 6(b).

Generally speaking, the decomposition techniques in [39, 18, 19, 41, 60, 61, 62, 63, 64, 65, 42, 66] are most useful if the feedback loop involves few variables. This is equivalent to asking that the graph G associated to the system be close to consistent, in the sense of the CD of G being small. This view of systems as monotone systems —which have strong stability properties, as discussed next,— with negative-feedback regulatory loops around them is very appealing from a control engineering perspective as well.

Dynamical Behavior of Monotone Systems

Continuous-time monotone systems have convergent behavior. For example, they cannot admit any possible stable oscillations [67, 68, 38]. When there is only one steady state, a theorem of Dancer [69] shows –under mild assumptions regarding possible constraints on the values of the variables, and boundedness of solutions–that every solution converges to this unique steady state (monostability). When, instead, there are multiple steady-states, the Hirsch Generic Convergence Theorem [51, 52, 37, 38] is the fundamental result. A *strongly* monotone system is one for which the an initial perturbation $z_i(0) > x_i(0)$ on the concentration of any species propagates as a strict up or down perturbation: $z_j(t) > x_j(t)$ for all $t > 0$ and all indices j for which $\sigma_j = \sigma_i$, and $z_j(t) < x_j(t)$ for all $t > 0$ and all j for which $\sigma_j = -\sigma_i$. Observe that this requirement is stronger (hence the terminology) than merely weak inequalities: $z_j(t) \geq x_j(t)$ or $z_j(t) \leq x_j(t)$ respectively as in Kamke's Theorem. A sufficient condition for strong monotonicity is that the Jacobian matrices must be irreducible for all x, which basically amounts to asking that the graph G must be strongly connected and that every non-identically zero Jacobian entry be everywhere nonzero. Even though they may have arbitrarily large dimensionality, monotone systems behave in many ways like one-dimensional systems: Hirsch's Theorem asserts that generic bounded solutions of strongly monotone differential equation systems must converge to the set of steady states. ("Generic" means "every solution except for a measure-zero set of initial conditions.") In particular, no "chaotic" or other "strange" dynamics can occur. For discrete-time strongly monotone systems, generically also stable oscillations are allowed besides convergence to equilibria, but no more complicated behavior.

The ordered behavior of monotone systems is robust with respect to spatial localization effects as well as signaling delays (such as those arising from transport, transcription, or translation). Moreover, their stability character does not change much if some inconsistent connections are inserted, but only provided that these added connections are weak ("small gain theorem") or that they operate at a comparatively fast time scale [70].

The intuition behind the convergence results is easy to explain in the very special case of just two interacting species, described by a two-dimensional system with variables $x(t)$ and $y(t)$:

$$\frac{dx}{dt} = f(x, y)$$
$$\frac{dy}{dt} = g(x, y)\,.$$

A system like this is monotone if either (a) the species are mutually activating (or, as is said in mathematical biology, "cooperative"), (b) they are mutually inhibiting ("competitive"), or (c) either x does not affect y, y does not affect x, or neither affects the other. Let us discuss the mutually activating case (a). (Case (b) is similar, and case (c) is easy, since the systems are partially or totally decoupled.) We want to argue that there cannot be any periodic orbit. Suppose that there would be a periodic orbit in which the motion is counterclockwise, as shown in Figure 9(a). We then pick two points in this orbit with identical x coordinates, as indicated by (x, y) and (x, y') in Figure 9(a). These points correspond to the concentrations at two times t_0, t_1, with $x(t_0) = x(t_1)$ and $y(t_0) < y(t_1)$. Since $y(t_1)$ is larger than $y(t_0)$, x is at the same concentration,

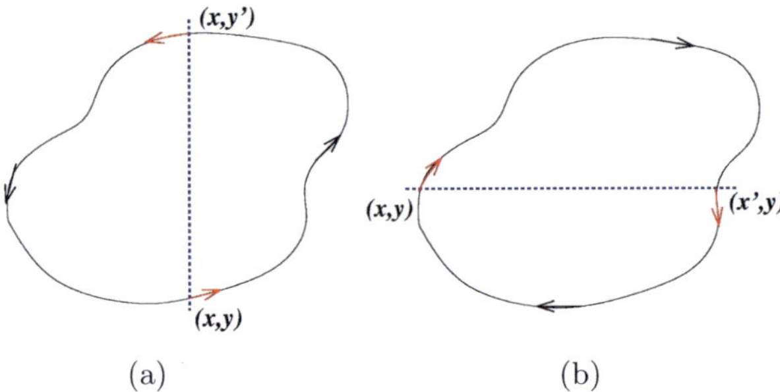

Fig. 9. Impossible (a) counterclockwise and (b) clockwise periodic orbits in planar cooperative system, each drawn in the (x, y)-plane

and the species are mutually activating, it follows that the rate of change in the concentration x should be comparatively larger at time t_1 than at time t_0, that is, $f(x, y') \geq f(x, y)$. However, this contradicts the fact that $x(t)$ is increasing at time t_0 $(f(x, y) \geq 0)$ but is decreasing at time t_1 $(f(x, y') \leq 0)$. The contradiction means that there cannot be any counterclockwise-oriented curve. To show that there cannot be any clockwise-oriented curve, one may proceed by an entirely analogous argument, using two points (x, y) and (x', y) as in Figure 9(b). Of course, the power of monotone systems theory arises in the analysis of systems of higher dimension, since two-dimensional systems are easy to study by elementary phase plane methods.

For general, non-monotone systems, on the other hand, no dynamical behavior, including chaos, can be mathematically ruled out. This is in spite of the fact that some features of non-monotone systems are commonly regarded as having a stabilizing effect. For example, negative feedback loops confer robustness with regard to certain types of structural as well as external perturbations [71,72,73,74]. However, and perhaps paradoxically, the behavior of non-monotone systems may also be very fragile: for instance, they can be destabilized by delays in negative feedback paths. Nonetheless, we conjecture that systems that are close to monotone must be better-behaved, generically, than those that are far from monotone. Preliminary evidence (unpublished) for this has been obtained from the analysis of random Boolean networks, at least for discrete analogs of the continuous system, but the work is not yet definitive.

Directed Cycles

Intuition suggests that somewhat less than monotonicity should suffice for guaranteeing that no chaotic behavior may arise, or even that no stable limit cycles exist. Indeed, monotonicity amounts to requiring that no undirected negative-parity cycles be present in the graph, but a weaker condition, that no *directed* negative parity cycles exist, should be sufficient to insure these properties. For a strongly connected graph, the property that no directed negative cycles exist is equivalent to the property that no undirected negative cycles exist, because the same proof as given earlier, but applied to directed paths, insures that a consistent spin assignment exists (and hence there cannot be any undirected negative cycles). However, for non-strongly connected graphs, the properties are not the same. On the other hand, every graph can be decomposed as a cascade of graphs that are strongly connected. This means (aside from some technicalities having to do with Jacobian entries being not identically zero but vanishing on large sets) that systems having no directed negative cycles can be written as a cascade of strongly monotone systems. Therefore, it is natural to conjecture that such cascades have nice dynamical properties. Indeed, under appropriate technical conditions for the systems in the cascade, one may recursively prove convergence to equilibria in each component, appealing to the theory of asymptotically autonomous systems [75] and thus one may conclude global convergence of the entire system [76, 77]. For example, a cascade of the form $dx/dt = f(x)$, $dy/dt = g(x, y)$ where the x system is monotone and where the system $dy/dt = g(x_0, y)$ is monotone for each fixed x_0, cannot have any attractive periodic orbits (except equilibria). This is because the projection of such an orbit on the first system must be a point x_0, and hence the orbit must have the form $(x_0, y(t))$. Therefore, it is an attractive periodic orbit of $dy/dt = g(x_0, y)$, and by monotonicity of this latter system we conclude that $y(t) \equiv$ a constant as well. The argument generalizes to any cascade, by an inductive argument. Also, chaotic attractors cannot exist [78].

The condition of having no directed negative cycles is the weakest one that can be given strictly on the basis of the graph G, because for any graph G with a negative feedback loop there is a system with graph G which admits stable periodic orbits. (First find a limit cycle for the loop, and then use a small perturbation to define a system with nonzero entries as needed, which will still have a limit cycle.)

Positive Feedback and Stability

The strong global convergence properties of monotone systems mentioned above would seemingly contradict the fact that positive feedback, which tends to increase the direction of perturbations, is allowed in monotone systems, but negative feedback, which tends to stabilize systems, is not. One explanation for this apparent paradox is that the main theorems in monotone systems theory only guarantee that *bounded* solutions converge, but they do not make any assertions about unbounded solutions. For example, the system $dx/dt = -x + x^2$ has the property that every solution starting at an $x(0) > 1$ is unbounded, diverging to $+\infty$, a fact which does not contradict its monotonicity (every one-dimensional system is monotone). This is not as important a restriction as it may seem, because for biochemical systems it is often the case that all trajectories must remain bounded, due to conservation of mass and other constraints. A second explanation is that negative *self-loops* are not ruled out in monotone systems, and such loops, which represent degradation or decay diagonal terms, help insure stability.

Intuition on Why Negative Self-loops Do Not Affect Monotonicity

In the definition of the graph associated to a continuous-time system, self-loops (diagonal terms in the Jacobian of the vector field f) were ignored. The theory (Kamke's condition) does not require self-loop information in order to guarantee monotonicity. Intuitively, the reason for this is that a larger initial value for a variable x_i implies a larger value for this variable, at least for short enough time periods, independently of the sign of the partial derivative df_i/dx_i (continuity of flow with respect to initial conditions). For example, consider a degradation equation $dp/dt = -p$, for the concentration $p(t)$ of a protein P. At any time t, we have that $p(t) = e^{-t}p(0)$, where $p(0)$ is the initial concentration. The concentration $p(t)$ is positively proportional to $p(0)$, even though the partial derivative $\frac{\partial(-p)}{\partial p} = -1$ is negative. Note that, in contrast, for a difference equation, a jump may occur: for instance the iteration $p(t+1) = -p(t)$ has the property that the order of two elements is reversed at each time step. Thus, for difference equations, diagonal terms matter.

Multiple Time Scale Analysis May Make Systems Monotone

A system may fail to be monotone due to the effect of negative regulatory loops that operate at a faster time scale than monotone subsystems. In such a case, sometimes an approximate but monotone model may be obtained, by collapsing negative loops into self-loops. Mathematically: a non-monotone system might be a singular perturbation of a monotone system. A trivial linear example that illustrates this point is $dx/dt = -x-y$, $\varepsilon dy/dt = -y+x$, with $\varepsilon > 0$. This system is not monotone (with respect to any orthant cone). On the other hand, for $\varepsilon \ll 1$, the fast variable y tracks x, so the slow dynamics is well-approximated by $dx/dt = -2x$ (monotone, since every scalar system is).

More generally, one may consider $dx/dt = f(x,y)$, $\varepsilon dy/dt = g(x,y)$ such that the fast system $dy/dt = g(x,y)$ has a unique globally asymptotically stable steady state $y = h(x)$ for each x (and possibly a mild input to state stability requirement, as with the special case $\varepsilon dy/dt = -y + h(x)$), and the slow system $dx/dt = f(x,h(x))$ is (strongly) monotone. Then one may expect that the original system inherits global convergence properties, at least for all $\varepsilon > 0$ small enough. The paper [79] employs tools from geometric invariant manifold theory [80, 81], taking advantage of the existence of a manifold M_ε invariant for the dynamics, which attracts all near-enough solutions, and with an asymptotic phase property. The system restricted to the invariant manifold M_ε is a regular perturbation of the fast ($\varepsilon = 0$) system, and hence inherits strong monotonicity properties. So, solutions in the manifold will be generally well-behaved, and asymptotic phase implies that solutions track solutions in M_ε, and hence also converge to equilibria if solutions on M_ε do. However, the technical details are delicate, because strong monotonicity only guarantees generic convergence, and one must show that the generic tracking solutions start from the "good" set of initial conditions, for generic solutions of the large system.

Discrete-Time Systems

As discussed, for autonomous differential equations monotonicity implies that stable periodic behaviors will not be observed, and moreover, under certain technical assumptions, all trajectories must converge to steady states. This is not exactly true for difference equation models, but a variant does hold: for discrete-time monotone systems, trajectories must converge to either steady states *or periodic orbits*. In general, even the simplest difference equations may exhibit arbitrarily complicated (chaotic) behavior, as shown by the logistic iteration in one dimension $x(t+1) = kx(t)(1 - x(t))$ for appropriate values of the parameter k [82]. However, for monotone difference equations, a close analog of Hirsch's Generic Convergence Theorem is known. Specifically, suppose that the equations are point-dissipative, meaning that all solutions converge to a bounded set [83], and that the system is strongly monotone, in the sense that the Jacobian matrix $(\partial f_i / \partial x_j)$ is irreducible at all states. Then, a result of Tereščák and coworkers [84,85,86,87] shows that there is a positive integer m such that generic solutions (in an appropriate sense of genericity) converge to periodic orbits with period at most m. Results also exist under less than strong monotonicity, just as in the continuous case, for example when steady states are unique [69].

Difference equations allow one to study wider classes of systems. As a simple example, consider the nondimensionalized harmonic oscillator (idealized mass-spring system with no damping), which has equations

$$\frac{dx}{dt} = y$$
$$\frac{dy}{dt} = -x \,.$$

(For this example, we allow variables to be negative; these variables might indicate deviations of concentrations from some reference value.) This system is not monotone, since $v_1 \rightarrow v_2$ is negative and $v_1 \rightarrow v_2$ is positive, so that its graph has a negative loop. On the other hand, suppose that one looks at this system every Δt seconds, where $\Delta t = \pi$. The discrete-time system that results (using a superscript $^+$ to indicate time-stepping) is now:

$$x^+ = -y$$
$$y^+ = -x$$

(this is obtained by solving the differential equation on an interval of length π). This system is monotone (both $v_1 \rightarrow v_2$ and $v_1 \rightarrow v_2$ are negative). Every trajectory of this discrete system is, in fact, of period two: $(x_0, y_0) \rightarrow (-y_0, -x_0) \rightarrow (x_0, y_0) \rightarrow \dots$. This periodic property for the difference equation corresponds to the period-2π behavior of the original differential equation.

Oscillatory Behaviors

Stable periodic behaviors are ruled-out in autonomous monotone continuous-time systems. However, stable periodic orbits may arise through various external mechanisms. Three examples are (1) inhibitory negative feedback from some species into others in a monotone monostable system, (2) the generation of relaxation oscillations from a hysteresis parametric behavior by negative feedback on parameters by species in a monotone system, and (3) entrainment of external periodic signals. These general mechanisms are classical and well-understood for simple, one or two-dimensional, dynamics, and they may be generalized to the case where the underlying system is higher-dimensional but monotone.

Embeddings in Monotone Systems

As observed by Gouzé [88, 89], any n-dimensional system can be viewed as a subsystem of a $2n$-dimensional monotone system. The mathematical trick is to first duplicate every variable (species), introducing a "dual" species, and then to replace every inconsistent edge by an edge connecting the source species and the "dual" of its target (and vice-versa). The construction is illustrated in Fig. 10.

At first, this embedding result may seem paradoxical, since all monotone (or strongly monotone) systems have especially nice dynamical behaviors, such as not having any attractive periodic orbits or chaotic attractors, and of course non-monotone systems may admit such behaviors. However, there is no contradiction. A non-monotone subsystem of a monotone system may well have, say, a chaotic attractor or a stable periodic orbit: it is just that this attractor or orbit will be unstable when seen as a subset of the extended ($2n$-dimensional) state space. Not only there is no contradiction, but a classical construction of Smale [90] shows that indeed any possible dynamics can be embedded in a larger monotone system. More generally, the Hirsch Generic Covergence Theorem guarantees convergence to equilibria from *almost every* initial condition; applied to the above construction, in general the exceptional set of initial conditions would include the "thin" set corresponding to the embedded subsystem. Yet, one may ask what happens for example if the larger $2n$-system has a unique equilibrium. In that case, it is known [69] that every trajectory converges (not merely generic ones), so, in particular, the embedded subsystem must also be "well-behaved". Thus, systems that may be embedded by the above trick into monotone systems with unique equilibria will have global convergence to equilibria. This property amounts to the "small gain theorem" shown in [39], see [42] for a discussion and further results using this embedding idea.

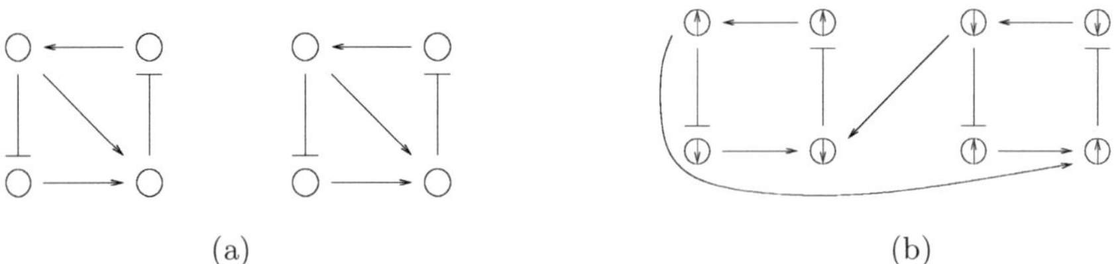

(a)   (b)

Fig. 10. (a) duplicated inconsistent graph, (b) replacing arrows and consistent assignment

Discrete Systems

We remark that one may also study difference equations for which the state components are only allowed to take values out of a finite set. For example, in Boolean models of biological networks, each variable $x_i(t)$ can only attain two values ($0/1$ or "on/off"). These values represent whether the ith gene is being expressed, or the concentration of the ith protein is above certain threshold, at time t. When detailed information on kinetic rates of protein-DNA or protein-protein interactions is lacking, and especially if regulatory relationships are strongly sigmoidal, such models are useful in theoretical analysis, because they serve to focus attention on the basic dynamical characteristics while ignoring specifics of reaction mechanisms $[91, 92, 93, 94, 95]$).

For difference equations over finite sets, such as Boolean systems, it is quite clear that all trajectories must either settle into equilibria or to periodic orbits, whether the system is monotone or not. However, cycles in discrete systems may be arbitrarily long and these might be seen as "chaotic" motions. Monotone systems, while also settling into steady states or periodic orbits, have generally shorter cycles. This is because periodic orbits must be anti-chains, i.e. no two different states can be compared; see $[96, 37]$. For example, consider a discrete-time system in which species concentrations are quantized to the k values $\{0, \ldots, k-1\}$; we interpret monotonicity with respect to the partial order: $(a_1, \ldots, a_n) \leq (b_1, \ldots, b_n)$ if every coordinate $a_i \leq b_i$. For non-monotone systems, orbits can have as many as k^n states. On the other hand, monotone systems cannot have orbits of size more than the width (size of largest antichains) of $P = \{0, \ldots, k-1\}^n$, which can be interpreted as the set of multisubsets of an n-element set, or equivalently as the set of divisors of a number of the form $(p_1 p_2 \ldots p_n)^{k-1}$ where the p_i's are distinct primes. The width of P is the number of possible vectors $(i_1, \ldots, i_n)$ such that $\sum i_j = \lfloor kn/2 \rfloor$ and each $i_j \in \{0, \ldots, k-1\}$. This is a generalization of Sperner's Theorem; see $[97]$. For example, for $n = 2$, periodic orbits in a monotone system evolving on $\{0, \ldots, k-1\}^2$ cannot have length larger than k, while non-monotone systems on $\{0, \ldots, k-1\}^2$ can have a periodic orbit of period k^2. As another example, arbitrary Boolean systems (i.e., the state space is $\{0, 1\}^n$) can have orbits of period up to 2^n, but monotone systems cannot have orbits of size larger than $\binom{n}{\lfloor n/2 \rfloor} \approx 2^n \sqrt{2/(n\pi)}$. These are all classical facts in Boolean circuit design $[96]$. It is worth pointing out that any anti-chain P_0 can be seen as a periodic orbit of a monotone system. This is proved as follows: we enumerate the elements of P_0 as $x_1, \ldots, x_\ell$, and define $f(x_i) = x_{i-1}$ for all i modulo ℓ. Then, f can be extended to all elements of the state space by defining $f(x) = (0, \ldots, 0)$ for every x which has the property that $x < x_i$ for some $x_i \in P_0$ and $f(x) = (k-1, \ldots, k-1)$ for every x which is not $\leq x_i$ for any $x_i \in P_0$. It is easy to see that this is a monotone map $[96, 98]$.

While on the subject of discrete and in particular Boolean systems, we mention a puzzling fact: any Boolean function may be implemented by using just two inverters, with all other gates being monotone. In other words, a circuit computing any Boolean rule whatsoever may be built so that its "consistency deficit" is just two. This is a well-known fact in circuit design [96,99]. Here is one solution, from [100]. One first shows how to implement the Boolean function that takes as inputs three bits A, B, C and outputs the vector of three complements $(\text{not}A, \text{not}B, \text{not}C)$, by using this sequence of operations:

$$2or3ones = (A \wedge B) \vee (A \wedge C) \vee (B \wedge C)$$
$$0or1ones = \text{not}(2or3ones)$$
$$1one = 0or1ones \wedge (A \vee B \vee C)$$
$$1or3ones = 1one \vee (A \wedge B \wedge C)$$
$$0or2ones = \text{not}(1or3ones)$$
$$0ones = 0or2ones \wedge 0or1ones$$
$$2ones = 0or2ones \wedge 2or3ones$$
$$\text{not}A = 0ones \vee (1one \wedge (B \vee C)) \vee (2ones \wedge (B \wedge C))$$
$$\text{not}B = 0ones \vee (1one \wedge (A \vee C)) \vee (2ones \wedge (A \wedge C))$$
$$\text{not}C = 0ones \vee (1one \wedge (A \vee B)) \vee (2ones \wedge (A \wedge B))$$

(the node labeled "2or3ones" computes the Boolean function "the input has exactly 2 or 3 ones" and so forth). Note that only two inverters have been used. If we now want to invert four bits A, B, C, D, we build the above circuit, but we implement the inversion of the three bits $(2or3ones, 1or3ones, D)$ by a subciruit with only two inverters. With a similar recursive construction, one may invert an arbitrary number of bits, using just two inverters.

3 I/O Monotone Systems

We next describe recent work on *monotone input/output systems* (*"MIOS"* from now on). Monotone i/o systems originated in the analysis of mitogen-activated protein kinase cascades and other cell signaling networks, but later proved useful in the study of a broad variety of other biological models. Their surprising breath of applicability notwithstanding, of course MIOS constitute a restricted class of models, especially when seen in the context of large biochemical networks. Indeed, the original motivation for introducing MIOS, in the 2003 paper [39], was to study an existing *non-monotone* model of negative feedback in MAPK cascades. The key breakthrough was the realization that this example, and, as it turned out, many others, can be profitably studied by *decompositions into MIOS*. In other words, a non-monotone system is viewed as an interconnection of monotone subsystems. Based on the architecture of the interconnections between the subsystems ("network structure"), one deduces properties of the original, non-monotone, system. (Later work, starting with [18], showed that even monotone systems can be usefully studied through this decomposition-based approach).

We review the basic notion from [39]. (For concreteness, we make definitions for systems of ordinary differential equations, but similar definitions can be given for abstract dynamical systems, including in particular reaction-diffusion partial differential equations and delay-differential systems, see e.g. [65].) The basic setup is that of an input/output system in the sense of mathematical systems and control theory [24], that is, sets of equations

$$\frac{dx}{dt} = f(x, u), \quad y = h(x) \,, \tag{1}$$

in which states $x(t)$ evolve on some subset $X \subseteq \mathbb{R}^n$, and input and output values $u(t)$ and $y(t)$ belong to subsets $U \subseteq \mathbb{R}^m$ and $Y \subseteq \mathbb{R}^p$ respectively. The coordinates $x_1, \ldots, x_n$ of states typically represent concentrations of chemical species, such as proteins, mRNA, or metabolites. The input variables, which can be seen as controls, forcing functions, or external signals, act as stimuli. Output variables can be thought of as describing responses, such as movement, or as measurements provided by biological reporter devices like GFP that allow a partial read-out of the system state vector $(x_1, \ldots, x_n)$. The maps $f : X \times U \to \mathbb{R}^n$ and $h : X \to Y$ are taken to be continuously differentiable. (Much less can be assumed for many results, so long as local existence and uniqueness of solutions is guaranteed.) An *input* is a signal $u : [0, \infty) \to U$ which is locally essentially compact (meaning that images of restrictions to finite intervals are compact), and we write $\varphi(t, x_0, u)$ for the solution of the initial value problem $dx/dt(t) = f(x(t), u(t))$ with $x(0) = x_0$, or just $x(t)$ if x_0 and u are clear from the context, and $y(t) = h(x(t))$. See [24] for more on i/o systems. For simplicity of exposition, we make the blanket assumption that solutions do not blow-up on finite time, so $x(t)$ (and $y(t)$) are defined for all $t \geq 0$. (In biological problems, almost always conservation laws and/or boundedness of vector fields insure this property. In any event, extensions to local semiflows are possible as well.)

Given three partial orders on X, U, Y (we use the same symbol $\preceq$ for all three orders), a monotone I/O system (MIOS), with respect to these partial orders, is a system (1) such that h is a monotone map (it preserves order) and: for all initial states x_1, x_2 for all inputs u_1, u_2, the following property holds: if $x_1 \preceq x_2$ and $u_1 \preceq u_2$ (meaning that $u_1(t) \preceq u_2(t)$ for all $t \geq 0$), then $\varphi(t, x_1, u) \preceq \varphi(t, x_2, u_2)$ for all $t > 0$. Here we consider partial orders induced by closed proper cones $K \subseteq \mathbb{R}^\ell$, in the sense that $x \preceq y$ iff $y - x \in K$. The cones K are assumed to have a nonempty interior and are pointed, i.e. $K \bigcap -K = \{0\}$. A *strongly* monotone system is one which satisfies the following stronger property: if $x_1 \preceq x_2$ and $u_1 \preceq u_2$, then the strict inequality $\varphi(t, x_1, u) \prec\prec \varphi(t, x_2, u_2)$ holds for all $t > 0$, where $x \prec\prec y$ means that $y - x$ is in the interior of the cone K.

The most interesting particular case is that in which K is an *orthant* cone in $\mathbb{R}^n$, i.e. a set S_ε of the form $\{x \in \mathbb{R}^n \,|\, \varepsilon_i x_i \geq 0\}$, where $\varepsilon_i = \pm 1$ for each i.

When there are no inputs nor outputs, the definition of monotone systems reduces to the classical one of monotone dynamical systems studied by Hirsch, Smith, and others [37]. This is what we discussed earlier, for the case of orthant cones. When there are no inputs, strongly monotone classical systems have especially nice dynamics. Not only is chaotic or other irregular behavior ruled out, but, in fact, almost all bounded trajectories converge to the set of steady states (Hirsch's generic convergence theorem [51,52]).

A useful test for monotonicity with respect to orthant cones, which generalizes Kamke's condition to the i/o case, is as follows. Let us assume that all the partial derivatives $\frac{\partial f_i}{\partial x_j}(x,u)$ for $i \neq j$, $\frac{\partial f_i}{\partial u_j}(x,u)$ for all i,j, and $\frac{\partial h_i}{\partial x_j}(x)$ for all i,j (subscripts indicate components) do not change sign, i.e., they are either always ≥ 0 or always ≤ 0. We also assume that X is convex (much less is needed.) We then associate a directed graph G to the given MIOS, with $n+m+p$ nodes, and edges labeled "+" or "−" (or ± 1), whose labels are determined by the signs of the appropriate partial derivatives (ignoring diagonal elements of $\partial f/\partial x$). One may define in an obvious manner undirected loops in G, and the *parity* of a loop is defined by multiplication of signs along the loop. (See e.g. [18,28] for more details.) Then, it is easy to show that a system is monotone with respect to *some* orthant cones in X, U, Y if and only if there are no negative loops in G. A sufficient condition for strong monotonicity is that, in addition to monotonicity, the partial Jacobians of f with respect to x should be everywhere irreducible. ("Almost-everywhere" often suffices; see [37,38]. See these references also for extensions to non-orthant cones in the case of no inputs and outputs, based on work of Schneider and Vidyasagar, Volkmann, and others [101,102,103,104]).

In inhibitory feedback, a chemical species x_j typically affects the rate of formation of another species x_i through a term like $h(x_j) = V/(K+x_j)$. The decreasing function $h(x_j)$ can be seen as the output of an *anti-monotone* system, i.e. a system which satisfies the conditions for monotonicity, except that the output map *reverses* order: $x_1 \preceq x_2 \Rightarrow h(x_2) \preceq h(x_1)$.

An interconnection of monotone subsystems, that is to say, an entire system made up of monotone components, may or may not be monotone: "positive feedback" (in a sense that can be made precise) preserves monotonicity, while "negative feedback" destroys it. Thus, oscillators such as circadian rhythm generators require negative feedback loops in order for periodic orbits to arise, and hence are not themselves monotone systems, although they can be decomposed into monotone subsystems (cf. [105]). A rich theory is beginning to arise, characterizing the behavior of non-monotone interconnections. For example, [39] shows how to preserve convergence to equilibria; see also the follow-up papers [61,62,65,42,66]. Even for monotone interconnections, the decomposition approach is very useful, as it permits locating and characterizing the stability of steady states based upon input/output behaviors of components, as described in [18]; see also the follow-up papers [19,63,64].

Moreover, a key point brought up in [39,18,41,60] is that new techniques for monotone systems in many situations allow one to characterize the behavior of an entire system, based upon the "qualitative" knowledge represented by general network topology and the inhibitory or activating character of interconnections, combined with only a relatively *small amount of quantitative* data. The latter data may consist of steady-state responses of components (dose-response curves and so forth), and there is no need to know the precise form of dynamics or parameters such as kinetic constants in order to obtain global stability conclusions and study global bifurcation behavior. We now discuss these issues.

Characteristics

The main results in [39, 18] were built around the study of *characteristics*, also called *step-input steady-state responses* or (nonlinear) *DC gains*. To explain this concept, we study the effect of a *constant* input $u(t) \equiv u_0, t \geq 0$ (in biological terms, the constant input may represent the extracellular concentration of a ligand during a particular experiment, for example). For each such constant input, we study the dynamical system $dx/dt = f(x, u_0)$. Let us assume that all the solutions of this system approach steady states, and let us call $K(u_0)$ the set of steady states that arises in this way. To each state x in this set $K(u_0)$, one may associate the corresponding output or measured quantity $h(x_0)$. Let $k(u_0)$ be the set of all output values that arise in this manner. The graph of the set-valued mapping $u_0 \mapsto k(u_0)$ is a subset of the cross product space $\mathbb{R}^m \times \mathbb{R}^p$, which may be though of as a curve when $m = p = 1$, and which describes the possible steady state output values for any given constant input.

Although many results may be given in more generality, we will assume for the remainder of this paper that these mappings are single-valued, not set-valued, in other words that the system is monostable. Thus, a (single-valued) characteristic is said to exist for the system if there is a unique steady state for the dynamical system $dx/dt = f(x, u_0)$, denoted $K(u_0)$, and this property is true for all possible constant levels u_0. We then define the (output) *characteristic* $k : U \to Y$ as the composition $h \circ K$. Under reasonable assumptions on X and boundedness, appealing to results from [106, 69] allows one to conclude that $K(u_0)$ is in fact a globally asymptotically stable ("GAS" from now on) state for $dx/dt = f(x, u_0)$, so that all trajectories (for this "frozen" value of the input u), converge to $K(u_0)$, and the output $y(t)$ converges to $k(u_0)$.

Characteristics (dose response curves, activity plots, steady-state expression of a gene in response to an external ligand, etc.) are frequently available from experimental data, especially in molecular biology and pharmacology (for instance, in the modeling of receptor-ligand interactions [107]). A goal of MIOS analysis is to *combine the numerical information provided by characteristics with the qualitative information given by (signed) network topology in order to predict global bifurcation behavior.* (See [60] for a longer discussion of this "qualitative-quantitative approach" to systems biology.) On the other hand, characteristics are also a very powerful tool for the purely mathematical analysis of existing models, as we show below. Monotone systems with well-defined characteristics constitute a very well-behaved set of building blocks for arbitrary systems. In particular, cascades of such systems inherit the same properties (monotone, monostable response). The original theorems, in the works [39, 18], dealt with systems obtained by interconnecting monotone (or anti-monotone) I/O systems with characteristics in *feedback*. Let us review them next.

Positive Feedback

The first basic theorem refers to a feedback interconnection of two MIOS

$$\frac{x_1}{dt} = f_1(x_1, u_1), \quad y_1 = h_1(x_1) \tag{2}$$

$$\frac{x_2}{dt} = f_2(x_2, u_2), \quad y_2 = h_2(x_2) \tag{3}$$

with characteristics denoted by "k" and "g" respectively. (A degenerate case, in which the second system is memory-free, that is, there are no state variables x_2 and y_2 is simply a static function $y_2(t) = g(u_2(t))$, is also allowed. In fact, the proof of the general case can be reduced to that of the degenerate case, simply by taking the first system as a cascade connection of the two systems.)

As in [18], we suppose that the inputs and outputs of both systems are scalar: $m_1=m_2=p_1=p_2= 1$ (see [63] for a generalization to high-dimensional inputs and outputs). The "positive feedback interconnection" of the systems (2) and (3) is defined by letting the output of each of them serve as the input of the other ($u_2=y_1=$"y" and $u_1=y_2=$"u"), as depicted in Figure 11(a). Such positive feedback systems may easily be multi-stable, even if the constituent pieces are monostable [23, 108, 14, 109]. Let us first discuss how multi-stability may arise in a very intuitive and simple example, and later present the general theorem.

Two typical steady-state responses are as follows. Suppose that P is a protein with Michaelis-Menten production rate and linear degradation: $dp/dt = V_{max}u/(k_m + u) - kp$, where u represents the concentration a substrate that is used in P's formation. The reporter variable is $y(t)=p(t)$. In this case, the steady state when $u(t)\equiv u_0$ is $p_0=k(u_0)=(V_{max}/k)u_0/(k_m + u_0)$, and we obtain a *hyperbolic* response, Fig. 12(a). The response in this example is *graded* ("light-dimmer"): it is proportional to the parameter u_0 on a large range of values until saturation. In contrast, a *sigmoidal* ("doorbell") response, Fig. 12(b), which may arise from $dp/dt = V_{max}u^r/(k_m^r + u^r) - kp$ with Hill coefficient (cooperativity index) $r>1$, implies that values of u_0 under some threshold will not

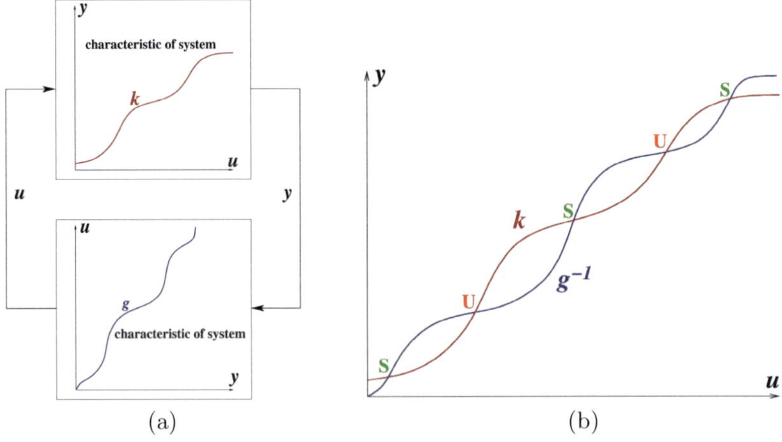

(a)   (b)

Fig. 11. (a) Positive feedback and (b) characteristics

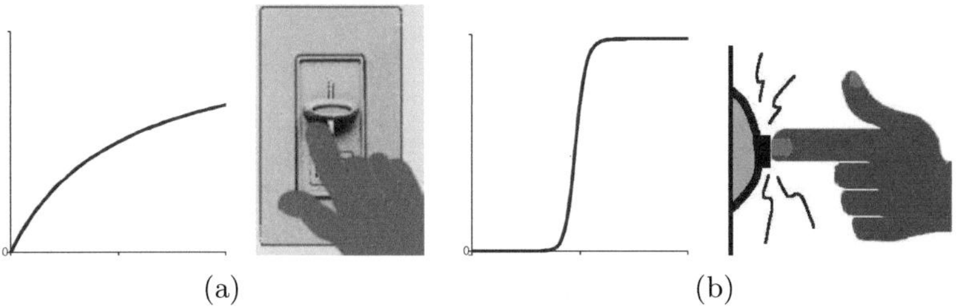

Fig. 12. (a) hyperbolic and (b) sigmoidal responses

result in an appreciable activity, while values over the threshold give an abrupt change (respectively, $p_0 \approx 0$ in steady state and $p_0 \approx V_{max}/k$ in steady state). Sigmoidal responses are critical e.g. if a cell must decide in a binary fashion whether a gene should be transcribed or not, depending on an extracellular signal [110,16,111,1,112,113,114,12,115,116,117,26,27,118,114,115,119]. Cascades of enzymatic reactions can be made to display such ultrasensitive response, if there is a Hill coefficient $r>1$ at each step [120].

Multiple attractors may appear if the output y (for example $y=p$ in the example) is fed-back as input u. The mechanism might be an autocatalytic process ($u=y$, e.g. if p helps promote its own transcription) or via a more complicated positive feedback pathway from p to u. Formally, substituting $u=p$ into $dp/dt = V_{max}u^r/k_m^r + u^r - kp$ (where $r=1$ or $r>1$), we obtain the closed-loop equation $dp/dt = V_{max}p^r/k_m^r + p^r - kp$. We plot in Fig. 13 both the first term (formation rate) and the second one (degradation), in cases where $r=1$ (left) or $r>1$ (right). For $r=1$, for small p the formation rate is larger than the degradation rate but for large p the opposite holds, so the concentration $p(t)$ converges to a unique intermediate value. For $r>1$, for small p the degradation rate is larger, so $p(t)$ converges to a low value, but for large p the formation rate is larger and $p(t)$ converges to a high value instead. Thus, two stable states are created, one low and one high, by this interaction of formation and degradation. (There is also an intermediate, unstable state.) *This reasoning is totally elementary, but it provides an intuition for the general result in [18], shown next, which represents a far-reaching generalization.* (The result can also be viewed as generalizing aspects of the papers [20,21,121,122,22,123,124,125,37,126], to arbitrary MIOS.)

We consider Fig. 11(b), where we have plotted together k and the inverse of g. It is quite obvious that there is a bijective correspondence between the steady states of the feedback system and the intersection points of the two graphs. With some mild technical conditions of transversality and "controllability" and "observability" (the recent papers [127, 128] show that even these mild conditions can be largely dispensed with), the following much less obvious facts are true. We first attach labels to the intersection points between the two graphs as follows: a label S (respectively, U) if the slope of k is smaller (respectively, larger) than the slope of g^{-1} at the intersection point. One can then conclude that "almost all" (in a measure-theoretic sense or in a Baire-category sense) bounded solutions of

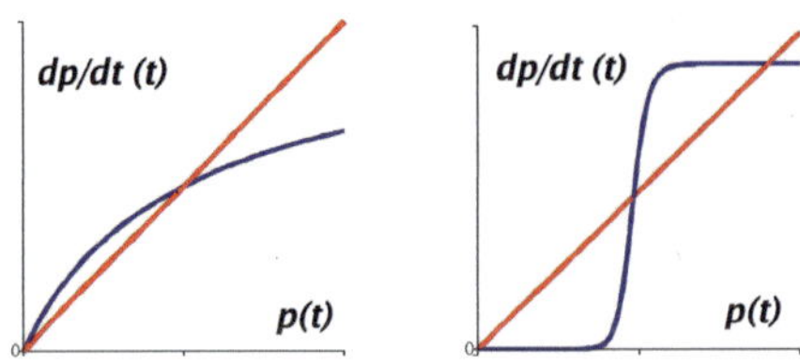

Fig. 13. intersections with degradation

the feedback system must converge to one of the steady states corresponding to intersection points labeled with an S. The proof reduces ultimately to an application of Hisrch's generic convergence theorem to the closed-loop system (the technical conditions insure strong monotonicity). However, the value-added is in the fact that stable states can be identified merely from the *one-dimensional plot* shown in Fig. 11(b). (If each subsystem would have dimension just one, one can also interpret the result in terms of a simple nullcline analysis; see the Supplementary Section of [19].) We remark that the theorems remain true even if arbitrary delays are allowed in the feedback loop and/or if space-dependent models are considered and diffusion is allowed (see [60] for a discussion). A new approach [129], based not on monotone theory but on a notion of "counterclockwise dynamics," extends in a different direction the range of applicability of this methodology.

We wish to emphasize the potential practical relevance of this result (and others such as [129]). The equations describing each of the systems are often poorly, or not at all, known. But, as long as we can assume that each subsystem is monotone and uni-stable, we can use the information from the planar plots in Fig. 11(b) to completely understand the dynamics of the closed-loop system, no matter how large the number of state variables. It is often said that the field of molecular systems biology is characterized by a *data-rich/data-poor* paradox: while on the one hand a huge amount of *qualitative* network (schematic modeling) knowledge is available for signaling, metabolic, and gene regulatory networks, on the other hand little of this knowledge is *quantitative*, at least at the level of precision demanded by most mathematical tools of analysis. On the other hand, input/output steady state data (from a signal such as a ligand, to a reporter variable such as the expression of a gene monitored by GFP, or the activity of a protein measured by a Western blot) is frequently available. The problem of exploiting qualitative knowledge, and effectively integrating relatively sparse quantitative data, is among the most challenging issues confronting systems biology. The MIOS approach provides one way to combine these two types of data. (For further discussion of this "data-rich/data-poor" issue, see [41,60].)

The theorem from [18] and its generalizations, as well as the negative feedback result discussed below, amount to a model-reduction approach. The bifurcation behavior of the complete closed-loop system is obtained from a low-order reduction (just to two one-dimensional systems, connected in feedback, when

$m = p = 1$) and information on the i/o behavior of the components. This model-reduction view is further developed in [63].

More Discussion Through an Example: MAPK Cascades

Mitogen-Activated Protein Kinase (MAPK) cascades are a ubiquitous "signaling module" in eukaryotes, involved in proliferation, differentiation, development, movement, apoptosis, and other processes [130, 131, 132]. There are several such cascades, sharing the property of being composed of a cascade of three kinases. The basic rule is that two proteins, called generically MAPK and MAPKK (the last K is for "kinase of MAPK," which is itself a kinase), are active when doubly phosphorylated, and MAPKK phosphorylates MAPK when active. Similarly, a kinase of MAPKK, MAPKKK, is active when phosphorylated. A phosphatase, which acts constitutively (that is, by default it is always active) reverses the phosphorylation. The biological model from [130, 19] is in Fig. 14(b), were we wrote $z_i(t), i = 1,2,3$ for MAPK, MAPK-P, and MAPK-PP concentrations and similarly for the other variables. The input represents an external signal to this subsystem (typically, the concentration of a kinase driving forward the reaction).

We make here the simplest assumptions about the dynamics, amounting basically to a quasi-steady state appproximation of enzyme kinetics. (For related results using more realistic, mass-action, models, see [48, 50, 49].) For example, take the reaction shown in the square in Fig. 14(a). As y_3 (MAPKK-PP) facilitates the conversion of z_1 into z_2 (MAPK to MAPK-P), the rate of change dz_2/dt should include a term $\alpha(z_1, y_3)$ (and dz_1/dt has a term $-\alpha(z_1, y_3)$) for some (otherwise unknown) function α such that $\alpha(0, y_3) = 0$ and $\frac{\partial \alpha}{\partial z_1} > 0$, $\frac{\partial \alpha}{\partial y_3} > 0$ when $z_1 > 0$. (Nothing happens if there is no substrate, but more enzyme or more substrate results in a faster reaction.) There will also be a term $+\beta(z_2)$ to reflect the phosphatase action. Similarly for the other species. The system as given would be represented by a set of seven ordinary differential equations (or reaction-diffusion PDE's, if spatial localization is of interest, or delay-differential equations, if appropriate).

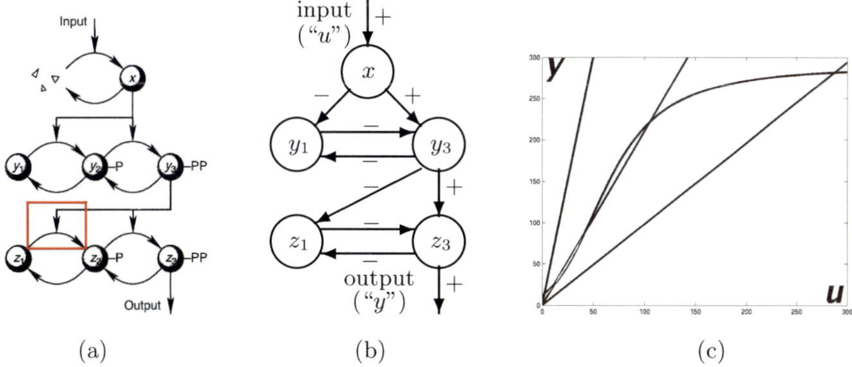

Fig. 14. (a) MAPK cascades; (b) graph; (c) characteristic

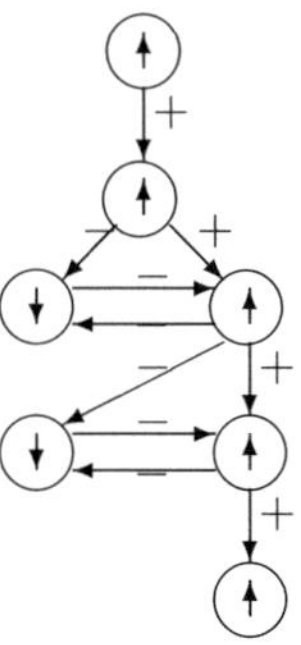

Fig. 15. Consistent assignment for simple MAPK cascade model

This system is not monotone (at least with respect to any orthant cone), as is easy to verify graphically. However, as with many other examples of biochemical networks, *the system is "monotone in disguise"*, so to speak, in the sense that a judicious change of variables allows one to apply MIOS tools. (Far more subtle forms of this argument are key to applications to signaling cascades. A substantial research effort, not reviewed here because of lack of space, addresses the search for graph-theoretic conditions that allow one to find such "monotone systems in disguise"; see [41, 60, 50] for references.)

In this example, which in fact was the one whose study initially led to the definition of MIOS, the following conservation laws: $y_1(t) + y_2(t) + y_3(t) \equiv y_{\mathrm{tot}}$ (total MAPKK) and $z_1(t) + z_2(t) + z_3(t) \equiv z_{\mathrm{tot}}$ (total MAPK) hold true, assuming no protein turn-over. This assumption is standard in most of the literature, because transcription and degradation occur at time scales much slower than signaling. (There is very recent experimental data that suggests that turn-over might be fast for some yeast MAPK species. Adding turn-over would lead to a different mathematical model.) These conservation laws allow us to eliminate variables. The right trick is to eliminate y_2 and z_2. Once we do this, and write $y_2 = y_{\mathrm{tot}} - y_1 - y_3$ and $z_2 = z_{\mathrm{tot}} - z_1 - z_3$, we are left with the variables x, y_1, y_3, z_1, z_3. For instance, the equations for z_1, z_3 look like:

$$\frac{dz_1}{dt} = -\alpha(z_1, y_3) + \beta(z_{\mathrm{tot}} - z_1 - z_3) \qquad \frac{dz_3}{dt} = \gamma(z_{\mathrm{tot}} - z_1 - z_3, y_3) - \delta(z_3)$$

for appropriate increasing functions $\alpha, \beta, \gamma, \delta$. The equations for the remaining variables are similar. The graph, ignoring, as usual, self-loops (diagonal of Jacobian), is shown in Fig. 14(b). This graph has no negative undirected loops, showing that the (reduced) *system is monotone.* A consistent spin assignment (including the top input node and the bottom output node) is shown in Figure 15. It is also true that this system *has a well-defined monostable state space response (characteristic)*; there is no space to discuss the proof here, so we refer the reader to the original papers [39, 28].

Positive and negative feedback loops around MAPK cascades have been a topic of interest in the biological literature. For example, see [114, 116] for positive feedback and [25, 133] for negative feedback. Since we know that the system is monotone and has a characteristic, MIOS theory as described here can indeed be applied to the example. We study next the effect of a positive feedback $u = gy$ obtained by "feeding back" into the input a scalar multiple g of the output. (This is a somewhat unrealistic model of feedback, since feedbacks act for example by enhancing the activity of a kinase. We pick it merely for illustration of the techniques).

The theorem does not require actual equations for its applicability. All that is needed is the knowledge that we have a MIOS, and a plot of its characteristic (which, in practice, would be obtained from interpolated experimental data). In order to illustrate the conclusions, on the other hand, it is worth discussing a particular set of equations. We take equations and parameters from [19, 41, 60]:

$$\frac{dx}{dt} = -\frac{v_2\,x}{k_2 + x} + v_0\,u + v_1$$

$$\frac{dy_1}{dt} = \frac{v_6\,(y_{\text{tot}} - y_1 - y_3)}{k_6 + (y_{\text{tot}} - y_1 - y_3)} - \frac{v_3\,x\,y_1}{k_3 + y_1}$$

$$\frac{dy_3}{dt} = \frac{v_4\,x\,(y_{\text{tot}} - y_1 - y_3)}{k_4 + (y_{\text{tot}} - y_1 - y_3)} - \frac{v_5\,y_3}{k_5 + y_3}$$

$$\frac{dz_1}{dt} = \frac{v_{10}\,(z_{\text{tot}} - z_1 - z_3)}{k_{10} + (z_{\text{tot}} - z_1 - z_3)} - \frac{v_7\,y_3\,z_1}{k_7 + z_1}$$

$$\frac{dz_3}{dt} = \frac{v_8\,y_3\,(z_{\text{tot}} - z_1 - z_3)}{k_8 + (z_{\text{tot}} - z_1 - z_3)} - \frac{v_9\,z_3}{k_9 + z_3}$$

with output z_3. Specifically, we will use the following parameters: $v_0 = 0.0015$, $v_1 = 0.09$, $v_2 = 1.2$, $v_3 = 0.064$, $v_4 = 0.064$, $v_5 = 5$, $v_6 = 5$, $v_7 = 0.06$, $v_8 = 0.06$, $v_9 = 5$, $v_{10} = 5$, $y_{\text{tot}} = 1200$, $z_{\text{tot}} = 300$, $k_2 = 200$, $k_3 = 1200$, $k_4 = 1200$, $k_5 = 1200$, $k_6 = 1200$, $k_7 = 300$, $k_8 = 300$, $k_9 = 300$, $k_{10} = 300$. (The units are: totals in nM (mol/cm^3), v's in nM·sec^{-1} and sec^{-1}, and k's in nM.)

With these choices, the steady state step response is the sigmoidal curve shown in Fig. 14(c), where y is the output z_3. We plotted in the same figure the inverse g^{-1} of the characteristic of the feedback system, in this case just the linear mapping $y = (1/g)u$, for three typical "feedback gains" ($g=1/0.98, 1/2.1, 1/6$).

For $g = 1/0.98$ (line of slope 0.98 when plotting y against u), there should be a unique stable state, with a high value of the output $y = z_3$, and trajectories should generically converge to it. Similarly, for $g = 1/2.1$ (line of slope 2.1) there should be two stable states, one with high and one with low $y = z_3$, with trajectories generically converging to one of these two, because the line intersects at three points, corresponding to two stable and one unstable state (exactly as in the discussion concerning the simple protein formation/degradation sigmoidal example in Fig. 13). Finally, for $g = 1/6$ (line of slope 6), only the low-y stable state should persist. Fig. 16(a-c) shows plots of the hidden variable $y_3(t)$ (MAPKK-PP) for several initial states, confirming the predictions. The same convergence results are predicted if there are delays in the feedback loop, or if concentrations depend on location in a convex spatial domain. Results for

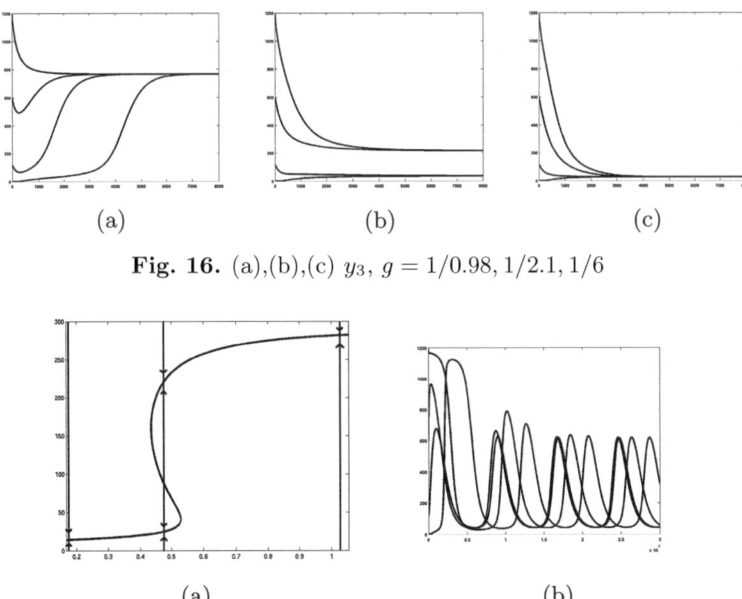

Fig. 16. (a),(b),(c) y_3, $g = 1/0.98, 1/2.1, 1/6$

Fig. 17. (a) bifurcation diagram and relaxation (b) oscillation (y_3)

reaction-diffusion PDE's and delay-differential systems are discussed in [60], and simulation results for this example are also provided there.

We may plot the steady state value of y, under the feedback $u = gy$, as the gain g is varied, Fig. 17(a). This resulting complete bifurcation diagram showing points of saddle-node bifurcation can be also completely determined just from the characteristic, with no need to know the equations of the system. Relaxation oscillations may be expected under such circumstances if a second, slower, feedback loop is used to negatively adapt the gain as a function of the output. Reasons of space preclude describing a very general theorem, which shows that indeed, relaxation oscillations can be guaranteed in this fashion: see [66] for technical details, and [60] for a more informal discussion. Fig. 17(b) shows a simulation confirming the theoretical prediction (details in [66, 60]).

Negative Feedback

A different set of results apply to inhibitory or negative feedback interconnections of two MIOS systems (2)-(3). A convenient mathematical way to define "negative feedback" in the context of monotone systems is to say that the orders on inputs and outputs are inverted (example: an inhibition term of the form $\frac{V}{K+y}$, as usual in biochemistry). Equivalently, we may incorporate the inhibition into the output of the second system (3), which is then seen as an *anti-monotone* I/O system, and this is how we proceed from now on. See Fig. 18(a). We emphasize that the closed-loop systems that result are *not* monotone, at least with respect to any known order.

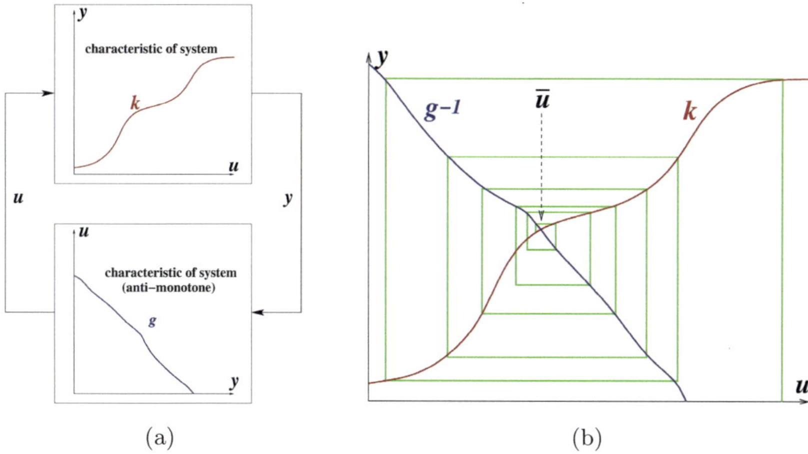

Fig. 18. (a) Negative feedback and (b) characteristics

The original theorem, from [39], is as follows. We assume once more that inputs and outputs are scalar ($m=p=1$; see [65] for generalizations). We once again plot together k and g^{-1}, as shown in Fig. 18(b). Consider the following discrete iteration:

$$u_{i+1} = (g \circ k)(u_i).$$

Then, if solutions of the closed-loop system are bounded and if this iteration has a globally attractive fixed point $\bar{u}$, as shown in Fig. 18(b), then the feedback system has a globally attracting steady state. (An equivalent condition, see [65], is that the iteration have no nontrivial period-two orbits.) We call this result a *small gain theorem* ("SGT"), because of its analogy to concepts in control theory.

It is easy to see that *arbitrary delays* may be allowed in the feedback loop. In other words, the feedback could be of the form $u(t) = y(t - h)$, and such delays (even with $h = h(t)$ time varying or even state-dependent, as long as $t - h(t) \to \infty$ as $t \to \infty$) *do not destroy global stability of the closed loop*. In [42], we have now shown also that *diffusion* does not destroy global stability either. In other words, a reaction-diffusion system (Neumann boundary conditions) whose reaction can be modeled in the shown feedback configuration, has the property that all solutions converge to a (unique) uniform in space solution. This is not immediately obvious, since standard parabolic comparison theorems do not immediately apply to the feedback system, which is not monotone.

Example: MAPK Cascade with Negative Feedback

As with the positive feedback theorem, an important feature is applicability to highly uncertain systems. As long as the component systems are known to be MIOS, the knowledge of I/O response curves and a planar analysis are sufficient to conclude GAS of the entire system, which may have an arbitrarily high dimension. For example, suppose we take a feedback like $u=a+b/(c+z_3)$, with a graph as shown in Fig. 19(a), which also shows the characteristic and a convergent

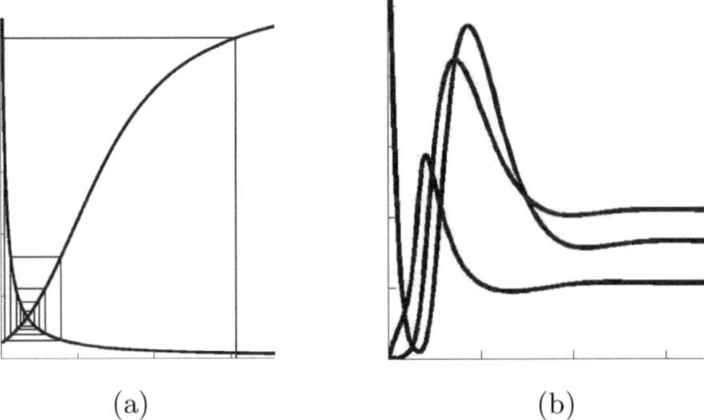

Fig. 19. inhibition: (a) spiderweb and (b) simulation

discrete 1-d iteration [60]. Then, we are guaranteed that all solutions of the closed-loop system converge to a unique steady state, as confirmed by the simulations in Fig.19(b), which shows the concentrations of the active forms of the kinases.

Example: Testosterone Model

This example is intended to show that even for a classical mathematical biology model, a very simple application of the result in [39] gives an interesting conclusion. The concentration of testosterone in the blood of a healthy human male is known to oscillate periodically with a period of a few hours, in response to similar oscillations in the concentrations of the luteinising hormone (LH) secreted by the pituitary gland, and the luteinising hormone releasing hormone (LHRH), normally secreted by the hypothalamus (see [134]). The well-known textbook [9] (and its previous editions) presents this process as an example of a biological oscillator, and proposes a model to describe it, introducing delays in order to obtain oscillations. (Since the textbook was written, the physiological mechanism has been much further elucidated, and this simple model is now known not to be correct. However, we want merely to illustrate a point about mathematical analysis.) The equations are:

$$\dot{R} = \frac{A}{K + T} - b_1 R$$
$$\dot{L} = g_1 R - b_2 L$$
$$\dot{T} = g_2 L(t - \tau) - b_3 T$$

(R, L, T = concentrations of hormones luteinising hormone releasing, luteinising, and testosterone, τ = delay; we use "$\dot{x}$" to denote time derivative). The system may be seen as the feedback connection of the MIOS system

$$\dot{R} = u - b_1 R$$
$$\dot{L} = g_1 R - b_2 L$$
$$\dot{T} = g_2 L - b_3 T$$

with the inhibitory feedback $u(t) = g(T - \tau) = A/(K + T(t - \tau))$ after moving the delay to the loop (without loss of generality). The characteristic is linear, $T = k(u) = \frac{g_1 g_2}{b_1 b_2 b_3} u$, so $g \circ k$ is a fractional transformation $S(u) = \frac{p}{q+u}$. Since such a transformation has no period-two cycles, global stability follows. (For arbitrary, even time-varying, delays.) This contradicts the existence of oscillations claimed in [9] for large enough delays. (See [135], which also explains the error in [9].)

Example: Lac Operon

The study of *E. Coli* lactose metabolism has been a topic of research in mathematical biology since Jacob and Monod's classical work which led to their 1995 Nobel Prize. For this example, we look at the subsystem modeled in [136]. The lac operon induces production of permease and β-gal, permease makes the cell membrane more permeable to lactose, and genes are activated if lactose present; lactose is digested by the enzyme β-gal, and the other species are degraded at fixed rates. (In this model from [136], lactose and isolactose are identified, and catabolic repression by glucose via cAMP is ignored.) Delays arise from translation of permease and β-gal. The equations are:

$$\begin{aligned}
\dot{x}_1(t) &= g(x_4(t - \tau)) - b_1 x_1(t) && \text{lac operon mRNA} \\
\dot{x}_2(t) &= x_1(t) - b_2 x_2(t) && \beta\text{-galactoside permease} \\
\dot{x}_3(t) &= r x_1(t) - b_3 x_3(t) && \beta\text{-galactosidase} \\
\dot{x}_4(t) &= S x_2(t) - x_3(t) x_4(t) && \text{lactose}
\end{aligned}$$

with $g(x) := (1 + K x^\rho)/(1 + x^\rho)$, $K > 1$, and the Hill exponent ρ representing a cooperativity effect. (All delays have been lumped into one.) We view this system as a negative feedback loop, where $u = x_1$, $v = x_4$, of a MIOS system (details in [65]). Since there are two inputs and outputs, now we must study the two-dimensional iteration

$$(u, v) \ \mapsto \ (g \circ k)(u, v) = \left[\frac{g(v)}{b}, \frac{S b_1 b_3 u}{r b_2 g(v)} \right].$$

Based on results on rational difference equations from [137], one concludes that there are no nontrivial 2-periodic orbits, provided that $\rho < (\sqrt{K}+1)/(K-1)$, for arbitrary b_1, b_2, b_3, r, S. Hence, by the theorem, there is a unique steady state of the original system, which is GAS, even when arbitrary delays are present,

These and other conditions are analyzed in [65], where it is also shown that the results from [136] are recovered as a special case. Among other advantages of this approach, besides generalizing the result and giving a conceptually simple proof, we have (because of [42]) the additional conclusion that also for the corresponding reaction-diffusion system, in which localization is taken account of, the same globally stable behavior can be guaranteed.

Example: Circadian Oscillator

As a final example of the negative feedback theorem, we pick Goldbeter's [138,7] original model of the molecular mechanism underlying circadian rhythms in

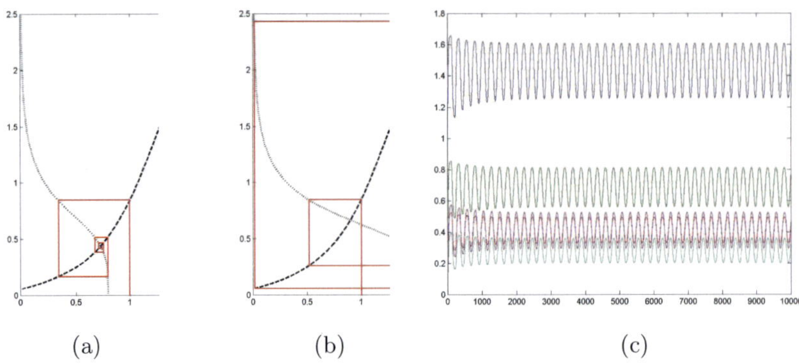

(a)   (b)   (c)

Fig. 20. (a) convergent. (b) divergent iterations. (c) oscillations.

Drosophila. (In this oversimplified model, only *per* protein is considered; other players such as *tim* are ignored.) PER protein is synthesized at a rate proportional to its mRNA concentration. Two phosphorylation sites are available, and constitutive phosphorylation and dephosphorylation occur with saturation dynamics, at maximum rate v_i's and with Michaelis constants K_i. Doubly phosphorylated PER is degraded, also satisfying saturation dynamics (with parameters v_d, k_d), and it is translocated to the nucleus with rate constant k_1. Nuclear PER inhibits transcription of the *per* gene, with a Hill-type reaction of cooperativity degree n and threshold constant K_I, and mRNA is produced. and translocated to the cytoplasm, at a rate determined by a constant v_s. Additionally, there is saturated degradation of mRNA (constants v_m and k_m). The model is ($P_i = per$ phosphorylated at i sites, $P_N =$ nuclear *per*, $M = per$ mRNA):

$$\dot{M} = v_s K_I/(K_I+P_N^n) - v_m M/(k_m+M)$$
$$\dot{P_0} = k_s M - V_1 P_0/(K_1+P_0) + V_2 P_1/(K_2+P_1)$$
$$\dot{P_1} = V_1 P_0/(K_1+P_0) - V_2 P_1/(K_2+P_1) - V_3 P_1/(K_3+P_1) + V_4 P_2/(K_4+P_2)$$
$$\dot{P_2} = V_3 P_1/(K_3+P_1) - V_4 P_2/(K_4+P_2) - k_1 P_2 + k_2 P_N - v_d P_2/(k_d+P_2)$$
$$\dot{P_N} = k_1 P_2 - k_2 P_N.$$

Parameters are chosen exactly as in Goldbeter's original paper, except that the rate v_s of mRNA translocation to the cytoplasm is taken as a bifurcation parameter. The value $v_s = 0.76$ from [138] gives oscillatory behavior. On the other hand, we may break up the system into the M and P_i, P_N subsystems. Each of these can be shown to be MIOS and have a characteristic. (The existence of a characteristic for the P-subsystem is nontrivial, and involves the application of Smillie's Theorem [139] for strongly monotone tridiagonal systems, and more precisely, repeated application of a proof technique in [139] involving "eventually monotonicity" of state variables.) When $v_s{=}0.4$, the discrete iteration is graphically seen to be convergent (see Fig. 20(a)), so the theorem guarantees global asymptotic stability even when arbitrary delays are introduced in the feedback. Bifurcation analysis on delay length and v_s indicates that local stability will fail for somewhat larger values. Using again the graphical test, we observe that for $v_s=0.5$ there appears limit cycle for the *discrete* iteration on characteristics, see Fig. 20(b). This suggests that oscillations may exist in the full nonlinear differential equation, at least for appropriate delays lengths. Indeed, the simulation in Fig. 20(c) displays such oscillations (see [105, 28]).

Multivalued Characteristics

For simplicity, we have not discussed the case when characteristics are set-valued instead of single-valued. This general case can also be productively studied with the toolkit afforded by MIOS interconnection theory, see [127, 128] for positive feedback and [64] for negative feedback.

4   Conclusions

There is a clear need in systems biology to study robust structures and to develop robust analysis tools. The theory of monotone systems provides one such tool. Interesting and nontrivial conclusions can be drawn from (signed) network structure alone, which is associated to purely stoichiometric information about the system, and ignores fluxes.

Associating a graph to a given system, we may define spin assignments and consistency, a notion that may be interpreted also as non-frustration of Ising spin-glass models. Every species in a monotone system (one whose graph is consistent) responds with a consistent sign to perturbations at every other species. This property would appear to be desirable in biological networks, and, indeed, there is some evidence suggesting the near-monotonicity of some natural networks. Moreover, "near"-monotone systems might be "practically" monotone, in the sense of being monotone under disjoint environmental conditions.

Dynamical behavior of monotone systems is ordered and "non chaotic". Systems close to monotone may be decomposed into a small number of monotone subsystems, and such decompositions may be usefully employed to study nonmonotone dynamics as well as to help detect bifurcations even in monotone systems, based only upon sparse numerical data, resulting in a sometimes useful model-reduction approach.

Acknowledgements

Much of the author's work on I/O monotone systems was done in collaboration with David Angeli, as well as Patrick de Leenheer, German Enciso, Bhaskar Dasgupta, and Hal Smith. The author also wishes to thank Moe Hirsch, Reka Albert, Tom Knight, Avi Maayan, Alex van Oudenaarden, and many others, for useful comments and suggestions regarding the material discussed here.

This work was supported in part by NSf Grants DMS-0504557 and DMS-0614371.

References

1. A. Novic and M. Weiner. Enzyme induction as an all-or-none phenomenon. *Proc. Natl. Acad. Sci. U.S.A.*, 43:553–566, 1957.
2. J. Monod and F. Jacob. Teleonomic mechanisms in cellular metabolism, growth and differentiation,. *Cold Spring Harb. Symp. Quant. Biol.*, 26:389–401, 1961.
3. J. Lewis, J.M. Slack, and L. Wolpert. Thresholds in development. *J. Theor. Biol.*, 65:579–590, 1977.
4. L.A. Segel. *Modeling dynamic phenomena in molecular and cellular biology.* Cambridge University Press, Cambridge, 1984.
5. D.L. DeAngelis, W.M. Post, and C.C. Travis. *Positive Feedback in Natural Systems.* Springer-Verlag, New York, 1986.
6. R. Thomas and R. D'ari. *Biological feedback.* CRC Press, Boca Raton, 1990.
7. A. Goldbeter. *Biochemical Oscillations and Cellular Rhythms.* Cambridge University Press, Cambridge, 1996.
8. J.P. Keener and J. Sneyd. *Mathematical Physiology.* Springer-Verlag, New York, 1998.
9. J.D. Murray. *Mathematical Biology, I, II: An introduction.* Springer-Verlag, New York, 2002.
10. R. Milo, S. Shen-Orr, S. Itzkovitz, N. Kashtan, D. Chklovskii, and U. Alon. Network motifs: Simple building blocks of complex networks. *Science*, 298:824–827, 2002.
11. L. Edelstein-Keshet. *Mathematical Models in Biology.* SIAM, Philadelphia, 2005.
12. M. Ptashne. *A Genetic Switch: Phage λ and Higher Organisms.* Cell Press and Blackwell Scientific Publications, Cambridge MA, 1992.
13. E. Plahte, T. Mestl, and W.S. Omholt. Feedback circuits, stability and multistationarity in dynamical systems. *J. Biol. Sys.*, 3:409–413, 1995.
14. O. Cinquin and J. Demongeot. Positive and negative feedback: striking a balance between necessary antagonists. *J. Theor. Biol.*, 216:229–241, 2002.
15. J.L. Gouze. Positive and negative circuits in dynamical systems. *J. Biol. Sys.*, 6:11–15, 1998.
16. R. Thomas and M. Kaufman. Multistationarity, the basis of cell differentiation and memory. i. structural conditions of multistationarity and other nontrivial behavior. *Chaos*, 11:170–179, 2001.
17. E. Remy, B. Mosse, C. Chaouiya, and D. Thieffry. A description of dynamical graphs associated to elementary regulatory circuits. *Bioinformatics*, 19 (Suppl 2):ii172ii178, 2003.
18. D. Angeli and E.D. Sontag. Multi-stability in monotone input/output systems. *Systems Control Lett.*, 51(3-4):185–202, 2004.
19. D. Angeli, J. E. Ferrell, and E.D. Sontag. Detection of multistability, bifurcations, and hysteresis in a large class of biological positive-feedback systems. *Proc Natl Acad Sci USA*, 101(7):1822–1827, February 2004. A revision of Suppl. Fig. 7(b) is here: http://www.math.rutgers.edu/(tilde)sontag/FTPDIR/nullclines-f-g-REV.jpg; and typos can be found here: http://www.math.rutgers.edu/(tilde)sontag/FTPDIR/angeli-ferrell-sontag-pnas04-errata.txt.
20. P.E. Rapp. A theoretical investigation of a large class of biochemical oscillations. *Math Biosciences*, 25:165–188, 1975.
21. S. Hastings, J. Tyson, and D. Webester. Existence of periodic solutions for negative feedback cellular control systems. *J. Diff. Eqs.*, 25:39–64, 1977.

22. J. Tyson and H.G. Othmer. The dynamics of feedback control circuits in biochemical pathways. *Progr. Theor. Biol.*, 5:1–60, 1978.

23. R. Thomas. On the relation between the logical structure of systems and their ability to generate multiple steady states or sustained oscillations. *Springer Ser. Synergetics*, 9:180–193, 1981.

24. E.D. Sontag. *Mathematical Control Theory. Deterministic Finite-Dimensional Systems*, volume 6 of *Texts in Applied Mathematics*. Springer-Verlag, New York, second edition, 1998.

25. B.N. Kholodenko. Negative feedback and ultrasensitivity can bring about oscillations in the mitogen-activated protein kinase cascades. *Eur. J. Biochem*, 267:1583–1588, 2000.

26. W. Sha, J. Moore, K. Chen, A.D. Lassaletta, C.S. Yi, J.J. Tyson, and J.C. Sible. Hysteresis drives cell-cycle transitions in xenopus laevis egg extracts. *Proc. Natl. Acad. Sci. USA*, 100:975–980, 2003.

27. J. R. Pomerening, E.D. Sontag, and J. E. Ferrell. Building a cell cycle oscillator: hysteresis and bistability in the activation of cdc2. *Nat Cell Biol*, 5(4):346–351, April 2003. Supplementary materials 2-4 are here: http://www.math.rutgers.edu/(tilde)sontag/FTPDIR/pomerening-sontag-ferrell-additional.pdf.

28. D. Angeli and E.D. Sontag. Interconnections of monotone systems with steady-state characteristics. In *Optimal control, stabilization and nonsmooth analysis*, volume 301 of *Lecture Notes in Control and Inform. Sci.*, pages 135–154. Springer, Berlin, 2004.

29. S. Mangan and U. Alon. Structure and function of the feed-forward loop network motif. *Proc. Natl. Acad. Sci. USA*, 110:11980–11985, 2003.

30. S. Mangan, A. Zaslaver, and U. Alon. The coherent feedforward loop serves as a sign-sensitive delay element in transcription networks. *J. Molec. Bio.*, 334:197–204, 2003.

31. B.L. Clarke. Stability of complex reaction networks. In I. Prigogine and S.A. Rice, editors, *Advances in Chemical Physics*, pages 1–215. John Wiley, New York, 1980.

32. F.J.M. Horn. The dynamics of open reaction systems. In *Mathematical aspects of chemical and biochemical problems and quantum chemistry (Proc. SIAM-AMS Sympos. Appl. Math., New York, 1974)*, pages 125–137. Amer. Math. Soc., Providence, 1974. SIAM-AMS Proceedings, Vol. VIII.

33. F.J.M. Horn and R. Jackson. General mass action kinetics. *Arch. Rational Mech. Anal.*, 49:81–116, 1972.

34. M. Feinberg and F.J.M. Horn. Dynamics of open chemical systems and algebraic structure of underlying reaction network. *Chemical Engineering Science*, 29:775–787, 1974.

35. M. Feinberg. Chemical reaction network structure and the stability of complex isothermal reactors - i. the deficiency zero and deficiency one theorems. *Chemical Engr. Sci.*, 42:2229–2268, 1987.

36. M. Feinberg. The existence and uniqueness of steady states for a class of chemical reaction networks. *Archive for Rational Mechanics and Analysis*, 132:311–370, 1995.

37. H. Smith. *Monotone dynamical systems: An introduction to the theory of competitive and cooperative systems, Mathematical Surveys and Monographs, vol. 41*. AMS, Providence, RI, 1995.

38. M. Hirsch and H.L. Smith. Monotone dynamical systems. In *Handbook of Differential Equations, Ordinary Differential Equations (second volume)*. Elsevier, Amsterdam, 2005.

39. D. Angeli and E.D. Sontag. Monotone control systems. *IEEE Trans. Automat. Control*, 48(10):1684–1698, 2003. Errata are here: http://www.math.rutgers.edu/ (tilde)sontag/FTPDIR/angeli-sontag-monotone-TAC03-typos.txt.

40. P. de Leenheer, D. Angeli, and E.D. Sontag. Monotone chemical reaction networks. *J. Math Chemistry*, to appear, 2006.

41. E.D. Sontag. Some new directions in control theory inspired by systems biology. *IEE Proc. Systems Biology*, 1:9–18, 2004.

42. G.A. Enciso, H.L. Smith, and E.D. Sontag. Non-monotone systems decomposable into monotone systems with negative feedback. *J. of Differential Equations*, 224:205–227, 2006.

43. M. Feinberg. Some recent results in chemical reaction network theory. In R. Aris, D. G. Aronson, and H. L. Swinney, editors, *Patterns and Dynamics in Reactive Media, IMA Vol. Math. Appl. 37*, page 43–70. Springer, Berlin, 1991.

44. V.N. Reddy, M.L. Mavrovouniotis, and M.N. Liebman. Petri net representations in metabolic pathways. *Proc. Int. Conf. Intell. Syst. Mol. Biol.*, 1:328–336, 1993.

45. I. Zevedei-Oancea and S. Schuster. Topological analysis of metabolic networks based on petri net theory. *In Silico Biol.*, 3, 2003.

46. G. Craciun and M. Feinberg. Multiple equilibria in complex chemical reaction networks: I. the injectivity property. *SIAM Journal on Applied Mathematics*, 65:1526–1546, 2005.

47. G. Craciun and M. Feinberg. Multiple equilibria in complex chemical reaction networks: II. the species-reactions graph. *SIAM Journal on Applied Mathematics*, 66:1321–1338, 2006.

48. D. Angeli and E.D. Sontag. Translation-invariant monotone systems, and a global convergence result for enzymatic futile cycles. *Nonlinear Analysis Series B: Real World Applications*, to appear, 2006.

49. D. Angeli, P. de Leenheer, and E.D. Sontag. A Petri net approach to the study of persistence in chemical reaction networks. (Submitted to Mathematical Biosciences, also arXiv q-bio.MN/068019v2, 10 Aug 2006), 2006.

50. D. Angeli, P. de Leenheer, and E.D. Sontag. On the structural monotonicity of chemical reaction networks. In *Proc. IEEE Conf. Decision and Control, San Diego, Dec. 2006*. IEEE, 2006. (to appear).

51. M. Hirsch. Differential equations and convergence almost everywhere in strongly monotone flows. *Contemporary Mathematics*, 17:267–285, 1983.

52. M. Hirsch. Systems of differential equations that are competitive or cooperative ii: Convergence almost everywhere. *SIAM J. Mathematical Analysis*, 16:423–439, 1985.

53. B. DasGupta, G.A. Enciso, E.D. Sontag, and Y. Zhang. Algorithmic and complexity aspects of decompositions of biological networks into monotone subsystems. *BioSystems*, to appear, 2006.

54. M. Goemans and D. Williamson. Improved approximation algorithms for maximum cut and satisfiability problems using semidefinite programming. *Journal of the ACM*, 42:1115–1145, 1995.

55. F. Barahona. On the computational complexity of Ising spin glass models. *J. Phys. A. Math. Gen.*, 15:3241–3253, 1982.

56. C. De Simone, M. Diehl, M. Junger, P. Mutzel, G. Reinelt, and G. Rinaldi. Exact ground states of Ising spin glasses: New experimental results with a branch and cut algorithm. *Journal of Statistical Physics*, 80:487–496, 1995.

57. S. Istrail. Statistical mechanics, three-dimensionality and np-completeness: I. universality of intractability of the partition functions of the Ising model across non-planar lattices. In *Proceedings of the 32nd ACM Symposium on the Theory of Computing (STOC00)*, pages 87–96. ACM Press, 2000.

58. http://www.weizmann.ac.il/mcb/UriAlon/Papers/networkMotifs/yeastData.mat.

59. Maria C. Costanzo, Matthew E. Crawford, Jodi E. Hirschman, Janice E. Kranz, Philip Olsen, Laura S. Robertson, Marek S. Skrzypek, Burkhard R. Braun, Kelley Lennon Hopkins, Pinar Kondu, Carey Lengieza, Jodi E. Lew-Smith, Michael Tillberg, and James I. Garrels. YPDTM, PombePDTM and WormPDTM: model organism volumes of the BioKnowledgeTM Library, an integrated resource for protein information. *Nucl. Acids Res.*, 29(1):75–79, 2001.

60. E.D. Sontag. Molecular systems biology and control. *Eur. J. Control*, 11(4-5):396–435, 2005.

61. D. Angeli, P. de Leenheer, and E.D. Sontag. A small-gain theorem for almost global convergence of monotone systems. *Systems Control Lett.*, 52(5):407–414, 2004.

62. P. de Leenheer, D. Angeli, and E.D. Sontag. On predator-prey systems and small-gain theorems. *Math. Biosci. Eng.*, 2(1):25–42, 2005.

63. G. Enciso and E.D. Sontag. Monotone systems under positive feedback: multistability and a reduction theorem. *Systems Control Lett.*, 54(2):159–168, 2005.

64. P. De Leenheer and M. Malisoff. A small-gain theorem for monotone systems with multivalued input-state characteristics. *IEEE Trans. Automat. Control*, 51:287–292, 2006.

65. G. A. Enciso and E.D. Sontag. Global attractivity, I/O monotone small-gain theorems, and biological delay systems. *Discrete Contin. Dyn. Syst.*, 14(3):549–578, 2006.

66. T. Gedeon and E.D. Sontag. Oscillations in multi-stable monotone systems with slowly varying feedback. *J. of Differential Equations*, page to appear, 2006.

67. K. Hadeler and D. Glas. Quasimonotone systems and convergence to equilibrium in a population genetics model. *J. Math. Anal. Appl.*, 95:297–303, 1983.

68. M.W. Hirsch. The dynamical systems approach to differential equations. *Bull. A.M.S.*, 11:1–64, 1984.

69. E.N. Dancer. Some remarks on a boundedness assumption for monotone dynamical systems. *Proc. of the AMS*, 126:801–807, 1998.

70. L. Wang and E.D. Sontag. Almost global convergence in singular perturbations of strongly monotone systems. In *Positive Systems*, pages 415–422. Springer-Verlag, Berlin/Heidelberg, 2006. (Lecture Notes in Control and Information Sciences Volume 341, Proceedings of the second Multidisciplinary International Symposium on Positive Systems: Theory and Applications (POSTA 06) Grenoble, France).

71. J.C. Doyle, B. Francis, and A. Tannenbaum. *Feedback Control Theory*. MacMillan publishing Co., 1990.

72. R. Sepulchre, M. Jankovic, and P.V. Kokotović. *Constructive Nonlinear Control*. Springer-Verlag, London, 1997.

73. E.D. Sontag. Stability and stabilization: discontinuities and the effect of disturbances. In *Nonlinear analysis, differential equations and control (Montreal, QC, 1998)*, volume 528 of *NATO Sci. Ser. C Math. Phys. Sci.*, pages 551–598. Kluwer Acad. Publ., Dordrecht, 1999.

74. H.K. Khalil. *Nonlinear Systems, Third Edition*. Prentice Hall, Upper Saddle River, NJ, 2002.

75. H.R. Thieme. Convergence results and a poincaré-bendixson trichotomy for asymptotically autonomous differential equations. *J. Math. Biol.*, 30:755–763, 1992.

76. M. Hirsch. Convergent activation dynamics in continuous-time networks. *Neural Networks*, 2:331–349, 1989.

77. H.L. Smith. Convergent and oscillatory activation dynamics for cascades of neural nets with nearest neighbor competitive or cooperative interactions. *Neural Networks*, 4:41–46, 1991.

78. D. Angeli, M. Hirsch, and E.D. Sontag. Remarks on cascades of strongly monotone systems. in preparation.

79. L. Wang and E.D. Sontag. A remark on singular perturbations of strongly monotone systems. In *Proc. IEEE Conf. Decision and Control, San Diego, Dec. 2006*. IEEE, 2006. (to appear).

80. N. Fenichel. Geometric singular perturbation theory for ordinary differential equations. *J. of Differential Equations*, 31:53–98, 1979.

81. C.K.R.T. Jones. Geometric singular perturbation theory. In *Dynamical Systems (Montecatini Terme, 1994), Lect. Notes in Math. 1609*. Springer-Verlag, Berlin, 1994.

82. R. Devaney. *An Introduction to Chaotic Dynamical Systems, 2nd ed.* Addison-Wesley, Redwood City, 1989.

83. J.K. Hale. *Asymptotic Behavior of Dissipative Systems.* Amer. Math. Soc., Providence, 1988.

84. P. Poláčik and I. Tereščák. Convergence to cycles as a typical asymptotic behavior in smooth strongly monotone discrete-time dynamical systems. *Arch. Rational Mech. Anal.*, 116:339–360, 1992.

85. P. Poláčik and I. Tereščák. Exponential separation and invariant bundles for maps in ordered banach spaces with applications to parabolic equations. *J. Dynam. Differential Equations*, 5:279–303, 1993.

86. P. Hess and P. Poláčik. Boundedness of prime periods of stable cycles and convergence to fixed points in discrete monotone dynamical systems. *SIAM J. Math. Anal.*, 24:1312–1330, 1993.

87. I. Tereščák. Dynamics of c^1 smooth strongly monotone discrete-time dynamical system. Technical report, Comenius University, Bratislava, 1996.

88. J.-L. Gouze. A criterion of global convergence to equilibrium for differential systems. application to lotka-volterra systems. Technical Report RR-0894, INRIA, 1988.

89. J.-L. Gouze and K. P. Hadeler. Order intervals and monotone flow. *Nonlinear World*, 1:23–34, 1994.

90. S. Smale. On the differential equations of species in competition. *Journal of Mathematical Biology*, 3:5–7, 1976.

91. S.A. Kauffman. Metabolic stability and epigenesis in randomly constructed genetic nets. *Journal of Theoretical Biology*, 22:437– 467, 1969.

92. S.A. Kauffman. Homeostasis and differentiation in random genetic control networks. *Nature*, 224:177–178, 1969.

93. S.A. Kauffman and K. Glass. The logical analysis of continuous, nonlinear biochemical control networks. *Journal of Theoretical Biology*, 39:103–129, 1973.

94. R. Albert and H.G. Othmer. The topology of the regulatory interactions predicts the expression pattern of the *drosophila* segment polarity genes. *J. Theor. Biol.*, 223:1–18, 2003.

95. M. Chaves, R. Albert, and E.D. Sontag. Robustness and fragility of Boolean models for genetic regulatory networks. *J. Theoret. Biol.*, 235(3):431–449, 2005.

96. E.N. Gilbert. Lattice theoretic properties of frontal switching functions. *Journal of Mathematics and Physics*, 33:57–67, 1954.

97. I. Anderson. *Combinatorics of Finite Sets*. Dover Publications, Mineola, N.Y., 2002.

98. J. Aracena, J. Demongeot, and E. Goles. On limit cycles of monotone functions with symmetric connection graph. *Theor. Comput. Sci.*, 322(2):237–244, 2004.

99. M.L. Minsky. *Computation: finite and infinite machines*. Prentice-Hall, Englewood Cliffs, N.J., 1967.

100. Clive Maxfield. How to invert three signals with only two not gates (and *no* xor gates). Technical report, http://www.mobilehandsetdesignline.com, 2006.

101. H. Schneider and M. Vidyasagar. Cross-positive matrices. *SIAM J. Numer. Anal.*, 7:508–519, 1970.

102. P. Volkmann. Gewohnliche differentialungleichungen mit quasimonoton wachsenden funktionen in topologischen vektorraumen. *Math. Z.*, 127:157–164, 1972.

103. S. Walcher. On cooperative systems with respect to arbitrary orderings. *Journal of Mathematical Analysis and Appl.*, 263:543–554, 2001.

104. W. Walter. *Differential and Integral Inequalities*. Springer-Verlag, Berlin, 1970.

105. D. Angeli and E.D. Sontag. An analysis of a circadian model using the small-gain approach to monotone systems. In *Proc. IEEE Conf. Decision and Control, Paradise Island, Bahamas, Dec. 2004, IEEE Publications*, pages 575–578, 2004.

106. J.F. Jiang. On the global stability of cooperative systems. *Bulletin of the London Math Soc*, 6:455–458, 1994.

107. M. Chaves, E.D. Sontag, and R. J. Dinerstein. Steady-states of receptor-ligand dynamics: A theoretical framework. *J. Theoret. Biol.*, 227(3):413–428, 2004.

108. E.H. Snoussi. Necessary conditions for multistationarity and stable periodicity. *J. Biol. Sys.*, 6:3–9, 1998.

109. J.J. Tyson, K. Chen, and B. Novak. Sniffers, buzzers, toggles, and blinkers: dynamics of regulatory and signaling pathways in the cell. *Curr. Opin. Cell. Biol.*, 15:221–231, 2003.

110. J.E. Ferrell Jr and W. Xiong. Bistability in cell signaling: How to make continuous processes discontinuous, and reversible processes irreversible. *Chaos*, 11:227–236, 2001.

111. J.E. Lisman. A mechanism for memory storage insensitive to molecular turnover: a bistable autophosphorylating kinase. *Proc. Natl. Acad. Sci. USA*, 82:3055–3057, 1985.

112. M. Laurent and N. Kellershohn. Multistability: a major means of differentiation and evolution in biological systems. *Trends Biochem. Sci.*, 24:418–422, 1999.

113. T.S. Gardner, C.R. Cantor, and J.J. Collins. Construction of a genetic toggle switch in escherichia coli. *Nature*, 403:339–342, 2000.

114. J.E. Ferrell Jr and E.M. Machleder. The biochemical basis of an all-or-none cell fate switch in xenopus oocytes. *Science*, 280:895–898, 1998.

115. C.P. Bagowski and J.E. Ferrell Jr. Bistability in the jnk cascade. *Curr. Biol.*, 11:1176–1182, 2001.

116. U.S. Bhalla, P.T. Ram, and R. Iyengar. Map kinase phosphatase as a locus of flexibility in a mitogen-activated protein kinase signaling network. *Science*, 297:1018–1023, 2002.

117. F.R. Cross, V. Archambault, M. Miller, and M. Klovstad. Testing a mathematical model of the yeast cell cycle. *Mol. Biol. Cell*, 13:52–70, 2002.

118. A. Becskei, B. Seraphin, and L. Serrano. Positive feedback in eukaryotic gene networks: cell differentiation by graded to binary response conversion. *EMBO J.*, 20:2528–2535, 2001.

119. C.P. Bagowski, J. Besser, C.R. Frey, and J.E. Ferrell Jr. The jnk cascade as a biochemical switch in mammalian cells: ultrasensitive and all-or-none responses. *Curr. Biol.*, 13:315–320, 2003.

120. J.E. Ferrell Jr. Tripping the switch fantastic: How a protein kinase cascade can convert graded inputs into switch-like outputs. *Trends Biochem. Sci.*, 21:460–466, 1996.

121. D.J. Allwright. A global stability criterion for simple control loops. *J. Math. Biol.*, 4:363–373, 1977.

122. H.G. Othmer. The qualitative dynamics of a class of biochemical control circuits. *J. Math. Biol.*, 3:53–78, 1976.

123. C.D. Thron. The secant condition for instability in biochemical feedback-control .1. The role of cooperativity and saturability. *Bull. Math. Biology*, 53:383–401, 1991.

124. J. Mallet-Paret and H.L. Smith. The poincaré-bendixson theorem for monotone cyclic feedback systems. *J. Dynamics and Diff. Eqns.*, 2:367–421, 1990.

125. T. Gedeon. Cyclic feedback systems. *Mem. Amer. Math. Soc.*, 134:1–73, 1998.

126. H.L. Smith. Oscillations and multiple steady states in a cyclic gene model with repression. *J. Math. Biol.*, 25:169–190, 1987.

127. G.A. Enciso and E.D. Sontag. A characterization of the stability of strongly monotone systems. in preparation, 2006.

128. G. Enciso and E.D. Sontag. A remark on multistability for monotone systems ii. In *Proc. IEEE Conf. Decision and Control, Seville, Dec. 2005, IEEE Publications*, pages 2957–2962, 2005.

129. D. Angeli. Systems with counterclockwise input-output dynamics. *IEEE Transactions on Automatic Control*, 51:1130– 1143, 2006.

130. C-Y.F. Huang and J.E. Ferrell Jr. Ultrasensitivity in the mitogen-activated protein kinase cascade. *Proc. Natl. Acad. Sci. USA*, 93:10078–10083, 1996.

131. A.R. Asthagiri and D.A. Lauffenburger. A computational study of feedback effects on signal dynamics in a mitogen-activated protein kinase (mapk) pathway model. *Biotechnol. Prog.*, 17:227–239, 2001.

132. C. Widmann, G. Spencer, M.B. Jarpe, and G.L. Johnson. Mitogen-activated protein kinase: Conservation of a three-kinase module from yeast to human. *Physiol. Rev.*, 79:143–180, 1999.

133. S.Y. Shvartsman, H.S. Wiley, and D.A. Lauffenburger. Autocrine loop as a module for bidirectional and context-dependent cell signaling. Technical report, MIT Chemical Engineering Department, 2000.

134. M. Cartwright and M.A. Husain. A model for the control of testosterone secretion. *J. Theor. Biol.*, 123:239–250, 1986.

135. G. Enciso and E.D. Sontag. On the stability of a model of testosterone dynamics. *J. Math. Biol.*, 49(6):627–634, 2004.

136. J. Mahaffy and E.S. Savev. Stability analysis for a mathematical model of the *lac* operon. *Quarterly of Appl. Math.*, LVII:37–53, 1999.

137. M.R.S. Kulenovic and G. Ladas. *Dynamics of Second Order Rational Difference Equations.* Chapman & Hall/CRC, New York, 2002.

138. A. Goldbeter. A model for circadian oscillations in the drosophila period protein (per). *Proc. Royal Soc. Lond. B.*, 261:319–324, 1995.

139. J. Smillie. Competitive and cooperative tridiagonal systems of differential equations. *SIAM J. Math. Anal.*, 15:530–534, 1984.

System and Control Theory Furthers the Understanding of Biological Signal Transduction

Eric Bullinger[1], Rolf Findeisen[2], Dimitrios Kalamatianos[1], and Peter Wellstead[1]

[1] The Hamilton Institute, National University of Ireland, Maynooth, Co. Kildare, Ireland

[2] Institute for Systems Theory and Automatic Control, University of Stuttgart, 70550 Stuttgart, Germany

Summary. This article discusses why novel modelling and analysis methods are required for biological systems, presents recent advances and outlines some future challenges. In this respect, the main focus is placed upon methods for parameter estimation and sensitivity analysis as they are encountered in systems biology.

1 Introduction

An essential characteristic of higher organisms is the communication of information by cellular signalling. Whether this communication is within or between cells, the communicating signals are usually transmitted using a range of chemical substances. For example, and depending upon the type of neural connection, both chemical neurotransmitters and ions pass information between brain cells. Within the cell, the different extracellular signals that are received are integrated with the intra-cellular state along signalling pathways that decide a cell's fate. Since typical cellular fates include proliferation, differentiation and death, it is clear that cell signalling must be tightly and correctly regulated. Happily this is almost always the case. However, where mis-regulation exists, it too often leads to fatal or debilitating conditions such as cancer (where cells resist signals telling them to die) or neurodegeneration (as in Parkinson's Disease), where internal signalling mechanisms are inhibited and healthy neurons die.

In general, the reasons and mechanisms for the malfunction of cell signal transduction are poorly understood. Where there is knowledge, it is usually of a qualitative and static nature. What is required are quantitative models of dynamic behaviour of the kind used in systems and control theory. With this motivation, our overall aim in this chapter is to indicate some ways in which systems and control theory might provide new insight in the mechanisms of cell signalling and thus help illuminate reasons for mis-regulation.

A motivation for our work is the examination of how a systems approach to such problems might assist. In addressing these problems however we are at pains to remind ourselves that biology is not simply a new area of application for known systems and control theory [74]. The existing body of systems theory can

I. Queinnec et al. (Eds.): Bio. & Ctrl. Theory: Current Challenges, LNCIS 357, pp. 123–135, 2007.
springerlink.com

form a base from which to begin. However, the true systems biology challenge in cell signal transduction lies in the many problem–specific difficulties that appear, resulting from the complexity of interactions, uncertainties, and the wide variability present in biological cells. Overcoming these challenges requires the development of novel theories and methods adapted to the purpose in hand. Some methods already exist and discussed by Sontag in his helpful overview articles [64, 65], in which he lists many research directions. In his review, Sontag consciously constrains his listing and there are many more challenges that would benefit from discussion. A complete examination of the challenges is however beyond our scope, too. Rather in this article we focus on the aspects of parameter identification and sensitivity analysis. For these fields we try to show how a control perspective allows the design of novel methods for modelling and analysis of biological systems.

Due to the diversity of methods and results reviewed, we have decided to follow a purely descriptive path, leaving out a detailed mathematical description, as such can be found in the specific references provided.

2 History and Motivation

At first glance, systems biology has the appearance of a development with roots in the 1960's. In particular, 1968 is often cited as a significant date, since it was then that Mesarović first used the term [51]. However, it was not until the 1990's that systems biology gained popular currency and scientific momentum through the pioneering work of Kitano and others [75, 27, 36, 37, 1, 35]. Since then, systems biology has become an important area in biology, with a growth pattern marked by several regular conferences, journals and books dedicated to the area. History is however more complex, and systems and modelling approaches to biology have been considered for well over sixty years. For instance, the Dublin lectures and the associated book "What is life" by Schrödinger [61] are famous for the way in which the concept of systems was introduced to the biological context.

Even though there has been continuous interest since Schrödinger's time, systems and modelling approaches have had a limited impact on biology. The main obstacle preventing systems approaches being widely applied in biology was the difficulty of obtaining the required data. One of the first success stories in this respect was the work by Hodgkin and Huxley who were able to explain the electrical activity of nerve cells by combining the measurement of membrane potentials with a corresponding quantitative mathematical model [25]. The modelling of other signalling networks has proven more difficult however, mainly due to experimental reasons associated with measurement. But things are changing, and recent years have seen huge progress. For example, the genome is now completely known for several organisms, and significant progress has been achieved for the quantification of intra-cellular concentrations. These can now be measured both as the average of a cell population using e.g. mass spectrometry, or within single cells using e.g. electro-chemical sensors [56] or fluorescence tagging. In particular,

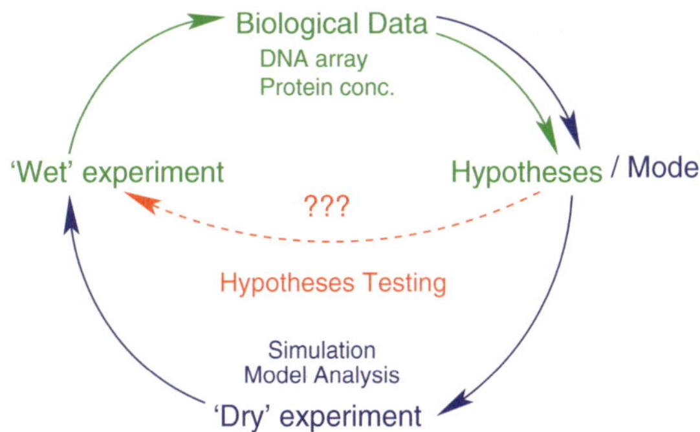

Fig. 1. The advancement of knowledge in systems biology can best be described by a cyclic process. Wet-lab experiments generate data such as mRNA expression profiles, protein concentrations or protein-protein interactions. These allow for simple cases to directly test hypotheses. In more complex cases, an intermediate step of modelling together with simulations and model analysis is required.

the last of these allows signalling to be quantified within unsynchronised cell populations as, for example, in cells undergoing programmed cell death [59].

The main goal of systems biology is that of gaining a holistic understanding of biology. In this spirit, systems biology builds on the knowledge of individual cellular components like genes and proteins as well as their interactions. The interactions are particularly important, since understanding systems properties (such as robustness) cannot be explained by the isolated study of individual parts of a cell [36, 37]. Rather, it requires us to study the dynamics and quantitative effects of the interaction of the systems' "pieces".

An essential tool for fitting the pieces together is mathematical modelling. Modelling allows the analysts to test hypotheses for their internal consistency and compare them with experimental data. It is particularly suited for understanding complex systems not accessible to purely observational description, such as they typically appear in systems biology. Furthermore, mathematical models have a capability that allows the analyst to make predictions, which can subsequently be tested experimentally [37].

From an organisational viewpoint, mathematical modelling is important since it provides a framework for integrating the work from many disciplines. This is vital to the systems approach, since systems biology requires the interdisciplinary cooperation and close interaction of experimentalists, data analysts, modellers and system analysts [37, 50]. The essential form of this interaction is shown in Figure 1 where, starting from experimental data, the first step consists of data filtering and data-mining. The information obtained from this process is then utilised to generate mathematical models, which in turn allow *in-vitro* simulations, predictions, as well as analytical studies of the models properties. The theoretical analysis of models is particularly important, since feedback from the analysis allows the design of better biological experiments with which to test hypotheses and validate (or falsify) models. This virtuous circle of experiment, modelling, hypothesis generation and validation is an essential feature of systems biology. It is in this way that systems biology currently facilitates and supports the understanding of biology, enabling the design of new medical therapies or the optimisation of biotechnological production control.

3 Importance of Dynamics and Quantitative Analysis

While most research in biology still focuses on static properties, like decoding the genome or finding possible protein-protein interaction, the dynamical characteristics of biological systems are increasingly recognised as essential for understanding the mechanisms of life [47]. For example, phenomena like the cellular responses to osmotic or heat shocks, or neuronal communication via spikes, clearly require analysis of signal dynamics (see [42], [16] and [46], respectively).

To understand biology at the system level, we must examine the structure and dynamics of cellular and organismal function, rather than the characteristics of isolated parts of a cell or organism. Properties of systems, such as robustness, emerge as central issues, and understanding these properties may have an impact on the future of medicine [37]. Metabolic analysis, sensitivity analysis, and dynamic analysis methods such as phase portrait and bifurcation analysis, play a significant role in understanding how a system behaves over time and under various conditions. Bifurcation analysis traces time-varying changes in the state of the system in a multidimensional space where each dimension represents a particular concentration of the biochemical factor involved. Developmental biologists and other scientists have proposed that cells behave as nonlinear dynamical systems, and so bifurcation phenomena play a major role [66].

Recently, it has become apparent that distinct spatio-temporal activation profiles of the same repertoire of signalling proteins result in different gene-expression patterns and diverse physiological responses [55, 26]. These observations indicate that pivotal cellular decisions, such as cytoskeletal reorganisation, cell-cycle checkpoints and cell death (apoptosis), depend on the precise temporal control and relative spatial distribution of activated signal transducers [33]. To understand how cells react to signals, it is therefore necessary to describe the signalling pathways with dynamic and quantitative models. Understanding the biological components and their interactions is a central part of the research area of systems biology, and an essential step in elucidating the control mechanisms within cells, organs and organisms [38, 76].

Although an impressive amount of qualitative network (schematic modelling) knowledge is available, little of this knowledge is quantitative at the level of precision demanded by most control and system theoretic analysis tools [66]. Molecular interactions have to be precisely described in mathematical formulas that reflect the right level of abstraction suitable for specific biology, and the necessary parameters, such as initial concentration of each component and kinetic constants, have to be estimated from a set of experiments. The problem of exploiting this qualitative knowledge, and effectively integrating relatively sparse quantitative data, is among the most challenging issues confronting systems biology [40].

Quantitative models that generate novel, experimentally testable hypotheses will have an increasingly important role in post-genomic biology. Future models will integrate data on the distinct spatio-temporal dynamics of signalling from different cellular compartments and provide new insights into the connection between external stimuli and the signalling outcome in terms of gene-expression responses. Challenges of the combinatorial complexity of signalling networks and experimental uncertainty in parameter values could be addressed by modular approaches, and stochastic and pattern-oriented modelling [33]. The goal of the pattern-oriented approach is to predict and explain dynamic patterns of cellular responses to a multitude of external cues and perturbations.

Since the communicating signals are themselves dynamic, it is clear that in order for understand how cells react to signals, it is also necessary to describe the signalling pathways in quantitative dynamic terms. In this context, the role of the biological components and their interactions in the dynamical response of a cell is a central part of systems biology research. In particular, the control functions of cells, organs and organisms are all crucially dependent upon dynamic phenomena [38, 76]. Thus, an understanding of cell dynamics is an essential precursor to understanding the control and regulation processes in and between cells.

A final cautionary note on models of dynamic features is needed before proceeding: the models currently used to represent dynamical features of cells are often approximations. For example, quasi-steady-state assumptions are common in modelling biological systems and are (for example) the theoretical reason behind Michaelis-Menten or Hill kinetic models [32]. Such approximations have been successfully used to represent biological sub-systems. However, the accumulation of a number of these approximations within a model of an overall system can significantly alter model behaviour [52] leading to erroneous results.

4 Mathematical Models

Modelling intra-cellular signalling transduction can be achieved in many different ways [11], though the most common model forms are systems of ordinary differential equations that are polynomial in the states. A further particularity of these models is that both states and parameters are positive. Usually, the structure of the model is fixed by the set of chemical reactions that are assumed to be essential to a signalling function. Although the structure can often be fixed with some degree of certainty, the values of the model parameters are often highly uncertain. Several system analysis methods take these model properties explicitly into account. One example is the Horn-Jackson-Feinberg modelling that allows, under certain structural conditions, to determine whether multiple steady states can occur [9].

The time and effort needed to assemble mathematical models, and determine the numerical values of their coefficients, has led to the sharing of models. As a result mathematical models of signalling networks can increasingly be found online. An overview of model databases can be found in [77], with particular examples:

the BioModels Database [3], JWS Online [29], the Physiome project [57] and the cellML model repository [6]. These online resources are of great value to the systems biologist, however there are issues that need addressing—the central one being the unreliability of the information concerning model parameter values. The key problem here is that the estimation of the parameters of biological models is intrinsically difficult. The peculiarities of model structures, the heterogeneous forms that measured data takes and the large level of noise often encountered, mean that classical systems identification algorithms are of limited value. In particular, the large body of traditional discrete-time systems identification approaches are of limited value compared to continuous-time identification [21]. These special aspects of biological systems present new problems in estimation theory and require new classes of quite distinct theoretical approaches. Some of these approaches are outlined in the following section.

5 Parameter Estimation

A common approach to parameter estimation in systems biology is the use of global search algorithms based on least square estimates or on evolutionary algorithms. Reference [23] gives an overview of these areas, and [53, 20] are surveys within systems biology. Typical methods proposed are genetic algorithms using multiple local optimisations that are solved as parallel problems. The solutions are then combined to generate a new starting point for local optimisation. As the problem is in general non-convex, parameter estimation becomes more and more difficult with increasing model size [58]. Less heuristic are branch-and-bound approaches, but they are usually applicable only to small systems [62].

Global estimation approaches usually require a high computational load, but even so, neither have a guarantee of convergence, nor a measure of how close their solution is to the optimal approximation. Also, they do not provide a conclusive answer as to whether parameters can be identified at all from the available measurements. There is therefore a strong need for new parameter estimation techniques that are designed for the particularities of biochemical reaction networks. These methods should go beyond (a) local identifiability analyses [80] (since these are problematic for nonlinear systems) or (b) algebraic approaches [2, 49, 78], which only work in very low state dimensions.

In mathematical models of signal transduction networks, flows usually follow the law of mass action, i.e. the flow is proportional to each of its substrates (see e.g. [32]). This results in the differential equation model being automatically linear in the parameters, which are the kinetic rate constants. Only few publications deal with this specificity of biological systems, e.g. [18] with the identifiability, while [31] shows that in the noise-free, full-state information case, identification can be solved using a least-square optimisation. This case is however far from realistic—an observation that underlines again the need for developing systems identification techniques, which address realistic biological model forms.

Under practical experimental conditions it is unusual for all concentrations to be measurable, at least not with a sufficiently fast sampling rate. Taking into account the special structure of biochemical reaction networks, we were able to derive a novel system description related to the Horn-Jackson-Feinberg formalism [18] where the states are concentrations and flows while the parameters are hidden in the initial conditions. A key component in deriving this system description is a factorisation of the stoichiometry and of the fluxes, similar to [22], and other publications on Horn-Jackson-Feinberg modelling. The main advantage of this novel system description is that it transforms the analysis of identifiability into an equivalent observability problem. This transformed form of the problem allows the reformulation of the parameter estimation task into a state estimation problem, which can be combined with the estimation of unmeasured states in a straight-forward way.

6 Sensitivity and Robustness

Biological systems belong to the important class of systems that involve high complexity, and within this class the systems concept of robustness is a key feature [4, 68]. In this context, a classical example of a biological signalling pathway for the study of robustness is chemotaxis. Chemotaxis is the mechanism responsible for cells navigating in the direction of an attractive chemical, such as sugar for a bacteria. It has been shown that robustness in the movement is achieved via feedback control [79] and can be used to discriminate between different models [44]. Other examples of systems where robustness has been used for model discrimination are the *Xenopus* cell cycle [54] or the apoptotic signalling system [14].

Biological systems, even more than complex technical systems, are inherently robust against changes in their environment. They also seem to be very robust against perturbations of kinetic parameters, see e.g. [72]. However, as cells also need to be able to react to certain changes, they cannot be completely robust. By analogy with corresponding control theory terminology [13], Carlson and Doyle called this principle "robust, yet fragile" and showed it to be an essential property of complex systems (see also [4]). Robustness can be achieved in different ways. Kitano lists four basic mechanisms: systems control, redundancy, modularity and decoupling [41]. The last three can be achieved by appropriate network structures. Best known are feedback loops for robustness, combined with feedforward loops to increase performance [12], as for example in the heat shock response system [17, 45] or the Tryptophan regulation in *E. coli* [71].

Negative feedback requires that the system parameters are within certain ranges. The right sign of the feedback is not enough as negative feedback can be destabilising and lead to oscillations, in particular if delays are present, see e.g. [60, 63, 73]. Fragility is a trade-of for achieving robustness, but can also be used in itself in a positive sense, for example in the design of novel drug targets [8]. In the same spirit, and as an essential property of biological systems, robustness can be used for model discrimination, i.e. for choosing the best suited mathematical models among a set of possible models [54, 39]. For example, *E. coli* uses the smallest sufficient robust network structure for chemotaxis [44].

The study of robustness with respect to parameter variations is commonly known as sensitivity analysis. Understanding robustness, in particular in the case of inherent control mechanisms, requires system analysis of a control theoretic kind [69, 67]. A now classical sensitivity analysis approach for the cellular metabolism is Metabolic Control Analysis (MCA) [30, 24, 10]. Sensitivity analysis is not only useful for model discrimination. Classical applications are metabolic engineering where the parameters usually correspond to enzyme concentrations [19]. A systems approach allows the investigator to uncover not only single targets, but also multiple targets [43]. While metabolic control analysis has mostly been applied in biotechnology, the same framework can also be used to look for possible drug targets [5].

For biological systems, robustness with respect to parameter changes is usually referred to as sensitivity. In signalling models, the most uncertain parts are the kinetic parameters. Knowledge of the precise value of some of them is important, while others are less important due to internal control mechanisms. Sensitivity analysis tries to uncover the important ones, and in systems biology, it is mostly linear sensitivity analysis that is used. MCA relates infinitesimal parameter changes to the resulting steady-state offset. For the analysis of signalling, such local analyses are often not sufficient and theoretical extensions are required. Possible extensions are sensitivity analyses along trajectories [28] or global analyses, see e.g. [14]. Alternatively, μ-analysis can be used to measure sensitivity with respect to multiple parameters around an operating point or limit cycle, see e.g. [34].

Recent developments show that systems and control can inspire new robustness analysis methods. For example, MCA's sensitivity analysis can also be expressed as the trace of the cross Gramian in a way that allows a nonlinear extension [70]. These new approaches are, as yet, few in number and there is a need for further methods, in particular for non-local robustness analysis. Non-local robust analysis is a relevant example, as different sensitivity analysis methods do not necessarily give similar results [15].

7 Closing Remarks

This article discusses why novel modelling and analysis methods are required for biological systems and presents recent advances as well as future challenges. In this context, it is sure that many improvements are destined to occur in systems biology, concerning both the experimental and the theoretical side of this field. The biggest challenge is smoothing the interplay between biology and systems sciences. This requires biologists to become familiar with systems approaches and mathematical modelling [7] and for control and systems scientists to learn the methods and "language" of biology. An important way to facilitate this process is the development of more and easier to use methods for feeding the results of theoretical studies back to the experimental biologist. The more seamless the transfer of information the more harmonious is the communication between biologist and systems analyst. For example, model-based experimental design that helps to choose the right sampling time, sampling instant and type of measurements, is of enormous value to the experimentalist and motivates a respect for systems analysis.

On the theoretical side, important challenges concern modelling and the analysis of mathematical models. At first sight, systems biology seems to be just another application area of systems and control. However, biological systems themselves, as well as the type of experiments that are possible, have many peculiarities which make the straight-forward transfer from classical systems theory to biology difficult, if not impossible [65]. To name just a few differences, measurements in biology are usually substantially noisier and the systems more uncertain than in areas such as mechanical or electrical engineering. Also, the number of samples (in the sense of different time points or the number of repeated experiments) is often very low. It is therefore necessary to adapt system theoretical results, or even develop new methodologies that are designed particularly for biological systems.

At present, published mathematical models of cellular signalling only describe very small parts of what is going on in cells. For example, a mathematical model might describe one signalling pathway or the cross-talk between two pathways. In the future, there will be the need to model many more pathways simultaneously as well as the combination of genomic and proteomic parts of the pathways. Other sources of model complexity are spatio-temporal dynamics, time-delays and heterogeneous cell populations. While the model structure is often relatively well known, and the concentrations of the components can, at least in part, be measured, the kinetic parameters such as protein degradation rates are often not known. This lack of knowledge may even extend to their order of magnitude. In engineering applications, numerical values for kinetic parameters can be extracted directly from time series data using methods from systems identification. These are however not directly applicable to biological systems. Thus, systems identification tools that are specifically tailored to systems biology are needed. They should both be optimal for the class of mathematical models used to describe biochemical reactions and for the type of experimental data. Usually, biological experiments provide few, but noisy data points and system identification methods need to be adapted to this class of application [48].

Complex mathematical models are difficult to understand. Running a few simulations can give a very biased view, and extensive simulation studies are intractable for the complex and uncertain mathematical models that do arise in systems biology. It is therefore necessary to extract key model features using systems analysis tools. Several methods for analysing systems are already applied in systems biology. The main challenges lie in the analysis of dynamic, non-steady-state systems such as those commonly found in cell signalling. Also, there should be methods that allow us to quantify how sensitive the dynamic properties of a mathematical model are to the parameters. In particular, the analysis of nonlinear phenomena like bifurcation theory or dynamic sensitivity analysis are important to advance the understanding of mathematical models of signalling pathways.

References

1. A. Agrawal. New institute to study systems biology. *Nat Biotechnol*, 17(8):743–744, 1999.
2. S. Audoly, G. Bellu, L. D'Angi'o, M. P. Saccomani, and C. Cobelli. Global identifiability of nonlinear models of biological systems. *IEEE T Bio-med Eng*, 48(1):55–65, 2001.
3. Biomodels database. www.biomodels.net.
4. J. M. Carlson and J. Doyle. Complexity and robustness. *Proc Natl Acad Sci USA*, 99(Suppl 1):2538–2545, 2002.
5. M. Cascante, L. G. Boros, B. Comin-Anduix, P. de Atauri, J. J. Centelles, and P. W.-N. Lee. Metabolic Control Analysis in drug discovery and disease. *Nat Biotechnol*, 20(3):243–249, Mar. 2002.
6. cellML model repository. www.cellml.org/examples/repository/.
7. K.-H. Cho and O. Wolkenhauer. Analysis and modelling of signal transduction pathways in systems biology. *Biochem Soc T*, 31(6):1503–1509, 2003.
8. A. Citri and Y. Yarden. EGF-ERBB signalling: Towards the systems level. *Nat Rev Mol Cell Biol*, 7(7):505–516, 2006.
9. C. Conradi, J. Saez-Rodriguez, E.-D. Gilles, and J. Raisch. Using chemical reaction network theory to d a kinetic mechanism hypothesis. *IEE Proc Syst Biol*, 152(4):243–248, 2005.
10. A. Cornish-Bowden. *Fundamentals of Enzyme Kinetics*. Portland Press, 3rd edition, 2004.
11. E. J. Crampin, S. Schnell, and P. E. McSharry. Mathematical and computational techniques to deduce complex biochemical reaction mechanisms. *Prog Biophys Mol Biol*, 86(1):77–112, 2004.
12. M. E. Csete and J. C. Doyle. Reverse engineering of biological complexity. *Science*, 295(5560):1664–1669, 2002.
13. P. Dorato. Non-fragile controller design: an overview. In *Proc. of the 1998 American Control Conf*, pages 2829–2831, 1998.
14. T. Eißing, F. Allgöwer, and E. Bullinger. Robustness properties of apoptosis models with respect to parameter variations and intrinsic noise. *IEE Proc Syst Biol*, 152(4):221–228, 2005.
15. T. Eißing, S. Waldherr, F. Allgöwer, P. Scheurich, and E. Bullinger. Response to bistability in apoptosis: Roles of Bax, Bcl-2, and mitochondrial permeability transition pores. *Biophys. J.*, 2007. In press.
16. H. El-Samad, H. Kurata, J. C. Doyle, C. A. Gross, and M. Khammash. Surviving heat shock: control strategies for robustness and performance. *Proc Natl Acad Sci USA*, 102(8):2736–2741, 2005.
17. H. El-Samad, H. Kurata, J. C. Doyle, C. A. Gross, and M. Khammash. Surviving heat shock: Control strategies for robustness and performance. *Proc Natl Acad Sci USA*, 102(8):2736–2741, 2005.
18. M. Farina, R. Findeisen, E. Bullinger, S. Bittanti, F. Allgöwer, and P. Wellstead. Results towards identifiability properties of biochemical reaction networks. In *Proc. of the 45th IEEE Conf on Decision and Control, San Diego, USA*, pages 2104–2109, 2006.
19. D. A. Fell. Increasing the flux in metabolic pathways: A Metabolic Control Analysis perspective. *Biotechnol Bioeng*, 58(2–3):121–124, 1998.
20. X.-J. Feng, S. Hooshangi, D. Chen, G. Li, R. Weiss, and H. Rabitz. Optimizing genetic circuits by global sensitivity analysis. *Biophys J*, 87(4):2195–2202, 2004.

21. H. Garnier, M. Mensler, and A. Richard. Continuous-time model identification from sampled data implementation issues and performance evaluation. *Int J Control*, 76(13):1337–1357, 2003.

22. K. Gatermann and B. Huber. A family of sparse polynomial systems arising in chemical reaction systems. *J Symb Comput*, 33:275–305, 2002.

23. P. Gray, W. Hart, L. Painton, C. Phillips, M. Trahan, and J. Wagner. A survey of global optimization methods. Technical report, Sandia National Laboratories, 1997.

24. R. Heinrich and T. A. Rapoport. A linear steady state treatment of enzymatic chains. general properties, control and effector strength. *European J of Biochemistry*, 42:89–95, 1974.

25. A. L. Hodgkin and A. F. Huxley. A quantitative description of membrane current and its application to conduction and excitation in nerve. *J Physiol.*, 117(4):500–544, 1952.

26. A. Hoffmann, A. Levchenko, M. L. Scott, and D. Baltimore. The IκB-NF-κB signaling module: temporal control and selective gene activation. *Science*, 298:1241–1245, 2002.

27. T. Ideker, T. Galitski, and L. Hood. A new approach to decoding life: Systems biology. *Annu Rev Genomics Hum Genet*, 2:343–372, 2001.

28. B. P. Ingalls and H. M. Sauro. Sensitivity analysis of stoichiometric networks: an extension of metabolic control analysis to non-steady state trajectories. *J Theor Biol*, 222(1):23–36, 2003.

29. JWS Online. `jjj.biochem.sun.ac.za`.

30. II. Kacser and J. A. Burns. The control of flux. *Symposia of thc Socicty for Experimental Biology*, 27:65–104, 1973.

31. A. V. Karnaukhov and E. V. Karnaukhova. Application of a new method of nonlinear dynamical system identification to biochemical problems. *Biochemistry (Moscow)*, 68(3):253–259, 2003. Translated from Biokhimiya, Vol. 68, No. 3, 2003, pp. 309–317.

32. J. Keener and J. Sneyd. *Mathematical Physiology*, volume 8 of *Interdisciplinary Applied Mathematics*. Springer-Verlag, New York, second edition, 2001.

33. B. N. Kholodenko. Cell-signalling dynamics in time and space. *Nat. Rev. Mol. Cell Biol.*, 7:165–176, 2006.

34. J. Kim, D. Bates, I. Postlethwaite, L. Ma, and P. Iglesias. Robustness analysis of biochemical network models. *IEE Proc Syst Biol*, 153(3):96–104, 2006.

35. H. Kitano. Perspectives on systems biology. *New Generation Computing*, 18(3):199–216, 2000.

36. H. Kitano. Preface. In H. Kitano, editor, *Foundations of Systems Biology*, pages xiii–xv. MIT Press, Cambridge/MA, 2001.

37. H. Kitano. Systems biology: a brief overview. *Science*, 295(5560):1662–1664, 2002.

38. H. Kitano. Systems biology: A brief overview. *Science*, 295:1662–1664, 2002.

39. H. Kitano. Biological robustness. *Nat Rev Genet*, 5:826–837, 2004.

40. II. Kitano. International alliances for quantitative modeling in systems biology. *Mol Syst Biol*, 1:2005.0007, 2005.

41. H. Kitano. Robustness from top to bottom. *Nat Genet*, 38(2):133–133, 2006.

42. E. Klipp, B. Nordlander, R. Krüger, P. Gennemark, and S. Hohmann. Integrative model of the response of yeast to osmotic shock. *Nat Biotechnol*, 23(8):975–982, 2005.

43. M. Koffas and G. Stephanopoulos. Strain improvement by metabolic engineering: Lysine production as a case study for systems biology. *Curr Opin Biotechnol*, 16(3):361–366, 2005.

44. M. Kollmann, L. Lovdok, K. Bartholomé, J. Timmer, and V. Sourjik. Design principles of a bacterial signalling network. *Nature*, 438(7067):504–507, 2005.

45. H. Kurata, H. El-Samad, R. Iwasaki, H. Ohtake, J. C. Doyle, I. Grigorova, C. A. Gross, and M. Khammash. Module-based analysis of robustness tradeoffs in the heat shock response system. *PLoS Comput Biol*, 2(7):e59, 2006.

46. R. Legenstein, C. Naeger, and W. Maass. What can a neuron learn with spike-timing-dependent plasticity? *Neural Comput*, 17(11):2337–2382, 2005.

47. A. Levchenko. Dynamical and integrative cell signaling: challenges for the new biology. *Biotechnol Bioeng*, 84(7):773–782, 2003.

48. L. Ljung. Bode lecture: Challenges of non-linear identification. In *39th IEEE Conf on Decision and Control, Maui, Hawaii, USA*, 2003.

49. L. Ljung and T. Glad. On global identifiability for arbitrary model parametrization. *Automatica*, 30(2):265–276, 1994.

50. J. O. McInerney. Bioinformatics in a post-genomics world—the need for an inclusive approach. *Pharmacogenomics J*, 2(4):207–208, 2002.

51. M. D. Mesarović. Systems theory and biology—view of a theoretician. In M. D. Mesarović, editor, *Systems Theory and Biology*, pages 59–87. Springer Verlag, 1968.

52. T. Millat, E. Bullinger, J. Rohwer, and O. Wolkenhauer. Approximations and their consequences for dynamic modelling of signal transduction pathways. *Math Bioscience*, 2006. In press.

53. C. G. Moles, P. Mendes, and J. R. Banga. Parameter estimation in biochemical pathways: a comparison of global optimization methods. *Genome Res*, 13(11):2467–2474, 2003.

54. M. Morohashi, A. E. Winn, M. T. Borisuk, H. Bolouri, J. Doyle, and H. Kitano. Robustness as a measure of plausibility in models of biochemical networks. *J Theor Biol*, 216:19–30, 2002.

55. L. O. Murphy, S. Smith, R. H. Chen, D. C. Fingar, and J. Blenis. Molecular interpretation of ERK signal duration by immediate early gene products. *Nature Cell Biol.*, 4:556–564, 2002.

56. R. D. O'Neill, J. P. Lowry, and M. Mas. Monitoring brain chemistry in vivo: Voltammetric techniques, sensors, and behavioral applications. *Crit Rev Neurobiol*, 12(1–2):69–127, 1998.

57. Physiome project models. `www.physiome.org/Models/`.

58. P. K. Polisetty, E. O. Voit, and E. P. Gatzke. Identification of metabolic system parameters using global optimization methods. *Theor Biol Med Model*, 3:4, 2006.

59. M. Rehm, H. Dussmann, R. U. Janicke, J. M. Tavare, D. Kogel, and J. H. M. Prehn. Single-cell fluorescence resonance energy transfer analysis demonstrates that caspase activation during apoptosis is a rapid process. role of caspase-3. *J Biol Chem*, 277(27):24506–24514, July 2002.

60. J. C. Schöning and D. Staiger. At the pulse of time: Protein interactions determine the pace of circadian clocks. *FEBS Lett*, 579(15):3246–3252, 2005.

61. E. Schrödinger. *What is Life? The physical aspect of the living cell.* Cambridge University Press, Cambridge, 1944.

62. A. B. Singer, J. W. Taylor, P. I. Barton, and W. H. Green. Global dynamic optimization for parameter estimation in chemical kinetics. *J Phys Chem A*, 110(3):971–976, 2006.

63. P. Smolen, D. A. Baxter, and J. H. Byrne. Modeling circadian oscillations with interlocking positive and negative feedback loops. *J Neurosci*, 21(17):6644–6656, 2001.

64. E. Sontag. Some new directions in control theory inspired by systems biology. *Syst Biol*, 1(1):9–18, 2004.

65. E. Sontag. Molecular systems biology and control. *European Journal of Control*, 11:396–435, 2005.

66. E. D. Sontag. Molecular systems biology and control: A qualitative-quantitative approach. In *Proc. of the 44th IEEE Conf on Decision and Control and European Control Conf, ECC'05, Seville, Spain*, pages 2314–2319, 2005.

67. J. Stelling, E. Gilles, and F. J. Doyle, III. Robustness properties of circadian clock architectures. *Proc Natl Acad Sci USA*, 101(36):13210 – 13205, 2004.

68. J. Stelling, U. Sauer, Z. Szallasi, F. J. Doyle, and J. Doyle. Robustness of cellular functions. *Cell*, 118(6):675–685, 2004.

69. J. Stelling, U. Sauer, Z. Szallasi, I. Francis J. Doyle, and J. Doyle. Robustness of cellular functions. *Cell*, 118:675–685, 2004.

70. S. Streif, R. Findeisen, and E. Bullinger. Relating cross Gramians and sensitivity analysis in systems biology. In *Proc. of the 17th International Symposium on Mathematical Theory of Networks and Systems, 24-28 July, Kyoto, Japan*, pages 437–442, 2006.

71. K. V. Venkatesh, S. Bhartiya, and A. Ruhela. Multiple feedback loops are key to a robust dynamic performance of tryptophan regulation in *escherichia coli*. *FEBS Lett*, 563(1–3):234–240, 2004.

72. G. von Dassow, E. Meir, E. M. Munro, and G. M. Odell. The segment polarity network is a robust developmental module. *Nature*, 406(6792):188–192, 2000.

73. R. Wang, Z. Jing, and L. Chen. Modelling periodic oscillation in gene regulatory networks by cyclic feedback systems. *Bull Math Biol*, 67(2):339–367, 2005.

74. O. Wolkenhauer. Systems biology: the reincarnation of systems theory applied in biology? *Brief Bioinform*, 2(3):258–270, 2001.

75. O. Wolkenhauer. Systems biology: The reincarnation of systems theory applied in biology? *Brief Bioinform*, 2(3):258–270, 2001.

76. O. Wolkenhauer, S. N. Sreenath, P. Wellstead, M. Ullah, and K.-H. Cho. A systems- and signal-oriented approach to intracellular dynamics. *Biochem Soc T*, 33(Pt 3):507–515, 2005.

77. www.systems-biology.org. `www.systems-biology.org/001/`.

78. X. Xia and C. H. Moog. Identifiability of nonlinear systems with application to HIV/AIDS models. *IEEE T Automat Contr*, 48(2):330–336, 2003.

79. T.-M. Yi, Y. Huang, M. Simon, and J. Doyle. Robust perfect adaptation in bacterial chemotaxis through integral feedback control. *Proc Natl Acad Sci USA*, 97(9):4649–4653, 2000.

80. D. E. Zak, G. E. Gonye, J. S. Schwaber, and F. J. Doyle III. Importance of input perturbations and stochastic gene expression in the reverse engineering of genetic regulatory networks: Insights from an identifiability analysis of an in silico network. *Genome Res*, 13(11):2396–2405, 2003.

Piecewise-Linear Models of Genetic Regulatory Networks: Theory and Example

Frédéric Grognard[1], Hidde de Jong[2], and Jean-Luc Gouzé[1]

[1] COMORE INRIA, Unité de recherche Sophia Antipolis, 2004 route des Lucioles, BP 93, 06902 Sophia Antipolis, France
`frederic.grognard@inria.fr, gouze@sophia.inria.fr`
[2] HELIX INRIA, Unité de recherche Rhône-Alpes, 655 avenue de l'Europe, Montbonnot, 38334 Saint Ismier Cedex, France
`Hidde.de-Jong@inrialpes.fr`

Summary. The experimental study of genetic regulatory networks has made tremendous progress in recent years resulting in a huge amount of data on the molecular interactions in model organisms. It is therefore not possible anymore to intuitively understand how the genes and interactions together influence the behavior of the system. In order to answer such questions, a rigorous modeling and analysis approach is necessary. In this chapter, we present a family of such models and analysis methods enabling us to better understand the dynamics of genetic regulatory networks. We apply such methods to the network that underlies the nutritional stress response of the bacterium *E. coli.*

The functioning and development of living organisms is controlled by large and complex networks of genes, proteins, small molecules, and their interactions, so-called *genetic regulatory networks*. The study of these networks has recently taken a qualitative leap through the use of modern genomic techniques that allow for the simultaneous measurement of the expression levels of all genes of an organism. This has resulted in an ever growing description of the interactions in the studied genetic regulatory networks. However, it is necessary to go beyond the simple description of the interactions in order to understand the behavior of these networks and their relation with the actual functioning of the organism. Since the networks under study are usually very large, an intuitive approach for their understanding is out of question. In order to support this work, mathematical and computer tools are necessary: the unambiguous description of the phenomena that mathematical models provide allows for a detailed analysis of the behaviors at play, though they might not exactly represent the exact behavior of the networks.

In this chapter, we will be mostly interested in the modeling of the genetic regulatory networks by means of *differential equations*. This classical approach allows precise numerical predictions of deterministic dynamic properties of genetic regulatory networks to be made. However, for most networks of biological interest the application of differential equations is far from straightforward. First, the biochemical reaction mechanisms underlying the interactions are usually not

I. Queinnec et al. (Eds.): Bio. & Ctrl. Theory: Current Challenges, LNCIS 357, pp. 137–159, 2007.

or incompletely known, which complicates the formulation of the models. Second, quantitative data on kinetic parameters and molecular concentrations is generally absent, even for extensively-studied systems, which makes standard numerical methods difficult to apply. In practice, the modeler disposes of much weaker information on the network components and their interactions. Instead of details on the mechanisms through which a protein regulates a gene, we typically only know whether the protein is an activator or an inhibitor. And even if it had been shown, for example, that the protein binds to one or several sites upstream of the coding region of the gene, numerical values of dissociation constants and other parameters are rarely available. At best, it is possible to infer that the regulatory protein strongly or weakly binds to the DNA, with a greater affinity for one site than for another.

Due to those uncertainties, we cannot hope to build a model that is guaranteed to reproduce the exact behavior of the considered genetic regulatory network. No model will be *quantitatively* accurate. It is therefore necessary to concentrate on the construction of models that reproduce the *qualitative dynamical properties* of the network, that is, dynamical properties that are invariant for a range of parameter values and reaction mechanisms. The qualitative properties express the intimate connection between the behavior of the system and the structure of the network of molecular interactions, independently from the quantitative details of the latter.

Consequently, qualitative approaches have been developed for the modeling, analysis, and simulation of genetic regulatory networks and other networks of biological interactions: Boolean networks [20, 30], Petri nets [22, 27], process algebras [28], qualitative differential equations [17], hybrid automata [11],... In this chapter, we concentrate on one particular class of qualitative models of genetic regulatory networks, originally proposed by Glass and Kauffman [12]: *piecewise-linear (PL) differential equations*. In Section 1, we describe this family of models and give a small example. In Section 2, we show qualitative results that have been obtained for the analysis of such systems. We then illustrate these models on the nutritional stress response of *E. coli* in Section 3, before discussing remaining challenges for the analysis and control of such models in Section 4.

1 Models of Genetic Regulatory Networks

Among the many emerging families of models (see [5]), a class of piecewise-linear (PL) models, originally proposed by Glass and Kauffman [12], has been widely used in modeling genetic regulatory networks. The variables in the piecewise-linear differential equation (PLDE) models are the concentrations of proteins encoded by the genes, while the differential equations describe the regulatory interactions in the network by means of step functions. The use of step functions is motivated by the switch-like behavior of many of the interactions in genetic regulatory networks [26], but it leads to some mathematical difficulties. The vector field for the PLDE model is undefined when one of the variables assumes a value where the step function is discontinuous, referred to as a threshold value. Recent work by Gouzé and Sari [13] uses an approach due to Filippov to define the solutions on the threshold hyperplanes. The approach involves extending the PLDE to a piecewise-linear differential inclusion (PLDI). As is well known, such discontinuities can lead to sliding modes. The definitions and results of this section are mainly taken from [3].

The family of PL-models is best illustrated with an example: the schematic diagram in Figure 1 describes a simple genetic regulatory network. In this example, the genes a and b code for the proteins A and B, which in turn control the expression of the two genes a and b. Protein A inhibits gene a and activates gene b above certain threshold concentrations, which are assumed to be different. Similarly protein B inhibits gene b and activates gene a above different threshold concentrations. This two-gene regulatory network is simple but represents many features of regulation found in real networks: auto-regulation, cross-regulation and inhibition/activation. Such a two-gene network could be found as a module of a more complex genetic regulatory network from a real biological system.

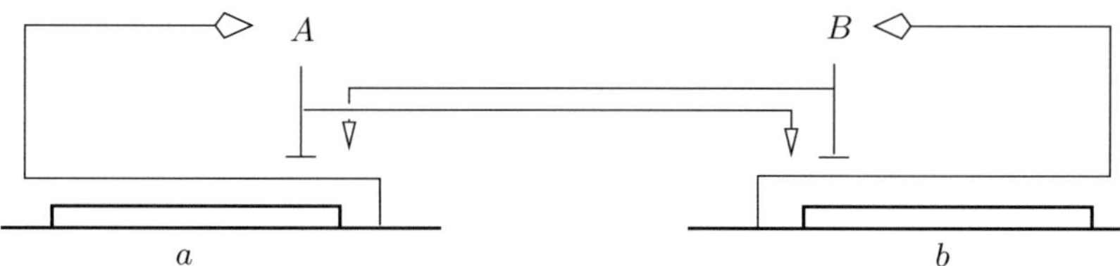

Fig. 1. Example of a genetic regulatory network of two genes (a and b), each coding for a regulatory protein (A and B)

The equations modeling the example network in Figure 1 can be written down as

$$\begin{cases} \dot{x}_a = \kappa_a s^+(x_b, \theta_b^1)s^-(x_a, \theta_a^2) - \gamma_a x_a \\ \dot{x}_b = \kappa_b s^+(x_a, \theta_a^1)s^-(x_b, \theta_b^2) - \gamma_b x_b \end{cases} \tag{1}$$

where $s^+(x_s, \theta_s)$ is equal to 0 when $x_s < \theta_s$ and equal to 1 when $x_s > \theta_s$ and $s^-(x_s, \theta_s) = 1 - s^+(x_s, \theta_s)$. In this model, gene a is expressed at a rate κ_a if the concentration x_b of protein b is above the threshold θ_b^1 and the concentration x_a of protein A is below the threshold θ_a^2. Similarly, gene b is expressed at a rate κ_b if the concentration x_a of protein A is above the threshold θ_a^1 and the concentration x_b of the protein B is below the threshold θ_b^2. Degradation of both proteins is assumed to be proportional to their own concentrations, so that the expression of the genes a and b is modulated by the degradation rates $\gamma_a x_a$ and $\gamma_b x_b$ respectively.

Such a model is readily generalized to models containing both expression and degradation terms for each gene:

$$\dot{x}_i = f_i(x) - \gamma_i x_i$$

where $f_i(x)$ represents the expression rate of gene i, depending on the whole state $x = (x_1, \cdots, x_n)^T$ and $\gamma_i x_i$ is the degradation rate. However, the expression rates of (1) have the additional property of being constant for values of x_a and x_b belonging to intervals that do not contain thresholds values θ_i^j. This can be rewritten by detailing $f_i(x)$ as follows:

$$f_i(x) = \sum_{l=1}^{L_i} \kappa_{il} b_{il}(x)$$

where $b_{il}(x)$ is a combination of step-functions $s^{\pm}(x_r, \theta_r^j)$ and $\kappa_{il} > 0$ is a rate parameter. The generalized form of (1) is a piecewise linear model

$$\dot{x} = f(x) - \gamma x \tag{2}$$

where the model is linear within hyper-rectangles of the state-space.

The dynamics of the piecewise-linear system (2) can be studied in the n-dimensional state-space $\Omega = \Omega_1 \times \Omega_2 \times \cdots \times \Omega_n$, where each Ω_i is defined by $\Omega_i = \{x \in \mathbb{R}_+ \mid 0 \leq x_i \leq max_i\}$ for some positive parameter $max_i > \max_{x \in \Omega} \left(\frac{f_i(x)}{\gamma_i} \right)$. A protein encoded by a gene will be involved in different interactions at different concentration thresholds, so for each variable x_i, we assume there are p_i ordered thresholds $\theta_i^1, \cdots, \theta_i^{p_i}$ (we also define $\theta_i^0 = 0$ and $\theta_i^{p_i+1} = max_i$). The $(n-1)$-dimensional hyperplanes defined by these thresholds partition Ω into hyper-rectangular regions we call *domains*. Specifically, a domain $D \subset \Omega$ is defined to be a set $D = D_1 \times \cdots \times D_n$, where D_i is one of the following:

$$D_i = \{x_i \in \Omega_i \mid 0 \leq x_i < \theta_i^1\}$$
$$D_i = \{x_i \in \Omega_i \mid \theta_i^j < x_i < \theta_i^{j+1}\} \quad \text{for } j \in \{1, \cdots, p_i - 1\}$$
$$D_i = \{x_i \in \Omega_i \mid \theta_i^{p_i} < x_i \leq max_i\}$$
$$D_i = \{x_i \in \Omega_i \mid x_i = \theta_i^j\} \quad \text{for } j \in \{1, \cdots, p_i\}$$

A domain $D \in \mathcal{D}$ is called a regulatory domain if none of the variables x_i has a threshold value in D. In contrast, a domain $D \in \mathcal{D}$ is called a switching domain of order $k \leq n$ if exactly k variables have threshold values in D [25]. The corresponding variables x_i are called switching variables in D. For convenience, we denote the sets of regulatory and switching domains by $\mathcal{D}_r$ and $\mathcal{D}_s$ respectively. It is also useful to define the concept of a supporting hyperplane for a domain.

Definition 1. *For every domain $D \in \mathcal{D}_s$ of order $k \geq 1$, define supp(D) to be the $(n-k)$-dimensional hyperplane containing D. If $D \in \mathcal{D}_r$ then we define supp(D) to be equal to Ω.*

1.1   Solutions in Regulatory Domains

For any regulatory domain $D \in \mathcal{D}_r$, the function $f(x)$ is constant for all $x \in D$, and it follows that the piecewise-linear system (2) can be written as a linear vector field

$$\dot{x} = f^D - \gamma x \tag{3}$$

where f^D is constant in D. Restricted to D, this is a classical linear ordinary differential equation. From (3), it is clear that all solutions in D monotonically converge towards the corresponding equilibrium $\phi(D)$, which is defined by $\gamma \phi(D) = f^D$. If $\phi(D)$ belongs to the closure of D, all solutions initiated in D converge towards $\phi(D)$; otherwise, all solutions reach the boundary of D in finite time (which means that they exit D).

Definition 2. *Given a regulatory domain $D \in \mathcal{D}_r$, the point $\phi(D) = \gamma^{-1} f^D \in \Omega$ is called the focal point for the flow in D.*

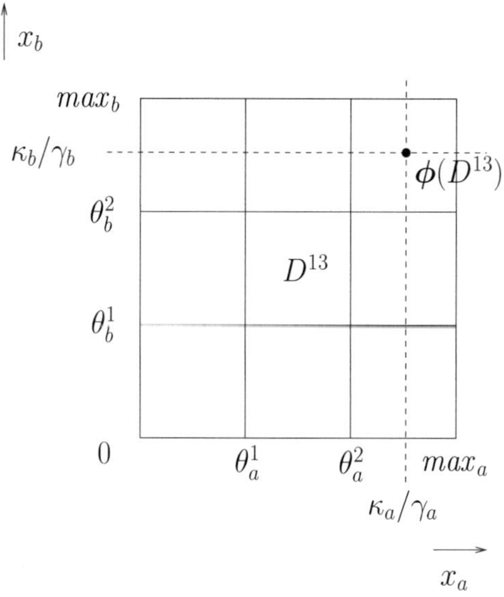

Fig. 2. Illustration of the focal point $\phi(D^{13})$ of a domain D^{13} in example (1)

In Figure 2, example (1) is used to illustrate this concept: the considered regulatory domain D is $\{x_a \in \Omega_a | \theta_a^1 < x_a < \theta_a^2\} \times \{x_b \in \Omega_b | \theta_b^1 < x_b < \theta_b^2\}$, so that system (1) becomes

$$\begin{cases} \dot{x}_a = \kappa_a - \gamma_a x_a \\ \dot{x}_b = \kappa_b - \gamma_b x_b \end{cases}$$

and the corresponding focal point is $\left(\frac{\kappa_a}{\gamma_a}, \frac{\kappa_b}{\gamma_b}\right)$. In the figure, this focal point is supposed to be outside of D: every solution starting in D therefore exits this domain in finite time.

1.2   Solutions in Switching Domains

In switching domains, the PL system (2) is not defined, since in a switching domain of order $k \geq 1$, k variables assume a threshold value. If solutions do not simply go through a switching domain, it is necessary to give a definition of what a solution can be on that domain. Classically, this is done by using a construction originally proposed by Filippov [10] and recently applied to PL systems of this form [13, 7].

The method consists of extending the system (3) to a differential inclusion,

$$\dot{x} \in H(x), \tag{4}$$

where H is a set-valued function (i.e. $H(x) \subseteq I\!R^n$). If D is a regulatory domain, then we define H simply as

$$H(x) = \{f^D - \gamma x\}, \tag{5}$$

for $x \in D$. If D is a switching domain, for $x \in D$, we define $H(x)$ as

$$H(x) = \overline{co}(\{f^{D'} - \gamma x \mid D' \in R(D)\}), \tag{6}$$

where $R(D) = \{D' \in \mathcal{D}_r \mid D \subseteq \partial D'\}$ is the set of all regulatory domains with D in their boundary, and $\overline{co}(X)$ is the closed convex hull of X. For switching domains, $H(x)$ is generally multi-valued so we define solutions of the differential inclusion as follows.

Definition 3. *A solution of (4) on $[0, T]$ in the sense of Filippov is an absolutely continuous function (w.r.t. t) $\xi_t(x_0)$ such that $\xi_0(x_0) = x_0$ and $\dot{\xi}_t \in H(\xi_t)$, for almost all $t \in [0, T]$.*

In order to more easily define these Filippov solutions, it is useful to define a concept analogous to the focal points defined for regulatory domains, extended to deal with switching domains.

Definition 4. *Let $D \in \mathcal{D}_s$ be a switching domain of order k. Then its focal set $\Phi(D)$ is*

$$\Phi(D) = supp(D) \cap \overline{co}(\{\phi(D') \mid D' \in R(D)\}). \tag{7}$$

Hence $\Phi(D)$ for $D \in \mathcal{D}_s$ is the convex hull of the focal points $\phi(D')$ of all the regulatory domains D' having D in their boundary, as defined above, intersected with the threshold hyperplane $supp(D)$ containing the switching domain D (Figure 3).

We have shown that

$$H(x) = \gamma(\Phi(D) - x) \tag{8}$$

which is a compact way of writing that $H(x) = \{y \in I\!R^n \mid \exists \phi \in \Phi(D)$ such that $y = \gamma(\phi - x)\}$. The Filippov vector field is defined by means of the focal set.

If $\Phi(D) = \{\ \}$, with D a switching domain, solutions will simply cross D; otherwise, sliding mode is possible and convergence takes place "in the direction" of $\Phi(D)$. If $\Phi(D) \cap D = \{\ \}$, solutions eventually leave D. In the case where $\Phi(D) \cap D$ is not empty, it can be assimilated to an equilibrium set within D towards which all solutions will converge in the following sense.

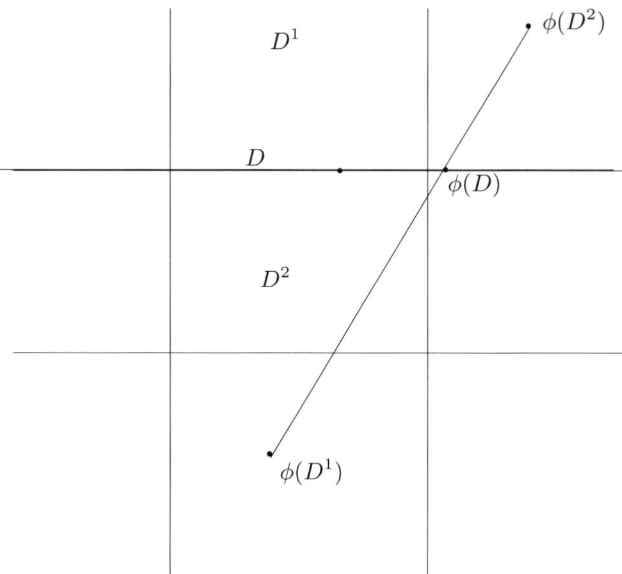

Fig. 3. Illustration of the definition of the focal set on a switching surface D according to the Filippov definition of solutions. The convex hull of the points $\phi(D^1)$ and $\phi(D^2)$ is simply the segment that links them, so that (7) implies that $\phi(D)$ is the intersection of this segment with $\mathrm{supp}(D)$.

Lemma 1. *[3] For every regulatory domain $D \in \mathcal{D}_r$, all solutions ξ_t in D monotonically converge towards the focal set $\Phi(D)$. For every switching domain $D \in \mathcal{D}_s$, the non-switching component $(\xi_t)_i$ of the solution ξ_t in D monotonically converges towards the closed interval*

$$\pi_i(\Phi(D)) = \{\phi_i \in \Omega_i \mid \phi \in \Phi(D)\},$$

the projection of $\Phi(D)$ onto Ω_i, if $(\xi_0)_i \notin \pi_i(\Phi(D))$. Every switching component $(\xi_t)_i$ of the solution ξ_t in D is a constant $(\xi_t)_i = \pi_i(\Phi(D)) = \theta_i^{q_i}$.

Basically, this means that convergence does not take place towards $\Phi(D)$, but towards the smallest hyper-rectangle that contains $\Phi(D)$. Indeed, if $\Phi(D)$ is neither empty, nor a singleton, and ξ_{t_0} belongs to $\Phi(D)$, the Filippov vector field at this point is defined as $H(\xi_{t_0}) = \gamma(\Phi(D) - \xi_{t_0})$ and there is no guarantee that no element of $H(\xi_{t_0})$ points outside of $\Phi(D)$ (we know however that a solution stays at ξ_{t_0}). Due to the structure of the differential equations, it is on the other hand certain that the transient solution does not leave the smallest hyper-rectangle containing $\Phi(D)$. This phenomenon is illustrated in Figure 4.

We then have the following corollary

Corollary 1. *[3] All solutions ξ_t in D converge towards $\Pi(D)$, if $\xi_0 \notin \Pi(D)$. For all solutions ξ_t in D, $\Pi(D)$ is invariant.*

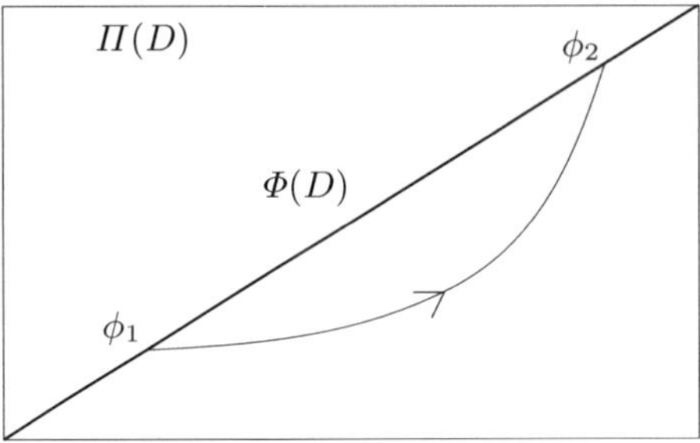

Fig. 4. Illustration of the non invariance of $\Phi(D)$: solutions with initial condition on $\Phi(D)$ stay inside the box $\Pi(D)$ but do not necessarily stay in $\Phi(D)$

Adding the following assumption

Assumption 1. *For all domains* $D \in \mathcal{D}$,

$$\Phi(D) \cap supp(D') = \{\}, \ \forall D' \subseteq \partial D. \tag{9}$$

it has been possible to develop stability results for this family of systems.

2   Stability and Qualitative Properties of PL Models

The stability analysis of the various equilibria is a direct consequence of the analysis in the previous section. It is easily seen that equilibria $\bar{x}_r$ in some $D \in \mathcal{D}_r$ are asymptotically stable. Indeed, they are the focal points of the domains in which they are contained, so that the convergence that was described in the previous section, leads to asymptotic stability. The more difficult part consists in defining and handling the stability of Filippov equilibria that lie in switching surfaces.

In a switching domain $D \in \mathcal{D}_s$, recall that solutions are defined by considering the differential inclusion $H(x)$. We say that a point $y \in \Omega$ is an equilibrium point for the differential inclusion if

$$0 \in H(y), \tag{10}$$

where H is computed using the Filippov construction in (6). In other words, there is a solution in the sense of Filippov, ξ_t, such that $\xi_t(y) = y$, $\forall t > 0$. We call such a point a *singular equilibrium point*. It is easily seen that, for y to be an equilibrium point inside D, it must belong to $\Phi(D)$. Also, since Assumption 1 prevents $\Phi(D)$ from intersecting the border of D, we then have that $\Phi(D) \subset D$. Every element ϕ of $\Phi(D)$ is then an equilibrium when $\Phi(D) \subset D$ so that, for every $\phi \in \Phi(D)$, there exists a solution $\xi_t(\phi) = \phi$ for all t.

One of the interesting results of [3] concerns the link between the configuration of the state transition graph and the stability of an equilibrium. This discrete, qualitative description of the dynamics of the PL system that underlies the qualitative simulation of genetic regulatory networks was originally due to Glass.

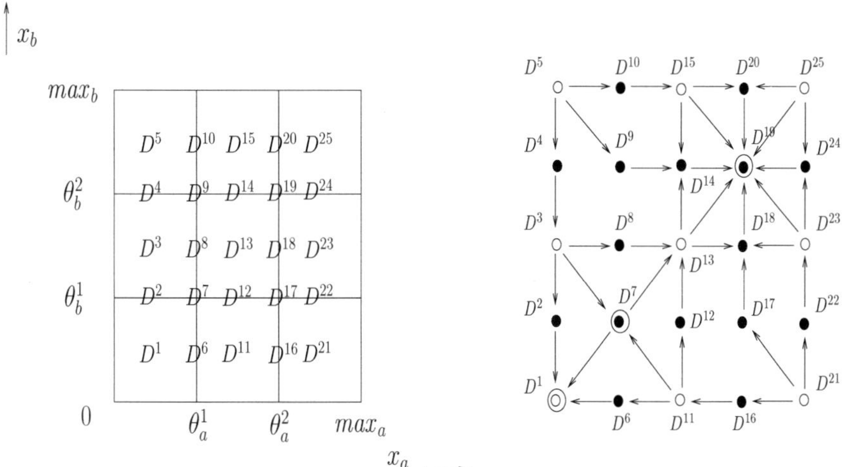

Fig. 5. Subdivision of the state-space in 25 domains and transition graph of system (1)

It indicates the passages between the different domains making up the phase space. A state transition graph is a directed graph whose vertices are the domains of the system and whose edges are the possible transitions between these domains (easily determined by examining the PL model [3]). The transition graph of system (1) is illustrated in Figure 5.

For a two-dimensional system, we show how this graph indicates the stability of singular equilibria:

Theorem 1. *[3] Let the dimension of the PL model be 2, and let D be a switching domain containing a singular equilibrium point $\phi(D)$. If for all regulatory domains $D' \in R(D)$ (that is, adjacent to D), there exists a transition from D' to D in the state transition graph, then $\phi(D)$ is asymptotically stable.*

This result is purely qualitative: the actual value of the parameters is not needed. It can be directly applied to show that the singular equilibrium $(x_a, x_b) = (\theta_a^2, \theta_b^2)$, corresponding to D^{19} on Figure 5, is asymptotically stable because there are transitions to D^{19} from D^{13}, D^{15}, D^{23} and D^{25}, the regulatory domains adjacent to D^{19}.

A generalization, but in a weaker form, of this theorem to dimension n is also available.

Theorem 2. *Assume $\Omega \subset \mathbb{R}^n$. Let $D \in \mathcal{D}_s$ be a switching domain of order $p \geq 1$ containing a singular equilibrium set $\Phi(D)$ that satisfies Assumption 1. If for all $D' \in R(D)$, there is a transition from D' to D in the state transition graph, then $\Pi(D)$ is asymptotically stable.*

These results are very helpful for the qualitative analysis of the genetic regulatory networks. However, some stable equilibria cannot be identified through those criteria. Some less restrictive criteria are therefore under development.

Besides this method, we can discover stable equilibria that would not have been directly identified by our criteria, through a rigorous simplification of the model. This can be done through model reduction or identification of regions of the state-space that cannot be reached by the solutions (maybe after some finite time). We will illustrate the kind of things that can be done on an example in the following section. In that section, since the resulting models are very simple, we do not need to go back to transition graph analysis at the end of the reduction procedure, but we could have done so, and it will be necessary to do so if the model reduction procedure does not yield very small models.

3   Carbon Starvation Response of *E. coli*

We will present a specific model reduction and stability analysis for the model in dimension 6 of the carbon starvation response of *E. coli* of Ropers et al. [29]. In their natural environment, bacteria like *Escherichia coli* rarely encounter conditions allowing continuous, balanced growth. While nutrients are available, *E. coli* cells grow quickly, leading to an exponential increase of their biomass, a state called *exponential phase*. However, upon depletion of an essential nutrient, the bacteria are no longer able to maintain fast growth rates, and the population consequently enters a non-growth state, called *stationary phase* (Figure 6). During the transition from exponential to stationary phase, each individual *E. coli* bacterium undergoes numerous physiological changes, concerning among other things the morphology and the metabolism of the cell, as well as gene expression [19]. These changes enable the cell to survive prolonged periods of starvation and be resistant to multiple stresses. This *carbon starvation response* can be reversed and growth resumed, as soon as carbon sources become available again.

On the molecular level, the transition from exponential phase to stationary phase is controlled by a complex genetic regulatory network integrating various environmental signals [18, 24, 32]. The molecular basis of the adaptation of the growth of *E. coli* to carbon starvation conditions has been the focus of extensive studies for decades [18]. However, notwithstanding the enormous amount of

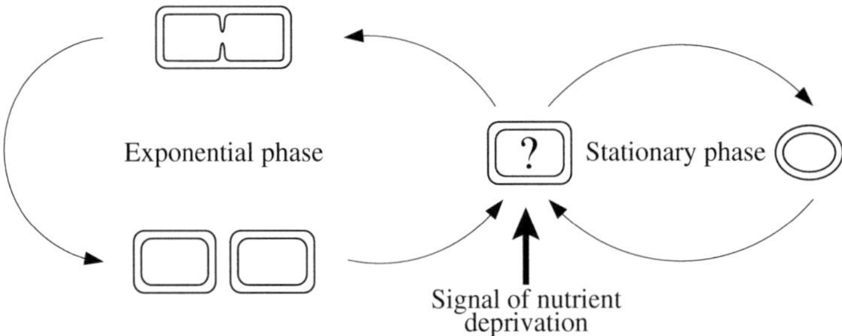

Fig. 6. Nutrient-stress response of bacteria during the transition from exponential to stationary phase

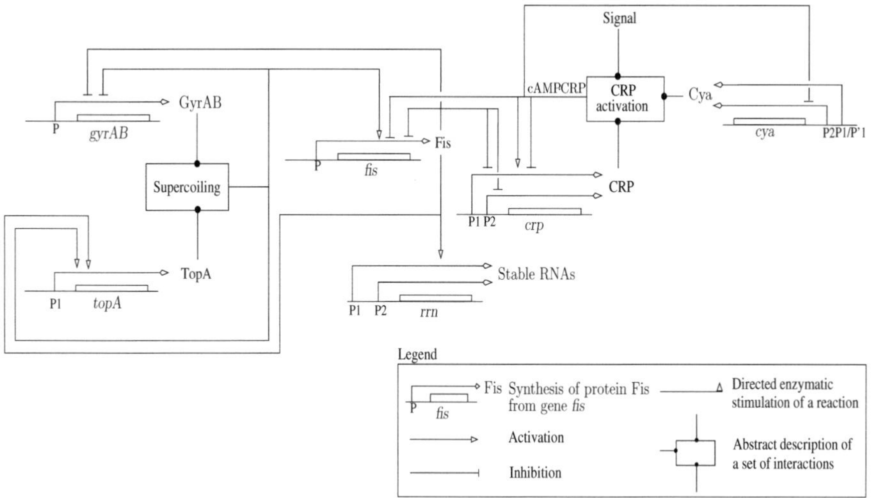

Fig. 7. Network of key genes, proteins, and regulatory interactions involved in the carbon starvation network in *E. coli*. The notation follows, in a somewhat simplified form, the graphical conventions proposed by Kohn [23]. The contents of the boxes labeled 'Activation' and 'Supercoiling' are detailed in [29].

information accumulated on the genes, proteins, and other molecules known to be involved in the stress adaptation process, there is currently no global understanding of how the response of the cell emerges from the network of molecular interactions. Moreover, with some exceptions [1, 16, 31], numerical values for the parameters characterizing the interactions and the molecular concentrations are absent, which makes it difficult to apply traditional methods for the dynamical modeling of genetic regulatory networks.

The above circumstances have motivated the qualitative analysis of the carbon starvation response network in *E. coli* [29]. The objective of the study was to simulate the response of an *E. coli* bacterium to the absence or presence of carbon sources in the growth medium. To this end, an initial, simple model of the carbon starvation response network has been built on the basis of literature data. It includes six genes that are believed to play a key role in the carbon starvation response (Figure 7). More specifically, the network includes genes encoding proteins whose activity depends on the transduction of the carbon starvation signal (the global regulator *crp* and the adenylate cyclase *cya*), genes involved in the metabolism (the global regulator *fis*), cellular growth (the *rrn* genes coding for stable RNAs), and DNA supercoiling, an important modulator of gene expression (the topoisomerase *topA* and the gyrase *gyrAB*).

3.1   Model of Carbon Starvation Response

The graphical representation of the network has been translated into a PL model supplemented with parameter inequality constraints. The resulting model

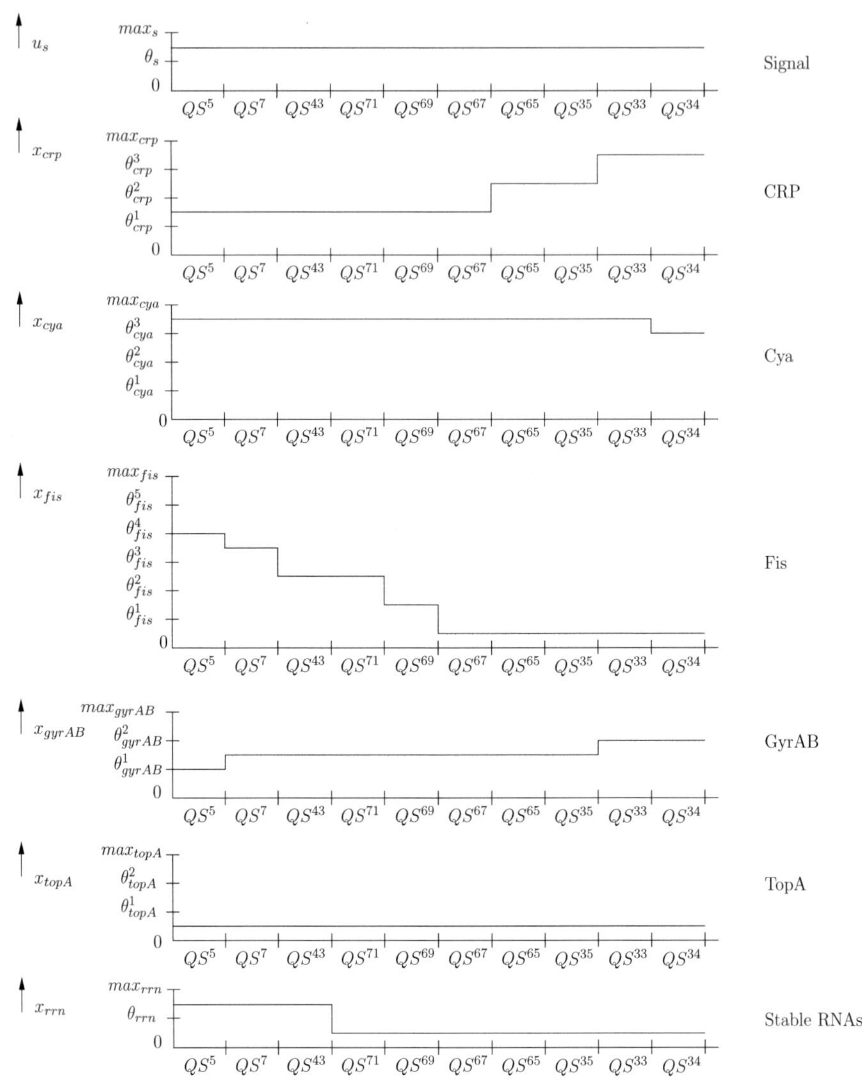

Fig. 8. Entry into stationary phase: qualitative temporal evolution of the proteins and stable RNA concentration in a depleted environment with the organisms being at the equilibrium of the exponential phase at the initial time. Convergence to one domain is detected (the domain where $x_c > \theta_c^3, x_y = \theta_y^3, x_f < \theta_f^1, x_g = \theta_g^2, x_t < \theta_t^1$ and $x_r < \theta_r$).

consists of seven variables, one concentration variable for the product of each of the six genes $((x_c, x_y, x_f, x_g, x_t, x_r)$ for (*crp, cya, fis, gyrAB, topA, rrn*)) and one input variable u_s representing the presence or absence of a carbon starvation signal [29]. The 38 parameters are constrained by 54 parameter inequalities, the choice of which is largely determined by experimental data.

The model of Ropers et al. is:

$$\begin{cases} \dot{x}_c = \kappa_c^1 + \kappa_c^2 s^-(x_f, \theta_f^2) s^+(x_c, \theta_c^1) s^+(x_y, \theta_y^1) s^+(u_s, \theta_s) \\ \qquad +\kappa_c^3 s^-(x_f, \theta_f^1) - \gamma_c x_c \\ \dot{x}_y = \kappa_y^1 + \kappa_y^2 \left(1 - s^+(x_c, \theta_c^3) s^+(x_y, \theta_y^3) s^+(u_s, \theta_s)\right) - \gamma_y x_y \\ \dot{x}_f = \left(\kappa_f^1 + \kappa_f^2 s^+(x_g, \theta_g^1) s^-(x_t, \theta_t^2)\right) \\ \qquad \left(1 - s^+(x_c, \theta_c^1) s^+(x_y, \theta_y^1) s^+(u_s, \theta_s)\right) s^-(x_f, \theta_f^5) - \gamma_f x_f \\ \dot{x}_g = \kappa_g \left(1 - s^+(x_g, \theta_g^2) s^-(x_t, \theta_t^1)\right) s^-(x_f, \theta_f^4) - \gamma_g x_g \\ \dot{x}_t = \kappa_t s^+(x_g, \theta_g^2) s^-(x_t, \theta_t^1) s^+(x_f, \theta_f^4) - \gamma_t x_t \\ \dot{x}_r = \kappa_r^1 s^+(x_f, \theta_f^3) + \kappa_r^2 - \gamma_r x_r \end{cases}$$

with $u_s = 0$ in the presence of carbon sources and $u_s = 1$ in a depleted environment (and $\theta_s = 0.5$). In order to uniquely determine the situation of the various focal points in the state-space, the following constraints on the parameters are needed:

$$\begin{cases} 0 < \theta_c^1 < \theta_c^2 < \theta_c^3 < max_c,\ \theta_c^1 < \frac{\kappa_c^1}{\gamma_c} < \theta_c^2,\ \theta_c^1 < \frac{(\kappa_c^1+\kappa_c^2)}{\gamma_c} < \theta_c^2, \\ \qquad\qquad\qquad\qquad\qquad \theta_c^3 < \frac{(\kappa_c^1+\kappa_c^3)}{\gamma_c} < max_c \\ 0 < \theta_y^1 < \theta_y^2 < \theta_y^3 < max_y,\ \theta_y^1 < \frac{\kappa_y^1}{\gamma_y} < \theta_y^2,\ \theta_y^3 < \frac{(\kappa_y^1+\kappa_y^2)}{\gamma_y} < max_y \\ 0 < \theta_f^1 < \theta_f^2 < \theta_f^3 < \theta_f^4 < \theta_f^5 < max_f,\ \theta_f^1 < \frac{\kappa_f^1}{\gamma_f} < \theta_f^2, \\ \qquad\qquad\qquad\qquad\qquad \theta_f^5 < \frac{(\kappa_f^1+\kappa_f^2)}{\gamma_f} < max_f \\ 0 < \theta_g^1 < \theta_g^2 < max_g,\ \theta_g^2 < \frac{\kappa_g}{\gamma_g} < max_g \\ 0 < \theta_t^1 < \theta_t^2 < max_t,\ \theta_t^2 < \frac{\kappa_t}{\gamma_t} < max_t \\ 0 < \theta_r < max_r,\ 0 < \frac{\kappa_r^2}{\gamma_r} < \theta_r,\ \theta_r < \frac{(\kappa_r^1+\kappa_r^2)}{\gamma_r} < max_r \end{cases}$$

A qualitative analysis of this model has been carried out in [29] by using GNA (*Genetic Network Analyzer* [6]), a computer tool that automatically generates the state-transition graph and possible trajectories in that graph, that is, qualitative solutions that are possible for this system. The following simulations are produced for the transition to the stationary phase (Figure 8) and to the exponential phase (Figure 9). In the first case, we see that the solution converges towards a single region of the state space, where we can guess that convergence towards an equilibrium takes place. In the second case, the behavior of the solution is not as clear: oscillations can be detected between various regions but it is impossible to say, based on the transition graph alone, if those oscillations are damped or not. Therefore, it is useful to try and analyze the model further to check what kind of oscillations take place (and in the same time if convergence actually takes place towards an equilibrium in the case of the entry in stationary phase).

3.2   Asymptotic Dynamics

Since the 6-dimensional model, with all its constraints, is too complex to handle directly, we first check if some kind of simplifications can be made. Independently

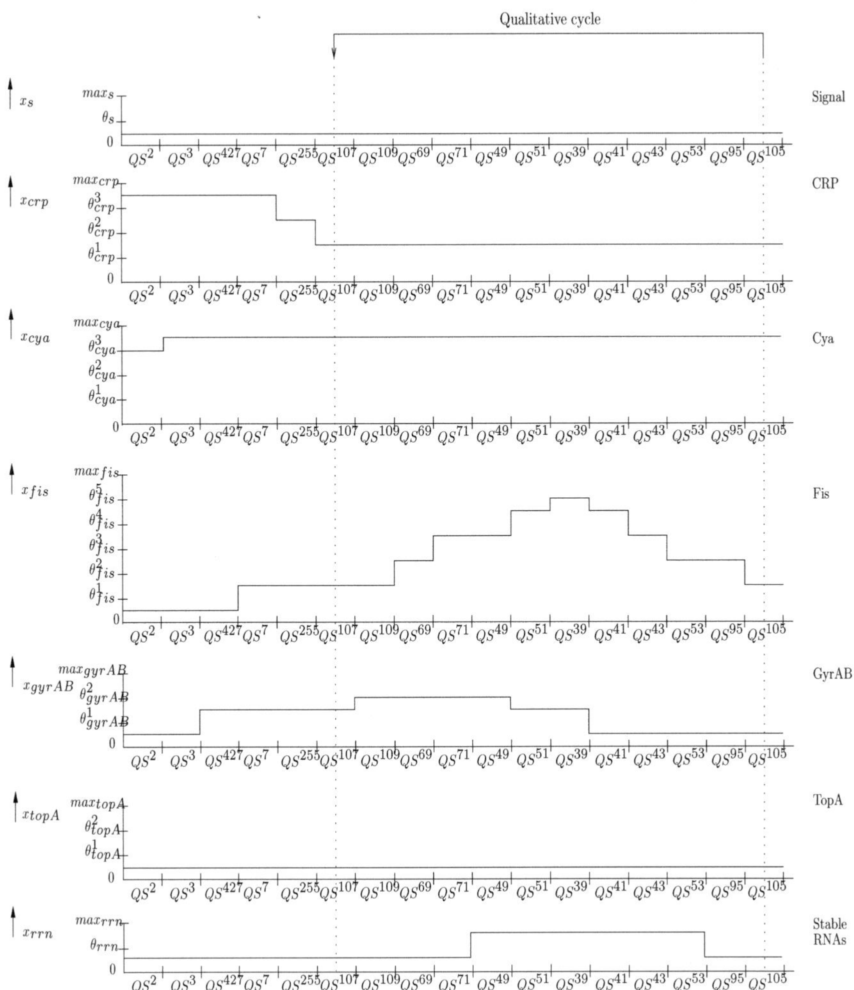

Fig. 9. Entry into exponential phase: qualitative temporal evolution of the proteins and stable RNA concentration in a rich environment with the organisms being at the equilibrium of the stationary phase at the initial time. Oscillations of the x_f and x_g states is detected.

of the case that we will study (stationary phase or exponential phase conditions), we notice that

- x_r is a variable whose evolution depends on, but does not influence the rest of the system. As a consequence, it can be removed from the analysis. Once the analysis of the remaining 5-dimensional system is completed, we will be able to easily identify the consequence of its behavior on the concentration of stable RNAs (x_r).

- There exists a finite time after which $x_t(t) \leq \theta_t^1$ since, as long as $x_t > \theta_t^1$, the x_t dynamics reduces to

$$\dot{x}_t = -\gamma_t x_t.$$

Once x_t reaches θ_t^1, we cannot a priori rule out a sliding mode along $x_t = \theta_t^1$. Since $\theta_t^1 < \theta_t^2$, this indicates that we can replace $s^-(x_t, \theta_t^2)$ with 1 for the purpose of our analysis. We simply consider that the aforementioned finite time has already occurred.
- Similar studies show that $x_c(t) \geq \theta_c^1$ and $x_y(t) \geq \theta_y^1$ after some finite time. We can then replace $s^+(x_c, \theta_c^1)$ and $s^+(x_y, \theta_y^1)$ with 1 in our analysis.

The system that we need to analyze has now become

$$\begin{cases} \dot{x}_c = \kappa_c^1 + \kappa_c^2 s^-(x_f, \theta_f^2) s^+(u_s, \theta_s) + \kappa_c^3 s^-(x_f, \theta_f^1) - \gamma_c x_c \\ \dot{x}_y = \kappa_y^1 + \kappa_y^2 \left(1 - s^+(x_c, \theta_c^3) s^+(x_y, \theta_y^3) s^+(u_s, \theta_s)\right) - \gamma_y x_y \\ \dot{x}_f = \left(\kappa_f^1 + \kappa_f^2 s^+(x_g, \theta_g^1)\right) s^-(u_s, \theta_s) s^-(x_f, \theta_f^5) - \gamma_f x_f \\ \dot{x}_g = \kappa_g \left(1 - s^+(x_g, \theta_g^2) s^-(x_t, \theta_t^1)\right) s^-(x_f, \theta_f^4) - \gamma_g x_g \\ \dot{x}_t = \kappa_t s^+(x_g, \theta_g^2) s^-(x_t, \theta_t^1) s^+(x_f, \theta_f^4) - \gamma_t x_t \end{cases}$$

The next simplification step consists in seeing that x_y does not influence the rest of the model, so that it can be removed, and that x_c does not influence the rest of the model either (except x_y) so that it can also be removed. These actions are in the same line of thought as the removal of x_r. As a consequence of these simplifications, we are able to see that the core of the long term dynamics is not really influenced by x_r, x_y and x_c. We now have the three-dimensional system:

$$\begin{cases} \dot{x}_f = \left(\kappa_f^1 + \kappa_f^2 s^+(x_g, \theta_g^1)\right) s^-(u_s, \theta_s) s^-(x_f, \theta_f^5) - \gamma_f x_f \\ \dot{x}_g = \kappa_g \left(1 - s^+(x_g, \theta_g^2) s^-(x_t, \theta_t^1)\right) s^-(x_f, \theta_f^4) - \gamma_g x_g \\ \dot{x}_t = \kappa_t s^+(x_g, \theta_g^2) s^-(x_t, \theta_t^1) s^+(x_f, \theta_f^4) - \gamma_t x_t \end{cases} \tag{11}$$

Once we have analyzed the behavior of the solutions of this model, we will be able to reconstruct what happens with x_c, x_y and x_r. For this analysis, we still suppose that $x_t \leq \theta_t^1$.

3.3   Asymptotic Dynamics in the Absence of Carbon Sources

The analysis of the case $u_s = 1$, the stationary phase solution in a depleted environment, is very straightforward. System (11) becomes

$$\begin{cases} \dot{x}_f = -\gamma_f x_f \\ \dot{x}_g = \kappa_g \left(1 - s^+(x_g, \theta_g^2) s^-(x_t, \theta_t^1)\right) s^-(x_f, \theta_f^4) - \gamma_g x_g \\ \dot{x}_t = \kappa_t s^+(x_g, \theta_g^2) s^-(x_t, \theta_t^1) s^+(x_f, \theta_f^4) - \gamma_t x_t \end{cases}$$

so that x_f goes to 0. It is then directly seen that, after a finite time (the time taken for x_f to fall below θ_f^4), we have

$$\dot{x}_t = -\gamma_t x_t$$

152   F. Grognard, H. de Jong, and J.-L. Gouzé

so that x_t also goes to zero. The x_g dynamics then reduce to

$$\dot{x}_g = \kappa_g s^-(x_g, \theta_g^2) - \gamma_g x_g$$

so that x_g reaches θ_g^2 in finite time. The three dimensional system thus has a very simple behavior: the state goes to $(x_f, x_g, x_t) = (0, \theta_g^2, 0)$.

Since the solutions of the 6-dimensional system are bounded, the behavior of the other three states can be deduced from the analysis of the corresponding equations with (x_f, x_g, x_t) approaching their equilibrium (so that $x_f < \theta_f^1, x_t < \theta_t^1$ and $x_g > \theta_g^1$). We then have:

$$\begin{cases} \dot{x}_c = \kappa_c^1 + \kappa_c^2 + \kappa_c^3 - \gamma_c x_c \\ \dot{x}_y = \kappa_y^1 + \kappa_y^2 \left(1 - s^+(x_c, \theta_c^3)s^+(x_y, \theta_y^3)\right) - \gamma_y x_y \\ \dot{x}_r = \kappa_r^2 - \gamma_r x_r \end{cases}$$

It is then directly seen that, once (x_f, x_g, x_t) is close to its equilibrium, the variables (x_c, x_r) exponentially converge towards $\left(\frac{\kappa_c^1 + \kappa_c^2 + \kappa_c^3}{\gamma_c}, \frac{\kappa_r^2}{\gamma_r}\right)$ while x_y reaches θ_y^3 in finite time.

3.4   Asymptotic Dynamics in the Presence of Carbon Sources

The case $u_s = 0$, the behavior of the model in an environment rich in carbon sources, is more intricate to analyze. System (11) becomes

$$\begin{cases} \dot{x}_f = \left(\kappa_f^1 + \kappa_f^2 s^+(x_g, \theta_g^1)\right) s^-(x_f, \theta_f^5) - \gamma_f x_f \\ \dot{x}_g = \kappa_g \left(1 - s^+(x_g, \theta_g^2)s^-(x_t, \theta_t^1)\right) s^-(x_f, \theta_f^4) - \gamma_g x_g \\ \dot{x}_t = \kappa_t s^+(x_g, \theta_g^2)s^-(x_t, \theta_t^1)s^+(x_f, \theta_f^4) - \gamma_t x_t \end{cases}$$

As stated earlier, we know that $x_t \leq \theta_t^1$ after some finite time; this does not help us for further simplifications of this model. In the following, we will show that, after some finite-time, we have $x_t < \theta_t^1$, which will help us eliminate the x_t equation. In order to do that, we first show that, after some finite time, $x_g \leq \theta_g^2$.

Indeed, if we suppose that $x_g > \theta_g^2$ for all times, system (11) would become

$$\begin{cases} \dot{x}_f = \left(\kappa_f^1 + \kappa_f^2\right) s^-(x_f, \theta_f^5) - \gamma_f x_f \\ \dot{x}_g = \kappa_g s^+(x_t, \theta_t^1)s^-(x_f, \theta_f^4) - \gamma_g x_g \\ \dot{x}_t = \kappa_t s^-(x_t, \theta_t^1)s^+(x_f, \theta_f^4) - \gamma_t x_t \end{cases}$$

which shows that x_f reaches θ_f^5 in finite time so that $\dot{x}_g$ becomes equal to

$$\dot{x}_g = -\gamma_g x_g$$

This leads to the convergence of x_g to 0 and thus to below θ_g^2, which is a contradiction. This shows that x_g should reach θ_g^2 in finite time when $x_g(0) > \theta_g^2$.

An ensuing case-by-case analysis shows that the region where $x_g \leq \theta_g^2$ is invariant [2].

We will now show that x_t is decreasing almost all of the time when $x_g \leq \theta_g^2$ and $x_t \leq \theta_t^1$, that is in a region which we have shown to be reached in finite time and invariant. Detailing three cases, we have:

$x_g < \theta_g^2$ or $x_f < \theta_f^4$: $\dot{x}_t = -\gamma_t x_t$.

$x_g = \theta_g^2$ and $x_f > \theta_f^4$: We have $\dot{x}_g = -\gamma_g x_g < 0$ at such a point and in a neighborhood surrounding each such point so that any solution directly enters the region where $x_g < \theta_g^2$ (and consequently $\dot{x}_t = -\gamma_t x_t$, as we have seen).

$x_g = \theta_g^2$ and $x_f = \theta_f^4$: We have $\dot{x}_f = \left(\kappa_f^1 + \kappa_f^2\right) - \gamma_f x_f > 0$ at this point and in a neighborhood surrounding it, so that any solution directly goes in one of the two previously described regions, where we have seen that x_t is decreasing.

For any solution of (11), x_t could only increase if x stayed in the second or third region, which we have shown not to be possible. We then have $\dot{x}_t = -\gamma_t x_t$ for almost all times in the region of interest. After elimination of x_t, we have to analyze the following system:

$$\begin{cases} \dot{x}_f = \left(\kappa_f^1 + \kappa_f^2 s^+(x_g, \theta_g^1)\right) s^-(x_f, \theta_f^5) - \gamma_f x_f \\ \dot{x}_g = \kappa_g s^-(x_g, \theta_g^2) s^-(x_f, \theta_f^4) - \gamma_g x_g \end{cases} \tag{12}$$

At first sight, this analysis is not straightforward because this is a second order piecewise linear system with two thresholds in each direction, which theoretically gives rise to 9 regions. However, as is illustrated on Figure 10, some of the regions have the same dynamics and can be grouped together, giving rise to six regions.

The behavior of the solutions along the thick black lines, where sliding modes are present, can be directly inferred from the Filippov construction. However, simple observations indicate what actually happens: along the line where $x_g = \theta_g^2$ and $x_f < \theta_f^4$, we have

$$\dot{x}_g = \kappa_g s^-(x_g, \theta_g^2) - \gamma_g x_g$$

with

$$\frac{\kappa_g}{\gamma_g} > \theta_g^2$$

so that the line is attractive (black wall). Moreover,

$$\dot{x}_f = \left(\kappa_f^1 + \kappa_f^2\right) - \gamma_f x_f > 0$$

so that x_f is increasing and all solutions reach the end-point $(x_f, x_g) = (\theta_f^4, \theta_g^2)$ in finite time. In some sense, each time the solution reaches this black wall, there is a *reset* taking place that sends the system to the end-point (θ_f^4, θ_g^2)

Along the line where $x_f = \theta_f^5$ and $x_g > \theta_g^1$, we have

$$\dot{x}_f = \left(\kappa_f^1 + \kappa_f^2\right) s^-(x_f, \theta_f^5) - \gamma_f x_f$$

so that this line also is a black wall (bearing in mind that $\frac{\kappa_f^1 + \kappa_f^2}{\gamma_f} > \theta_f^5$). In addition,

$$\dot{x}_g = -\gamma_g x_g$$

so that x_g is decreasing and all solutions reach the end-point $(x_f, x_g) = (\theta_f^5, \theta_g^1)$ in finite time.

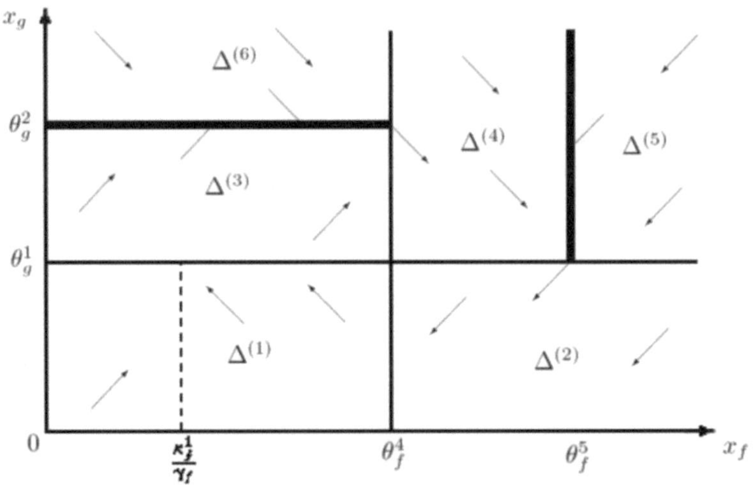

Fig. 10. Illustration of the vector field and the various regions for system (12). The thick black lines indicate where sliding modes can occur.

The observation of Figure 10 (as well as a detailed analysis of the linear systems in each of the regions) indicate that, eventually, the solutions oscillate around $(x_f, x_g) = (\theta_f^4, \theta_g^1)$. Whether this oscillation is damped, neutrally stable or unstable is still unclear. It is clear, though, that the oscillation is bounded, as it cannot go beyond the black walls.

In order to analyze the oscillations, we will compute the first return map from and to the segment that links (θ_f^4, θ_g^1) to (θ_f^4, θ_g^2). We will therefore consider some (θ_f^4, x) as initial condition and compute the function $f(x)$ such that $(\theta_f^4, f(x))$ is the image of (θ_f^4, x) on the segment after one cycle around (θ_f^4, θ_g^1). The computation of this first-return map can be handled in four steps, corresponding to the passages in the four regions surrounding (θ_f^4, θ_g^1).

The first step consists in computing the image of (θ_f^4, x), belonging to the initial segment, on the horizontal segment that links (θ_f^4, θ_g^1) to (θ_f^5, θ_g^1). The transition takes place in the region $\Delta^{(4)}$ so that (12) becomes

$$\begin{cases} \dot{x}_f = \kappa_f^1 + \kappa_f^2 - \gamma_f\, x_f \\ \dot{x}_g = -\gamma_g\, x_g \end{cases} \tag{13}$$

whose solution is

$$\begin{cases} x_f(t) = \theta_f^4\, e^{-\gamma_f t} + \frac{\kappa_f^1 + \kappa_f^2}{\gamma_f}\, (1 - e^{-\gamma_f t}) \\ x_g(t) = x\, e^{-\gamma_g t} \end{cases} \tag{14}$$

In a first computation, we will suppose the absence of the vertical black wall, and use the dynamics (13) for both regions $\Delta^{(4)}$ and $\Delta^{(5)}$: it is then straightforward to see that the solution impacts the target segment when $x_g(t) = \theta_g^1$, that is, at

$t = t_1(x) = \dfrac{ln(x) - ln(\theta_g^1)}{\gamma_g}$, so that

$$x_f(t_1(x)) = \theta_f^4 \, e^{-\gamma_f t_1(x)} + \frac{\kappa_f^1 + \kappa_f^2}{\gamma_f} \left(1 - e^{-\gamma_f t_1(x)}\right)$$

$$= \theta_f^4 \left(\frac{\theta_g^1}{x}\right)^{\frac{\gamma_f}{\gamma_g}} + \frac{\kappa_f^1 + \kappa_f^2}{\gamma_f} \left(1 - \left(\frac{\theta_g^1}{x}\right)^{\frac{\gamma_f}{\gamma_g}}\right)$$

However, we must account for the black-wall and it is possible that the actual solution hits this wall before reaching the target segment, so that the previously computed $x_f(t_1(x)) > \theta_f^5$. In that case, the actual solution stays on the vertical black wall until it reaches the point (θ_f^5, θ_g^1). Therefore the target of the point (θ_f^4, x) on the horizontal segment is

$$(f_1(x), \theta_g^1) = \left(\min\left(\theta_f^4 \left(\frac{\theta_g^1}{x}\right)^{\frac{\gamma_f}{\gamma_g}} + \frac{\kappa_f^1 + \kappa_f^2}{\gamma_f} \left(1 - \left(\frac{\theta_g^1}{x}\right)^{\frac{\gamma_f}{\gamma_g}}\right), \theta_f^5 \right), \theta_g^1 \right)$$

Similarly we can define $(\theta_f^4, f_2(x))$ as the image of (x, θ_g^1) (with $x \in [\theta_f^4, \theta_f^5]$) on the vertical segment below the equilibrium, $(f_3(x), \theta_g^1)$ as the image of (θ_f^4, x) (with $x \in [0, \theta_g^1]$) on the horizontal segment on the left of the equilibrium and $(\theta_f^4, f_4(x))$ as the image of (x, θ_g^1) (with $x \in [0, \theta_f^4]$) on the initial segment. This yields

$$f_1(x) - \min\left(0_f^4 \left(\frac{\theta_g^1}{x}\right)^{\frac{\gamma_f}{\gamma_g}} + \frac{\kappa_f^1 + \kappa_f^2}{\gamma_f} \left(1 - \left(\frac{\theta_g^1}{x}\right)^{\frac{\gamma_f}{\gamma_g}}\right), \theta_f^5 \right)$$

$$f_2(x) = \theta_g^1 \left(\frac{\theta_f^4 - \frac{\kappa_f^1}{\gamma_f}}{x - \frac{\kappa_f^1}{\gamma_f}} \right)^{\frac{\gamma_g}{\gamma_f}}$$

$$f_3(x) = \theta_f^4 \left(\frac{\theta_g^1 - \frac{\kappa_g}{\gamma_g}}{x - \frac{\kappa_g}{\gamma_g}} \right)^{\frac{\gamma_f}{\gamma_g}} + \frac{\kappa_f^1}{\gamma_f} \left(1 - \left(\frac{\theta_g^1 - \frac{\kappa_g}{\gamma_g}}{x - \frac{\kappa_g}{\gamma_g}} \right)^{\frac{\gamma_f}{\gamma_g}} \right)$$

$$f_4(x) = \min\left(\theta_g^1 \left(\frac{\theta_f^4 - \frac{\kappa_f^1 + \kappa_f^2}{\gamma_f}}{x - \frac{\kappa_f^1 + \kappa_f^2}{\gamma_f}} \right)^{\frac{\gamma_g}{\gamma_f}} + \frac{\kappa_g}{\gamma_g} \left(1 - \left(\frac{\theta_f^4 - \frac{\kappa_f^1 + \kappa_f^2}{\gamma_f}}{x - \frac{\kappa_f^1 + \kappa_f^2}{\gamma_f}} \right)^{\frac{\gamma_g}{\gamma_f}} \right), \theta_g^2 \right)$$

and $f(x) = f_4(f_3(f_2(f_1(x))))$ which has $x = \theta_g^1$ as a fixed point. It was then shown in [2] that $f'(x) < 1$ when $x > \theta_f^4$, so that the sequence $x_{n+1} = f(x_n)$, which represents the successive impacts on the initial segment converges to $x = \theta_f^4$. We can then conclude that the cyclic solutions that surround (θ_f^4, θ_g^1) are damped. This point is therefore a globally attractive equilibrium of (12) (cf. [9] that gives more general results in n dimensions for a negative feedback loop).

Having elucidated the dynamical behavior of the (x_f, x_g) subsystem, we can now deduce the behavior of all other states. From the moment that we have $x_g < \theta_g^2$, it comes from (11) that

$$\dot{x}_t = -\gamma_t x_t$$

so that x_t goes to 0. Once those three states are close to their equilibrium value, the remaining three equations become

$$\begin{cases} \dot{x}_c = \kappa_c^1 - \gamma_c x_c \\ \dot{x}_y = \kappa_y^1 + \kappa_y^2 - \gamma_y x_y \\ \dot{x}_r = \kappa_r^1 + \kappa_r^2 - \gamma_r x_r \end{cases}$$

so that convergence of (x_c, x_y, x_r) towards $\left(\frac{\kappa_c^1}{\gamma_c}, \frac{\kappa_y^1 + \kappa_y^2}{\gamma_y}, \frac{\kappa_r^1 + \kappa_r^2}{\gamma_r} \right)$ takes place.

3.5   Comparison of the Equilibria

It is interesting to compare both equilibria: we have

	x_c	x_y	x_f	x_g	x_t	x_r
$u_s = 1$	$\frac{\kappa_c^1 + \kappa_c^2 + \kappa_c^3}{\gamma_c}$	θ_y^3	0	θ_g^2	0	$\frac{\kappa_r^2}{\gamma_r}$
$u_s = 0$	$\frac{\kappa_c^1}{\gamma_c}$	$\frac{\kappa_y^1 + \kappa_y^2}{\gamma_y}$	θ_f^4	θ_g^1	0	$\frac{\kappa_r^1 + \kappa_r^2}{\gamma_r}$

We see that most genes settle at different levels depending on the absence or presence of carbon sources. The most illustrative of the difference between the two states (carbon starved or not) is x_r, which represents the concentration of stable RNAs and is a good indicator of the cellular growth. As expected, when carbon sources are depleted, the equilibrium level of x_r is smaller than when carbon sources are abundant: when carbon shortage occurs, x_r stays at a "house-keeping"-level whereas $\frac{\kappa_r^1 + \kappa_r^2}{\gamma_r}$, the equilibrium value in the presence of carbon sources, allows for fast cell growth. Also to be noted is the fact that $x_t = 0$ in both cases; this does not mean that *topA*, the gene corresponding to x_t, is useless. Indeed, when the carbon sources are either continuously present or absent, the effect of *topA* eventually dies down. However, in a time-varying environment, where nutrients are alternatively present and absent, an increase of the x_t concentration can occur whenever $x_g > \theta_g^2$ and $x_f > \theta_f^4$. TopA thus influences the transients.

3.6   Abstraction of the Reduction Method

We have seen that the preliminary model reduction has allowed for a simplification of the model analysis. Indeed, a global stability analysis of a 6-order model is no easy task, whereas there are various methods for the analysis of second order models. The reduction of the dimension of dynamical models is critical in the further development of the mathematical methods for genetic regulatory networks analysis because the networks typically are very large, so that it is rarely possible to study them directly. Classically, it has been attempted to apply time-scale separation methods, but these are mainly efficient for eliminating the fast metabolic components from mixed metabolic-genetic networks. Also balanced truncation methods have been introduced for genetic regulatory networks where inputs signals (action on the network) and output signals (measurements) are clearly identified ([15, 21]). In this example, we have exploited the *hierarchical* triangular structure of the model arising after a finite time (this finite time allowed us to get rid of some of the interactions interfering with the triangular structure). We notice from graph theory that the identification of such a structure in the graph corresponding to the network is equivalent to the search for the strongly connected components of the graph. There are efficient algorithms to do so on large graphs, so that this model reduction method is tractable for the huge graphs that represent genetic regulatory networks (preliminary work on that subject has been done in [4] with links to GNA). Combining this approach with thresholds elimination allows for a progressive simplification of the graphs.

4 Challenges in PL Models Analysis

One of the major challenges in the analysis of models of genetic regulatory networks lies in the difficulty of obtaining accurate parameters. Therefore, one has to develop methods to identify the qualitative behavior of the system: when the parameters are linked together through inequalities (instead of being fixed at given values), we would like to be able to say something about the stability of the equilibria. Some interesting results have been obtained on that subject in [3], as was shown in Section 2, and we would like to identify other cases where stability results can be deduced.

As we have seen in the analysis of *E. coli*, we are able to mathematically analyze PL models that are not trivial (dimension 6). However, actual genetic regulatory networks are much larger than that. It is therefore of paramount importance to develop methods that will help analyzing such large systems. Two major research directions are explored for that purpose: the model reduction approach (through balancing or through singular perturbations, in the linear case) and the separation of the original model into smaller, interconnected pieces that can be easily analyzed, as we have shown here.

Moreover, experimental techniques (e.g. gene deletion) are now available and allow to modify the production or degradation terms of some genes of the networks. This leads to problems of mathematical control of piecewise affine genetic networks, similar to more general problems for hybrid affine systems [14]. The global problem is to control the trajectories through some prescribed sequence of rectangular regions. Some preliminary results have been obtained in [8]. For example, we have shown that a simple two-gene inhibitor system with a single equilibrium can be controlled to a bistable switch. We believe that interesting and original control problems are still to be solved in this domain.

Acknowledgments

This research is supported by the HYGEIA project of the NEST-Adventure program of the European Union (NEST-004995) on Hybrid Systems for Biochemical Network Modeling and Analysis.

References

1. J. Botsford and J. Harman. Cyclic AMP in prokaryotes. *Microbiological reviews*, 56(1):100–122, 1992.
2. K. Bouraima Madjebi. Etude de modèles de réseau de régulation génique. Master's thesis, University of Orsay, 2005.
3. R. Casey, H. de Jong, and J.-L. Gouzé. Piecewise-linear models of genetic regulatory networks: Equilibria and their stability. *J. Math. Biol.*, 52:27–56, 2006.
4. D. Cristescu. Algorithmic study on genetic regulatory networks. Technical report, Automatic control and computer science faculty, Politechnica University of Bucharest, 2006. internship report.
5. H. de Jong. Modeling and simulation of genetic regulatory systems: a literature review. *J. Comput. Biol.*, 9:67–103, 2002.
6. H. de Jong, J. Geiselmann, C. Hernandez, and M. Page. Genetic Network Analyzer: Qualitative simulation of genetic regulatory networks. *Bioinformatics*, 19(3):336–344, 2003.
7. H. de Jong, J.-L. Gouzé, C. Hernandez, M. Page, T. Sari, and J. Geiselmann. Qualitative simulation of genetic regulatory networks using piecewise-linear models. *Bull. Math. Biol.*, 6:301–340, 2004.
8. E. Farcot and J.-L. Gouzé. How to control a biological switch: a mathematical framework for the control of piecewise affine models of gene networks. Research Report 5979, INRIA, 09 2006.
9. E. Farcot and J.-L. Gouzé. Periodic solutions of piecewise affine gene network models: the case of a negative feedback loop. Research Report 6018, INRIA, 11 2006.
10. A. F. Filippov. *Differential Equations with Discontinuous Righthand Sides*. Kluwer Academic Publishers, Dordrecht, 1988.
11. R. Ghosh and C. Tomlin. Symbolic reachable set computation of piecewise affine hybrid automata and its application to biological modelling: Delta-Notch protein signalling. *Systems Biology*, 1(1):170–183, 2004.
12. L. Glass and S. Kauffman. The logical analysis of continuous non-linear biochemical control networks. *J. Theor. Biol.*, 39:103–129, 1973.
13. J. Gouzé and T. Sari. A class of piecewise linear differential equations arising in biological models. *Dyn. Syst*, 17:299–316, 2002.
14. L. Habets and J. van Schuppen. A control problem for affine dynamical systems on a full-dimensional polytope. *Automatica*, 40:21 – 35, 2004.
15. H. M. Hardin and J. van Schuppen. System reduction of nonlinear positive systems by linearization and truncation. In C. Commault and N. Marchand, editors, *Positive systems - Proceedings of the Second Multidisciplinary Symposium on Positive Systems: Theory and Applications (POSTA 06)*, volume 341 of *Lecture Notes in Control and Information Sciences*, pages 431–438. Grenoble, France, 2006.
16. J. Harman. Allosteric regulation of the cAMP receptor protein. *Biochimica et Biophysica Acta*, 1547(1):1–17, 2001.
17. K. Heidtke and S. Schulze-Kremer. Design and implementation of a qualitative simulation model of λ phage infection. *Bioinformatics*, 14(1):81–91, 1998.
18. R. Hengge-Aronis. The general stress response in *Escherichia coli*. In G. Storz and R. Hengge-Aronis, editors, *Bacterial Stress Responses*, pages 161–177. ASM Press, Washington, DC, 2000.

19. G. Huisman, D. Siegele, M. Zambrano, and R. Kolter. Morphological and physiological changes during stationary phase. In F. Neidhardt, R. Curtiss III, J. Ingraham, E. Lin, K. Low, B. Magasanik, W. Reznikoff, M. Riley, M. Schaechter, and H. Umbarger, editors, *Escherichia coli and Salmonella: Cellular and Molecular Biology*, pages 1672–1682. ASM Press, Washington, DC, 2nd edition, 1996.

20. S. Kauffman. *The Origins of Order: Self-Organization and Selection in Evolution.* Oxford University Press, New York, 1993.

21. A. Keil and J.-L. Gouzé. Model reduction of modular systems using balancing methods. Technical report, Munich University of Technology, 2003.

22. I. Koch, B. Junker, and M. Heiner. Application of Petri net theory for modelling and validation of the sucrose breakdown pathway in the potato tuber. *Bioinformatics*, 2005. In press.

23. K. Kohn. Molecular interaction maps as information organizers and simulation guides. *Chaos*, 11(1):84–97, 2001.

24. A. Martinez-Antonio and J. Collado-Vides. Identifying global regulators in transcriptional regulatory networks in bacteria. *Current Opinion in Microbiology*, 6(5):482–489, 2003.

25. T. Mestl, E. Plahte, and S. Omholt. A mathematical framework for describing and analysing gene regulatory networks. *Journal of Theoretical Biology*, 176(2):291–300, 1995.

26. M. Ptashne. *A Genetic Switch: Phage λ and Higher Organisms.* Cell Press & Blackwell Science, Cambridge, MA, 2nd edition, 1992.

27. V. Reddy, M. Liebman, and M. Mavrovouniotis. Qualitative analysis of biochemical reaction systems. *Computers in Biology and Medicine*, 26(1):9–24, 1996.

28. A. Regev, W. Silverman, and E. Shapiro. Representation and simulation of biochemical processes using the π-calculus process algebra. In R. Altman, A. Dunker, L. Hunter, K. Lauderdale, and T. Klein, editors, *Pacific Symposium on Biocomputing, PSB'01*, volume 6, pages 459–470, Singapore, 2001. World Scientific Publishing.

29. D. Ropers, H. de Jong, M. Page, D. Schneider, and J. Geiselmann. Qualitative simulation of the carbon starvation response in E*scherichia coli. BioSystems*, 84:124–152, 2006.

30. R. Thomas and R. d'Ari. *Biological Feedback.* CRC Press, Boca Raton, FL, 1990.

31. J. Wang, E. Gilles, J. Lengeler, and K. Jahreis. Modeling of inducer exclusion and catabolite repression based on a PTS-dependent sucrose and non-PTS-dependent glycerol transport systems in *Escherichia* coli K-12 and its experimental verification. *Journal of Biotechnology*, 92(2):133–158, 2001.

32. L. Wick and T. Egli. Molecular components of physiological stress responses in *Escherichia coli. Advances in Biochemical Engineering/Biotechnology*, 89:1–45, 2004.

Modelling and Analysis of Cell Death Signalling

Thomas Eißing, Steffen Waldherr, and Frank Allgöwer

Institute for Systems Theory and Automatic Control, University of Stuttgart,
70550 Stuttgart, Germany
{eissing,waldherr,allgower}@ist.uni-stuttgart.de

1 Introduction

Biology has always been in close touch to other scientific disciplines which strongly contributed to its development. Quantitative reasoning based on mathematical considerations had strong and driving influences on biology [1]. However, with the emergence of molecular biology many exciting questions were raised and answered that did not require mathematical models to allow a qualitative understanding of many principle aspects that make up life. By now the wealth of information about molecular players and their interactions is becoming overwhelming and cannot be understood by merely drawing pictures and looking at them. Mathematical biology, for a long time rather a peripheral science, is emerging as the "post-omic" frontier. Whereas the "omic"-technologies produce large amounts of data, systems biology is promising to put the pieces back together.

Quantitative and dynamic modelling approaches describing aspects of, or on the long term, even the whole living cell or organisms will become essential to organize and understand biological complexity. The analysis of these models will allow to more rapidly test biological hypothesis and provides insight not easily accessible by classical experimentation. However, despite the ever increasing amounts of biological data, more, and more suitable, data will be needed. Also, the computational tools for dynamical systems not easily deal with very large and nonlinear models as easily encountered when entering biology [2].

In this contribution we would like to introduce selected aspects around our work to better understand apoptosis signalling using mathematical models. Apoptosis, also called programmed cell death, is a very important biological process that can eliminate selected cells for the benefit of the organisms as a whole. It is crucial during development and for cellular homoeostasis balancing cellular reproduction. Failures are implicated in severe diseases.

Motivated by this interesting biological process, we will introduce insights already gained, but also point to current and future challenges. Employing the apoptosis example, we will further discuss theoretical aspects of wider interest, such as bistability in cell signalling and robustness of biological systems. We will focus on rather small aspects of biology and mainly use ordinary differential

I. Queinnec et al. (Eds.): Bio. & Ctrl. Theory: Current Challenges, LNCIS 357, pp. 161–180, 2007.
springerlink.com

equations (ODEs) to model observed phenomena. In Section 2 we will introduce the necessary biological background. Section 3 overviews modelling approaches to describe apoptosis signalling. Thereby, we will introduce simple models to illustrate principle phenomena with a focus on viewing apoptosis as a bistable process. In Section 4, we will introduce robustness considerations for bistable systems, again employing the apoptosis example.

2   Apoptosis Biology

Apoptosis can be triggered externally, e.g. by certain cytokines binding to so-called death receptors, or internally, e.g. in response to DNA damage [3, 4, 5, 6]. At the heart of the apoptotic program are caspases, representing a number of aspartate directed cystein proteases [7]. Caspases are produced in an inactive pro-form and become activated through proteolytic cleavage. Initiator caspases sense apoptotic stimuli and propagate the signal to executioner caspases. These cleave many target proteins within the cell committed to die, thereby dismantling it and tagging the remainders for clearance [8]. A prominent example is the Inhibitor of CAD (ICAD), CAD being a Caspase Activated DNase which can

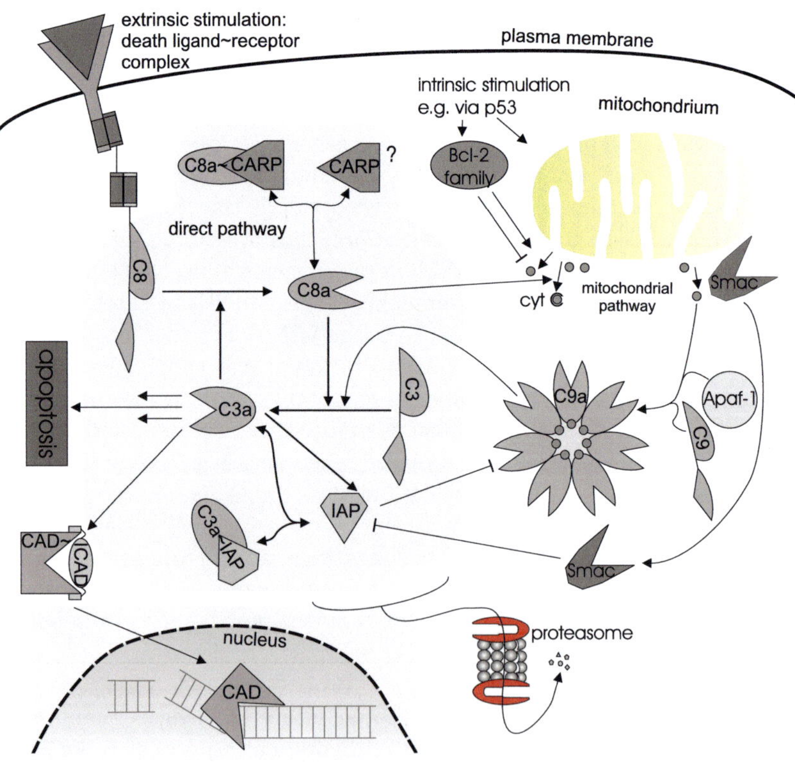

Fig. 1. Apoptosis pathways

cleave nuclear DNA. Internally triggered apoptosis proceeds via initiator caspase 9 activation through cytochrome c release from mitochondria. Caspase 9 then activates executioner caspases, most prominently caspase 3. Externally triggered apoptosis is initiated by the activation of receptor associated initiator caspases 8 and 10. The signalling process then proceeds either via the activation of the mitochondrial pathway (Type II cells) or by direct activation of caspase 3 (Type I cells) [9].

These steps are regulated at different levels. Inhibitor of Apoptosis Protein (IAP) family members directly inhibit activated caspases 3 and 9 and target them for proteasomal degradation. Bcl-2 family member proteins regulate the mitochondrial cytochrome c and Smac release, Smac being an inhibitor of IAPs, thus further contributing to the cytochrome c mediated caspase activation. In addition, various feedback loops complicate matters to yield a large and complex reaction system at the caspase level already [10, 11]. Also, many signalling pathways interact and influence apoptosis signalling and alternative pathways to cell death are known [12, 13, 14].

The physiological importance of apoptosis is evidenced by the fact that in adults about 10 billion cells per day commit apoptosis balancing those produced through cell division [4]. Disturbances to this delicate balance can lead to severe diseases. For example, hindered apoptosis is a hallmark of cancer whereas increased apoptosis is associated to neuro-degenerative diseases such as Alzheimer or Huntington [3, 7, 15].

3 Modelling Approaches to Apoptosis

As outlined in the last section, apoptosis is a complex but crucial phenomenon. Recently, several mathematical approaches to apoptosis signalling have been published. These range from large scale static models mainly used to interpret large amounts of experimental data [16, 17] to small scale dynamic models used to better understand selected aspects as outlined in the following.

One of the first dynamic models covers large parts of apoptosis signalling in a simplified manner and nicely describes several facets of both the externally and the internally triggered signalling pathways [18]. The hybrid model contains heuristic parts but is able to illustrate principle features and known regulatory mechanisms. A similar approach was taken in [19] and [20] to model death receptor induced apoptosis. There, also experimental time course data of relevant molecules on the population level are presented in order to identify the model parameters. However, the number of parameters is generally large compared to the amount and quality of data available, making the identification process difficult. Several theoretical studies focussed on the role of death receptor-ligand interactions as the signal initiating event and investigated the potential role of receptor clustering and the ligand trimer structure [21, 22]. Several other studies have focussed on the mitochondrial pathway of apoptosis or selected parts thereof [23, 24, 25, 26]. Rehm et al. [25] were able to closely link their mathematical model to own experiments and thereby experimentally verify their model predictions on the single cell level. This example nicely indicates the great potential of mathematical models to guide experimentation.

Especially on the level of the single cell, apoptosis is essentially an all-or-none process – the cell dies or continues to live. Several evidences indicate that this all-or-none behaviour is reflected on the executioner caspase level. ODE models can reflect such a behaviour by exhibiting bistability, i.e. two stable steady states. Consequently, several studies examined the steady state and stability properties of apoptosis models to reveal important ingredients to achieve bistability, indicate possibilities to shift the delicate balance of death and survival, and illustrate how pathological states of the cell can arise [23, 27, 28, 29, 24].

Following that idea, in Section 3.1 we will first illustrate how bistability can be realized in biochemical reaction systems at the example of simplified models resembling core processes of apoptosis signalling. In Section 3.2, we will then analyse a basic model of the direct pathway of apoptosis based on available literature data. As the analysis reveals deficiencies of the model, we will then present an extended model in Section 3.3 which captures important qualitative and quantitative aspects of this pathway and which is now supported by new experimental findings.

3.1   Bistability and Simple Apoptosis Models

Bistable behaviour is not restricted to apoptosis signalling but a reoccurring motif in biology. It is implicated in cell decision processes and cellular memory, as a bistable system can convert continuous input signals into discrete (all-or-none) output signals and thereby switch reversibly or irreversibly between two states. Motivated by mathematical model analysis, bistability has, for example, been experimentally demonstrated on the single cell level for the *lac* operon in *E. coli* and for the MAP kinase cascade in *Xenopus* oocytes [30, 31].

Several studies have outlined how bistability can be achieved by simple biochemical reaction systems and general theories about necessary requirements are emerging for special system classes [32, 33]. Generally, bistability requires two ingredients. One is positive feedback, which can be implemented by only positive (activatory) feedback or an even number of negative (inhibitory) interactions along the loop [32], and the second is an ultrasensitive reaction mechanism. Ultrasensitivity is defined as a system response that is more sensitive to changes in the component concentrations than is the normal hyperbolic response given by the Michaelis-Menten equation [34], e.g. a Hill type response [35]. However, both ingredients are only necessary but not sufficient to generate bistability. Whether the system is then bistable or not is strongly dependent on the parameter values as will be illustrated in the following.

Phase Plane Analysis of Simple Apoptosis Models

We described two alternative simple models capturing core properties of the caspase cascade at the heart of apoptosis, and more generally proteolytic reaction systems [28].

Following, we provide two alternative simple models of two mutually activating caspases similar to the illustration in Fig. 5. A detailed derivation is given in [28]. The ODE for the relative amount of activated initiator caspases common to both models is

$$\dot{X}_r = k_x \cdot (1 - X_r) \cdot Y_r - k_d \cdot X_r. \tag{1}$$

The ODE for the relative amount of activated executioner caspases for the cooperative model is

$$\dot{Y}_r = k_y \cdot (1 - Y_r) \cdot X_r^n - k_d \cdot Y_r, \tag{2}$$

and for the inhibitor model

$$\dot{Y}_r = k_y \cdot (1 - Y_r) \cdot X_r - k_d \cdot Y_r - \frac{k_f \cdot I_t \cdot Y_r \cdot (k_d + k_y \cdot X_r)}{(k_f \cdot Y_r + k_d + k_b) \cdot Y_t}. \tag{3}$$

The inhibitor model corresponds to the basic model (Fig. 5) without the rate v_4 and under the assumptions summarized in the text. The cooperative model is even simpler, additionally ignoring the rates v_3, v_7 and v_8 but with a modified version of v_1 introducing n.

Due to the normalization all parameters can be considered in one per unit of time. If not specified otherwise we use the following parameter values $k_x = k_y = 0.01$, $k_f = 1$, $k_b = 0.001$ and $k_d = 0.003$. The total amount of proteins are in concentration units and we assume $X_t = Y_t = 3 \cdot I_t = 1$.

Fig. 2. Simple apoptosis models

Two caspases X and Y, representing initiator and executioner caspases respectively, are considered. Their mutual activation constitutes a positive feedback loop. The two models are distinct in how they implement the ultrasensitive reaction mechanism. In the cooperative model, $n > 1$ molecules of activated X interact in order to activate Y, whereas in the inhibitor model (with $n = 1$) an inhibitor can bind to and inactivate activated Y. While caspase inhibitors are well known, especially the mitochondrial pathway of apoptosis also contains potential cooperative steps. Assuming the law of mass action and balancing the rates yields four and six ODEs for the cooperative and the inhibitor model respectively. Under some simplifying assumptions like equal degradation rates and a quasi steady state approximation for the inhibitor binding reaction, we can reduce both models to two ODEs describing the relative amounts of activated initiator (X_r) and executioner (Y_r) caspases respectively (Fig. 2).

Fig. 3 shows a classical phase plane analysis of the two models. Whereas the X_r nullcline (green) has a hyperbolic shape, the Y_r nullcline (red) has a sigmoidal shape. This sigmoidal shape reflects the ultrasensitive reaction mechanisms and results in three intersections of the nullclines corresponding to steady states. Generally, the exact form of the nullclines and therefore the number of intersections is parameter dependent (see below). However, two hyperbolic curves could at most yield two intersections. Stable steady states are marked green whereas unstable steady states are marked red. As can be seen, both systems are bistable

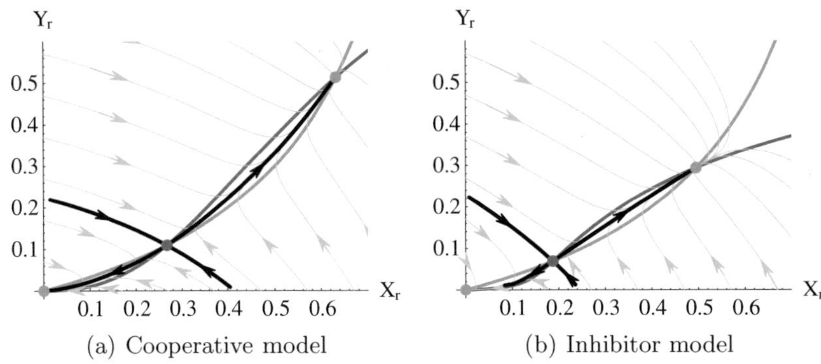

(a) Cooperative model   (b) Inhibitor model

Fig. 3. Phase plane analysis of two simple proteolytic models (similar to [28]). The green and red lines correspond to the X_r and Y_r nullcline respectively. Stable steady states are marked in green, unstable ones in red. The black lines show the stable manifold of the saddle separating the regions of attraction and the unstable manifold of the saddle connecting the three steady states. Representative time trajectories are depicted in grey.

and the regions of attraction are separated by the stable manifold of the saddle. Thus, depending on the initial conditions, the systems either converges to the steady state where both caspases are strongly activated or to the steady state where there are no activated caspases. Thereby, it is assumed that the initial conditions reflect processes not considered in the model itself.

Thus, ultrasensitivity can deform hyperbolic curves into sigmoidal curves enabling bistability. Ultrasensitivity can also be produced by other mechanisms such as saturation effects and the feedback can also be more indirect [28]. Latter was recently also indicated for the mitochondrial pathway of apoptosis where IAPs bind to both caspase 9 and 3 [24]. The binding to caspase 3 withdraws IAPs from caspase 9, which then can further contribute to caspase 3 activation.

Bifurcation Analysis of a Simple Inhibitor Model

As outlined above, both the simple cooperative and inhibitor model are bistable for the parameter sets shown. The number and stability properties of steady states changes when varying the parameters. This is commonly depicted in bifurcation diagrams such as shown in Fig. 4. For simple models, the steady states and their stability properties can be explicitly calculated in dependence of the parameters. Generally, they can be tracked in the parameter space employing continuation methods [36]. These are mainly restricted by numerical accuracy, computational power and our visual conception.

The bifurcation diagram of the simple inhibitor model for varying the activation rate constant k_x is shown in Fig. 4. The bistable behaviour is restricted to the left by a saddle node bifurcation – when the activation rate constant is too small a high caspase activation level cannot be sustained. To the right, the bistable behaviour is restricted by a transcritical bifurcation – a large activation constant will lead to a strong caspase activation for any small perturbation.

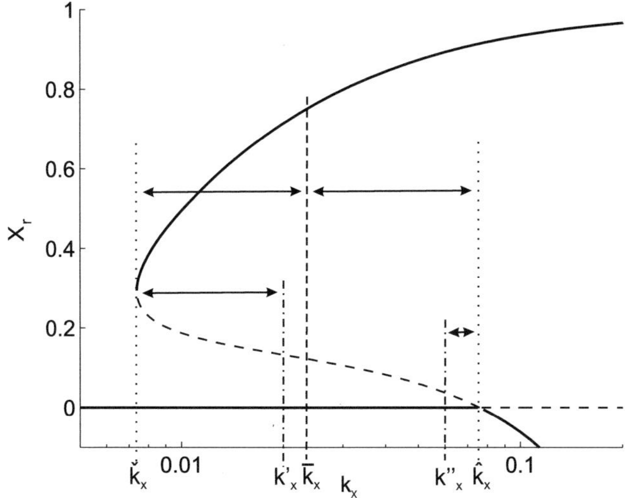

Fig. 4. Bifurcation analysis and the robustness measure according to Ma and Iglesias (Section 4.1). Stable steady states are drawn as solid lines, unstable steady state as dashed lines. The X_r axis is extended to biologically irrelevant negative values for clarity of drawing.

The bifurcation diagrams have similar properties for varying the other parameters and also for the cooperative model. However, unlike for the inhibitor model the Y_r nullcline of the cooperative model has a zero slope in the origin. Therefore, there is no transcritical bifurcation but the unstable steady state converges to zero for $k_x \to \infty$. Fig. 4 also illustrates a robustness measure, which will be discussed in Section 4.1.

3.2   Unstable "Life" in Basic Model for Direct Apoptotic Pathway

The inhibitor model introduced above closely resembles the direct pathway of apoptosis where we only consider one additional reaction, i.e. the cleavage of IAP molecules by caspase 3. However, we do not employ the simplifications and normalization as before. The model named "basic model" in the following is detailed in Fig. 5 and additional explanations are given in [27].

The qualitative behaviour of the basic model closely resembles that of the simple inhibitor model analysed above. However, similar bifurcation analysis indicate that bistability is only possible for parameter values far away from experimentally measured values reported in literature. This can be illustrated in different ways. In [27] we derived analytical conditions used for a plot in three parameter dimensions where several additional parameters are connected in a biologically meaningful way. Here we would like to illustrate this using an alternative approach we introduced in [37]. The methodology can also be used to define a robustness measure as discussed in Section 4.1.

For the direct pathway of receptor induced apoptosis, we consider a basic model based on literature knowledge (below) and an extended model (to the right) suggested by an extensive analysis of the basic model (compare figure and see text). For the basic model, we consider the rates

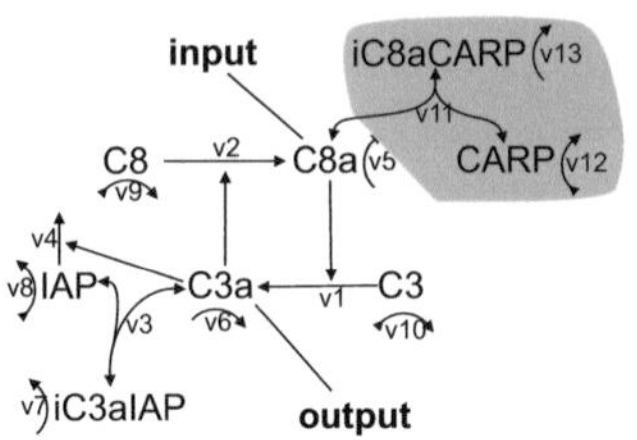

$$v_1 = k_1 \cdot [C8a] \cdot [C3]$$
$$v_2 = k_2 \cdot [C3a] \cdot [C8]$$
$$v_3 = k_3 \cdot [C3a] \cdot [IAP]$$
$$- k_{-3}[C3a{\sim}IAP]$$
$$v_4 = k_4 \cdot [C3a] \cdot [IAP]$$
$$v_5 = k_5 \cdot [C8a] \tag{4}$$
$$v_6 = k_6 \cdot [C3a]$$
$$v_7 = k_7 \cdot [C3a{\sim}IAP]$$
$$v_8 = k_8 \cdot [IAP] - k_{-8}$$
$$v_9 = k_9 \cdot [C8] - k_{-9}$$
$$v_{10} = k_{10} \cdot [C3] - k_{-10}.$$

Balancing (4) yields (without v_{11})

$$[\dot{C8}] = -v_2 - v_9$$
$$[\dot{C8a}] = v_2 - v_5(-v_{11})$$
$$[\dot{C3}] = -v_1 - v_{10}$$
$$[\dot{C3a}] = v_1 - v_3 - v_6 \tag{5}$$
$$[\dot{IAP}] = -v_3 - v_4 - v_8$$
$$[\dot{C3a{\sim}IAP}] = v_3 - v_7.$$

For the extended model we derive the additional rates

$$v_{11} = k_{11} \cdot [C8a] \cdot [CARP]$$
$$- k_{-11} \cdot [C8a{\sim}CARP]$$
$$v_{12} = k_{12} \cdot [CARP] - k_{-12} \tag{6}$$
$$v_{13} = k_{13} \cdot [C8a{\sim}CARP],$$

and extending the rates (4) by (6), the basic model (5) is extended by two variables

$$[\dot{CARP}] = -v_{11} - v_{12}$$
$$[\dot{C8a{\sim}CARP}] = v_{11} - v_{13}. \tag{7}$$

k_1	$5.8 \cdot 10^{-5}$	k_{11}	$5 \cdot 10^{-4}$
k_2	$1 \cdot 10^{-5}$	k_{12}	$1 \cdot 10^{-3}$
k_3	$5 \cdot 10^{-4}$	k_{13}	$1.16 \cdot 10^{-2}$
k_4	$3 \cdot 10^{-4}$	k_{-3}	0.21
k_5	$5.8 \cdot 10^{-3}$	k_{-8}	464
k_6	$5.8 \cdot 10^{-3}$	k_{-9}	507
k_7	$1.73 \cdot 10^{-2}$	k_{-10}	81.9
k_8	$1.16 \cdot 10^{-2}$	k_{-11}	0.21
k_9	$3.9 \cdot 10^{-3}$	k_{-12}	540
k_{10}	$3.9 \cdot 10^{-3}$		

For simplicity, we consider state variables (given in molecules per cell) as dimensionless. Then, the standard parameter values displayed in the table can all be considered in $[min^{-1}]$.

Fig. 5. Models of the direct apoptotic pathway (similar to [27, 37])

A Monte Carlo Approach to Evaluate Complex Behaviour

We proposed a Monte Carlo approach to evaluate complex behaviour such as bistability or limit cycles [37]. In this approach random parameter sets are generated from predefined ranges and evaluated for the property of interest, i.e. bistability here.

In Fig. 6 we show the distribution of hits (blue dots), i.e. parameter sets allowing for a bistable behaviour, and nonhits (yellow dots) in the parameter

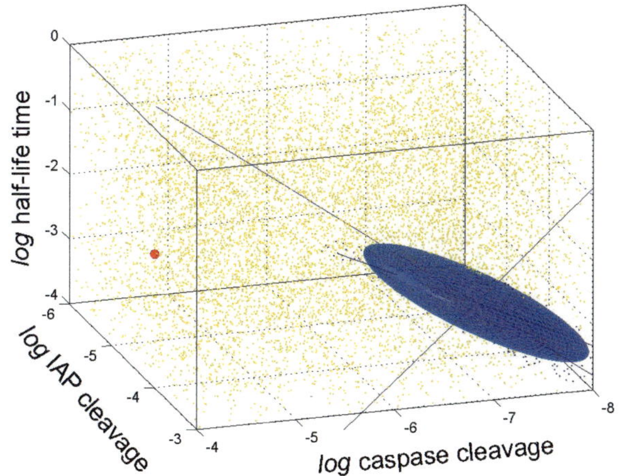

Fig. 6. Bistable parameter space as a evaluated by a Monte Carlo approach. Bistable parameter sets are depicted blue and yellow otherwise. The red dot indicates literature values. The blue dots were fitted by an 95 % ellipsoid whose principal components are depicted as blue lines. The following parameter ratios were chosen as axes, for caspase activation, $k_1 = 2 \cdot k_2$; for half-life time, $k_7 = k_8 = 2 \cdot k_5 = 2 \cdot k_6 = 4 \cdot k_9 = 4 \cdot k_{10}$; and for IAP cleavage, k_4; v_3 was fixed according to literature values.

space. Thereby, 95 % of the blue dots are contained in the ellipsoid shown. As can be seen, a bistable behaviour is favoured for slow cleavage kinetics and high turnovers. In addition, the IAP cleavage is crucial, which can be explained by the assumption of excess IAP molecules compared to caspase 3 (however, literature values are not consistent on these average amounts). The red dot indicates parameter values obtained from literature. Clearly, the values suggested in literature are far away from the region allowing for bistable behaviour. Literature values result in a fast activation of caspases for any initial condition that has some activated caspase (bound or free) included. The resulting biological interpretation would be that within a short time, every cell would undergo apoptosis, which is clearly not the case. Thus, if we assume that the measurements should not be that far away from the true values and if we ask for bistable behaviour, we have to modify our model. Additional simulation studies suggest that activated caspase 8 is a major cause as it is not controlled by an inhibitor like caspase 3 is and consequently minor amounts can directly activate caspase 3.

Recapitulating, the model analysis can reveal inconsistencies in the literature view of the direct pathway of apoptosis. The known interactions and parameter values do not fit to the observed behaviour. Thus, although more than 125000 studies relating to apoptosis have been published (PubMed search for "apoptosis"), an in-depth understanding of the core processes remains elusive.

3.3 Dynamic Behaviour of an Extended Model

Based on the above analysis, we proposed an extended model, which incorporates inhibitors binding to activated caspase 8 [27]. By now, this hypothesis is supported by high-throughput experiments in which IAP like molecules have been identified (named CARPs) that bind to receptor-associated initiator caspases including caspase 8. Further, suppression of these molecules has been implicated in tumour cell growth inhibition [38].

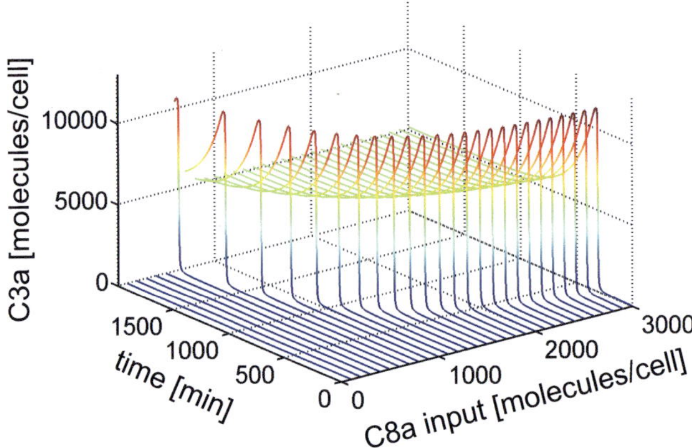

Fig. 7. Time trajectories for $[C3a]$ (output) in dependence on the initial $[C8a]$ concentration (input) (redrawn from [27])

The extended model is bistable with parameter values reasonably close to those reported in literature. The dynamic behaviour of caspase 3 for different initial concentrations of activated caspase 8 is depicted in Fig. 7 for one bistable parameter set (given in Fig. 5). The available data do not yet allow a systematic parameter identification. However, the model captures important qualitative and quantitative aspects of caspase activation [39]. Small amounts of activated caspases 8 do not result in a strong or sustained activation of caspase 3. For larger amounts of caspase 8, the majority of caspase 3 is, after a lag phase, activated within a short time frame after. The length of the lag phase is inversely dependent on the initial concentration of activated caspase 8.

The discrepancies between the fast caspase activation observed on the single cell level and the slower activation on the population level can be understood looking at the model behaviour – either a homogeneous population exposed to a distributed input or a heterogeneous population or a combination of both will result in different lengths of the lag phase for the individual cells. Thus, while each cell rapidly activates the majority of its caspases, the different timing blurs this behaviour on the population level. Interestingly, the distribution of initial concentrations of activated caspase 8 needed to reconcile population and single cell behaviour nicely matches measured distributions of death receptors which produce the activated caspase 8 [27]. Regarding our model, this can be viewed as a source of "extrinsic" noise [40]. In the following, we discuss additional influences on the single cell behaviour in the context of robustness.

4 Robustness Aspects

Robustness refers to the resistance of a function to perturbations in operating conditions and appears to be an inherent common property of biological systems [41, 42]. Clearly, many biological systems have to function reproducibly despite disturbances of internal and environmental conditions. This is especially important when considering processes deciding on the cell fate, as for example in programmed cell death. However, biological systems also need to respond to certain influences such as extra- and intracellular signals leading to a trade-off between robustness and performance. Further, this "robust yet fragile" character might be a more general consequence of the design of complex systems [43, 44] (see also other chapters).

Below, we will focus on robustness to parameter variations. However, biological systems also face stochastic influences. In fact, due to the small volume of cells and low number of molecules often involved, biological systems sometimes belie a deterministic and continuous description as assumed when employing ODEs [45, 46, 40]. The stochastic nature of the biochemical reactions can strongly influence behaviour. For example, in Fig. 7 it can be seen that the activated caspase 3 remains at very low concentrations for rather long times even for large inputs. Strong stochastic influences should be expected. However, we were able to show that stochastic effects due to the intrinsic stochasticity are negligible for a relevant and large range of inputs because the inhibitors buffer the activated caspases and thereby filter out noise [37].

In the following, we will first present two measures based on bifurcations and second, a measure based on control theoretical ideas.

4.1 Robustness Measures Based on Bifurcations

A robustness measure that is based on the distance of the nominal parameter value to a parameter value where a bifurcation happens has been introduced by Ma and Iglesias [47]. For dynamical behaviour like bistability, the relevant bifurcations are the ones where the unstable steady state turns stable. Precisely, Ma and Iglesias define the degree of robustness (DOR) of the dynamical behavior with respect to variations in the parameter k_j, for a given parameter vector K as

$$DOR_j(K) = 1 - \max\left\{ \frac{\check{k}_j(K)}{k_j}, \frac{k_j}{\hat{k}_j(K)} \right\},$$

where bifurcations occur at $\check{k}_j$ and $\hat{k}_j$, with $\check{k}_j < k_j < \hat{k}_j$, and which depend on the reference parameter values K. The definition is illustrated in Fig. 4. The parameter that is considered there is k_x. Bifurcations which destroy bistability appear for $k_x = \check{k}_x$ and $k_x = \hat{k}_x$. k'_x and k''_x indicate two possible parameter values for which $\check{k}_x$ and $\hat{k}_x$ are the relevant bifurcation values, respectively. The maximal robustness is obtained for $k_x = \bar{k}_x$, the geometric mean of the bifurcation values.

We computed the robustness measure according to Ma and Iglesias for parameter variations in the extended model ((5),(7)) of the direct apoptotic pathway. The results are shown in Fig. 10.

A similar method which also evaluates the boundaries of bistability in parameter space, is the Monte Carlo approach illustrated in Fig. 6. Especially for higher dimensional spaces, the relative frequency of occurrence of bistability provides an efficiently calculable estimate of the volume in the parameter space allowing for a bistable behaviour that can be used as a robustness measure. Another often advantageous feature is that it does not rely on a reference parameter set. For example, using this approach we were able to show that the extended model is much more robust than the basic model, indicating an additional advantage of the model extension [37]. Also, unlike reported otherwise, inhibitors as encountered in apoptosis signalling can allow a similar robust bistable performance as other mechanisms [48].

However, both measures introduced above generally require a global consideration of the model. Following, we will provide an alternative measure that can be obtained based on local considerations and which can interestingly but surprisingly be correlated to the global measure by Ma and Iglesias introduced above.

4.2   Robustness Measure Based on State Relevance

In a recent publication, Schmidt and Jacobsen [49] suggested a computational method to evaluate the relevance that individual state variables in a system contribute to dynamical behaviour like bistability or oscillations. Here, we summarize their method, apply it to the extended model of caspase activation (Fig. 5) and relate it to robustness of bistability with respect to parameter variations via the local sensitivity of the unstable steady state.

Computation of a State's Relevance to Bistability

The arguments in the previous sections show that the third, unstable steady state is required to generate bistability in the caspase activation model. Thus the unstable steady state has a crucial role in the computation of each state variable's relevance to bistability.

Based on a linear approximation of the model around this decision state, Schmidt and Jacobsen [49] perturb the influence of one state variable on the other variables of the system. In the unperturbed case, the linear approximation is unstable, since the decision state is unstable. One now searches for a perturbation that renders the decision state stable, which would roughly correspond to reaching a bifurcation point in the original nonlinear system. The magnitude of a perturbation found in this way is then a measure of the relevance to bistability of the perturbed variable with its connections to other variables.

Consider a system given by the differential equation

$$\frac{d}{dt}x = f(x),\tag{8}$$

with $x \in \mathbb{R}^n$. We assume that the system is bistable and has an unstable steady state x_d. Setting $\Delta x = x - x_d$, the linear approximation around the decision state is given by

$$\frac{d}{dt}(\Delta x) = A\Delta x,\tag{9}$$

with $A = \frac{\partial f}{\partial x}(x_d)$. Note that, by our assumption, the system (9) is unstable.

The computation of relevance is then done as follows: for each state variable x_i, a perturbation ϵ_i is introduced into the interactions of this variable (Fig. 8) towards other variables. This yields the system

$$L_i(\epsilon_i): \quad \frac{d}{dt}(\Delta x) = A\Delta x + A_i\epsilon_i\Delta x_i,$$

where A_i is the i-th column of A, but with a zero at the i-th component.

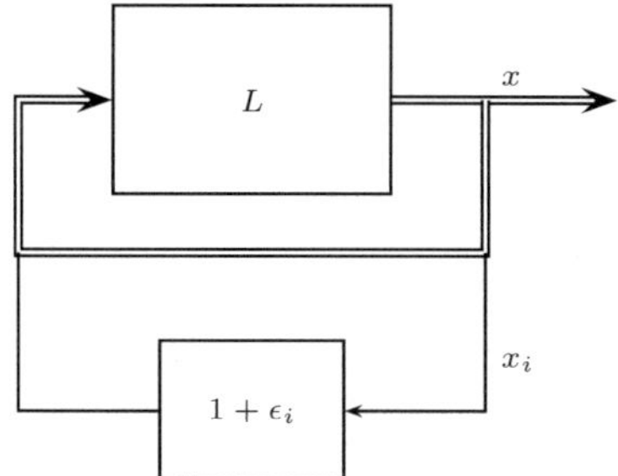

Fig. 8. Closed feedback loop with one feedback path perturbed

We are now searching for the minimal perturbation that will stabilize the system L_i and define the value $\bar{\epsilon}_i$ as

$$\bar{\epsilon}_i = \min\{\epsilon_i > 0 \mid L_i(\epsilon_i) \text{ or } L_i(-\epsilon_i) \text{ is stable}\}.\tag{10}$$

If the minimum does not exist, set $\bar{\epsilon}_i = \infty$. The higher the value of $\bar{\epsilon}_i$, the more difficult it becomes to perturb the connections among the considered state variable x_i and the remaining variables such that instability is lost, and the less relevant the variable x_i is to bistability. Formally, we use the following definition of relevance.

Definition 1. *The relevance R_i of the state variable x_i to the bistability of the system* (8) *is*

$$R_i = \frac{1}{\bar{\epsilon}_i},$$

where $\bar{\epsilon}_i$ is given by equation (10).

Table 1. Relevance of components to bistability in the direct apoptotic pathway

C8	C8a	C3	C3a	IAP	C3a_IAP	CARP	C8a_CARP
0.064	1210	0.466	833	0.712	758	0.759	1116

This computation has been implemented numerically in the Systems Biology toolbox for Matlab [50] and has been applied to the extended model (Fig. 5). The results given in Table 1 show that there is actually a large gap between the highly relevant variables C8a, C3a, C3a_IAP and C8a_CARP and the other variables, whose relevance to bistability is very low.

Steady State Sensitivity and Parameter Robustness

In this section, we show that the relevance of model components as introduced in the previous section can be interpreted as a robustness measure. This robustness measure is computed for the model of the direct apoptotic pathway.

The rationale behind the definition of relevance in the previous section is that bistability is due to an unstable steady state emerging from some bifurcation point more or less close to the actual point in parameter space that is considered. However, the bifurcation point is actually not used in the computation, but the analysis is done locally by a linearisation of the model at the unstable steady state.

Bistability is lost when the unstable steady state becomes stable by introducing perturbations in the feedback path of one component as shown in Fig. 8. Thus the relevance R_i as defined in Def. 1 can be interpreted as a measure of the sensitivity of the bistable behavior to disturbances in the influence of the considered state. R_i depends on the parameters of the model, and in this section, we are thus considering $R_i(K)$, where $K = (k_1, \ldots, k_{-12})$ is the vector of reference parameter values.

Since sensitivity can be considered as the inverse of robustness, we define the robustness of bistability with respect to variations in state i for the parameter values K as

$$RS_i(K) = \frac{1}{R_i(K)}.$$

When comparing the robustness measures DOR_j by Ma and Iglesias [47] and RS_i based on the relevance measure by Schmidt and Jacobsen [49], we note that the underlying principle is quite similar, but the form of robustness that is considered is different. The measure RS_i considers disturbances or variations in the influence of the state i on the other states, while the measure DOR_j regards variations in the value of the parameter j. Thus, to be able to compare these measures, we have to apply a relation between states and parameters of the model. To obtain such a relation, we use a local sensitivity analysis for the unstable steady state. The absolute value of the normalized local sensitivity of a variable's steady state value versus a model parameter is taken as an indication for the strength of the relation between the considered state variable and the parameter. See e.g. [51] and the references therein for details on the local sensitivity

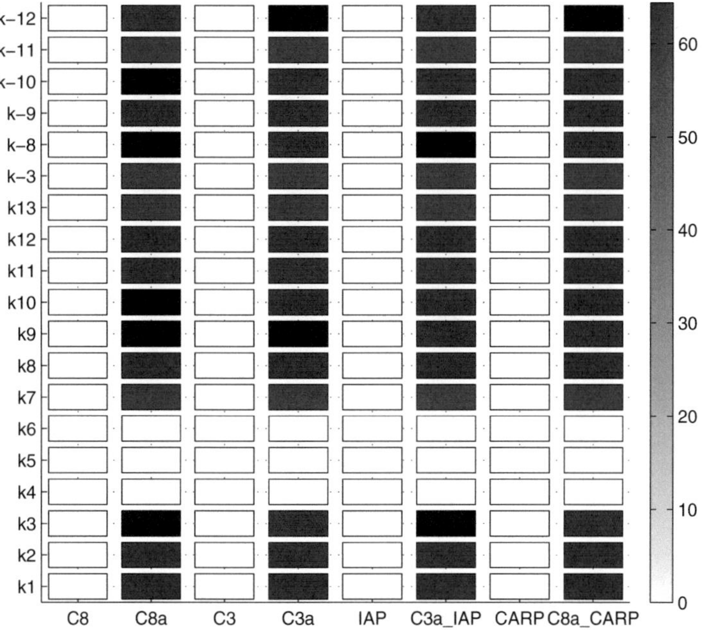

Fig. 9. Absolute values of normalized local parameter sensitivities of steady state values for the unstable steady state

analysis. The analysis was done numerically using the method implemented in the Systems Biology toolbox for Matlab [50].

The results of the local sensitivity analysis for the caspase activation model are shown in Fig. 9. They demonstrate that the local sensitivity values are highly structured: the components with high relevance have high local sensitivities with respect to all parameters but k_4, k_5 and k_6, while the other components are not sensitive with respect to any parameter.

Exploiting the sensitivity as a relation between states and parameters of the model, we can then define a robustness measure for bistability with respect to variations in the parameter k_j as

$$RP_j(K) = \sum_{i=1}^{n} \frac{1}{|S_{ij}|R_i(K)} \tag{11}$$

where K are reference parameters and S_{ij} is the local sensitivity of the state variable i versus the parameter j.

We have computed the robustness measure $RP_j(K)$ for each parameter in the extended model of the direct apoptotic pathway ((5),(7)). The results are shown in Fig. 10, normalized to range from 0 to 1 for better comparison with the robustness measure $DOR_j(K)$. Comparing the robustness measures RP_j and DOR_j, it is obvious from Fig. 10 that they give very similar results. For the extended model and using the reference parameter values, bistability is rather

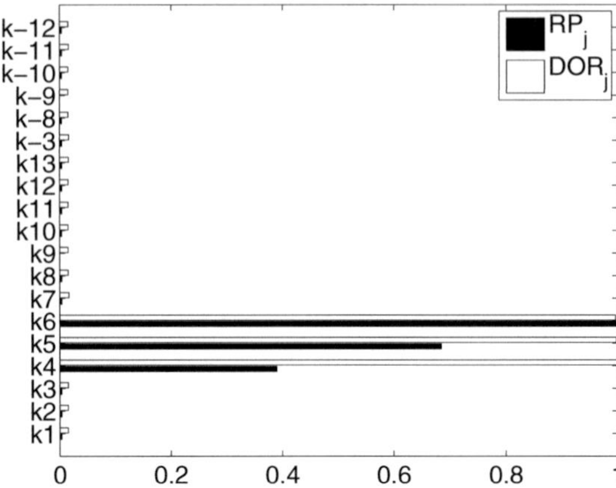

Fig. 10. Robustness of bistability against parameter variations in the extended model of the direct apoptotic pathway

sensitive with respect to variations in all parameters but k_4, k_5 and k_6. For this model, we also in detail analysed the effects on certain system characteristics when varying each single parameter over a wide range [29].

It is actually rather surprising that the conclusions from the two robustness measures are that similar. The robustness measure by Ma and Iglesias requires to compute the exact bifurcation points when each parameter in the model is varied, which may be a very hard computational task for a typical system. On the other hand, the robustness measure based on Schmidt and Jacobsen's state relevance uses only a linear approximation of the system at the unstable equilibrium point and is typically much easier to compute. This finding highlights again the important roles that the unstable steady state and the interactions among state variables take in the emergence of bistability.

5   Discussion and Outlook

In this chapter we described different simple models of apoptosis signalling introducing the idea of apoptosis as a bistable system. We then presented a model of the direct pathway of apoptosis capturing core properties of the signalling pathway. Using the apoptosis example, we outlined how parameter and robustness considerations can support or argue against certain biological hypotheses.

Clearly, the presented considerations need to be extended before apoptosis signalling as a whole can be understood. For example, many known reactions of the apoptosis network have been neglected (compare Fig. 1). Also, it is known that other pathways interact with the apoptosis pathways fine-tuning the final outcome as outlined in a nice set of experiments [16, 17]. Dealing with the complexity indicated in those experiments to derive a dynamical model closely resembling experimental data will require efficient computational tools, e.g. for identification and model analysis. Also, important phenomena such as bistability might not manifest on the population level, but only on the single cell level [25, 27]. Although the single cell level can be tackled with modern microscopy, many parameters needed for detailed modelling approaches cannot be measured on the single cell level to date. Combining data from different sources in a useful way will likely be crucial. Thereby, engineering competence cannot only contribute to modelling and model analysis but to data generation and processing [52] as well as automation, as was also crucial for the success of the human genome project.

When analysing larger models, the reduction of these models or the isolation of model parts most important to the behaviour of interest will be of major interest. For example, one might address the questions, which combination of proteins should be targeted to achieve a strong impact on an undesired behaviour while hardly influencing desired, or more generally other, behaviours? The framework of sensitivity and robustness analysis promises to be an important tool towards these goals. Of course, alternative measures of sensitivity and robustness than those introduced here can be considered. Other suitable measures for biological signalling networks are, for example, coefficients derived from dynamic sensitivities evaluated at different locations in the parameter space [53], overall coefficients developed in the framework of metabolic control analysis [54], direct Lyapunov exponents [55] or measures based on robust control ideas [47, 56]. Nevertheless, improved methodologies and an implementation easily accessible for the non-expert would be helpful [57].

The involved scales in biology and the inherent complexity are posing great challenges. However, it is hoped that behind the complexity, simple (design) principles will emerge [58]. Identifying and isolating those principles to allow a detailed analyses and understanding will move biology one step further and rationalize medicine. Thereby, dynamic models and ideas routed in systems and control theory are contributing to this important goal. Vice versa, systems and control theory can be inspired by biology [59].

References

1. Wingreen, N. and D. Botstein (2006). Back to the future: education for systems-level biologists. *Nat. Rev. Mol. Cell Biol.*, 7(11):829–832.
2. Kitano, H. (2002). Systems biology: a brief overview. *Science*, 295(5560):1662–4.
3. Danial, N. N. and S. J. Korsmeyer (2004). Cell death: critical control points. *Cell*, 116(2):205–219.
4. Heemels, M. T., R. Dhand, and L. Allen (2000). Apoptosis. *Nature*, 407(6805):769.
5. Hengartner, M. O. (2000). The biochemistry of apoptosis. *Nature*, 407(6805):770–6.
6. Rich, T., R. L. Allen, and A. H. Wyllie (2000). Defying death after DNA damage. *Nature*, 407(6805):777–783.
7. Wolf, B. B. and D. R. Green (1999). Suicidal tendencies: apoptotic cell death by caspase family proteinases. *J. Biol. Chem.*, 274(29):20049–20052.
8. Savill, J. and V. Fadok (2000). Corpse clearance defines the meaning of cell death. *Nature*, 407(6805):784–8.

9. Scaffidi, C., S. Fulda, A. Srinivasan, C. Friesen, F. Li, K. J. Tomaselli, K. M. Debatin, P. H. Krammer, and M. E. Peter (1998). Two CD95 (APO-1/Fas) signaling pathways. *EMBO J.*, 17(6):1675–1687.

10. Sohn, D., K. Schulze-Osthoff, and R. U. Jänicke (2005). Caspase-8 can be activated by interchain proteolysis without receptor-triggered dimerization during drug-induced apoptosis. *J. Biol. Chem.*, 280(7):5267–5273.

11. Stennicke, H. R. and G. S. Salvesen (1999). Catalytic properties of the caspases. *Cell Death Differ.*, 6(11):1054–1059.

12. Okada, H. and T. W. Mak (2004). Pathways of apoptotic and non-apoptotic death in tumour cells. *Nat. Rev. Cancer*, 4(8):592–603.

13. Karin, M. and A. Lin (2002). NF-kappaB at the crossroads of life and death. *Nat. Immunol.*, 3(3):221–7.

14. Leist, M. and M. Jaattela (2001). Four deaths and a funeral: from caspases to alternative mechanisms. *Nat. Rev. Mol. Cell Biol.*, 2(8):589–98.

15. Hanahan, D. and R. A. Weinberg (2000). The hallmarks of cancer. *Cell*, 100(1):57–70.

16. Janes, K. A., J. G. Albeck, S. Gaudet, P. K. Sorger, D. A. Lauffenburger, and M. B. Yaffe (2005). A systems model of signaling identifies a molecular basis set for cytokine-induced apoptosis. *Science*, 310(5754):1646–53.

17. Janes, K. A., S. Gaudet, J. G. Albeck, U. B. Nielsen, D. A. Lauffenburger, and P. K. Sorger (2006). The response of human epithelial cells to TNF involves an inducible autocrine cascade. *Cell*, 124(6):1225–1239.

18. Fussenegger, M., J. E. Bailey, and J. Varner (2000). A mathematical model of caspase function in apoptosis. *Nat. Biotechnol.*, 18(7):768–74.

19. Schoeberl, B., E. D. Gilles, and P. Scheurich (2001). A mathematical vision of TNF receptor interaction. In Yi, T., M. Hucka, M. Morohashi, and H. Kitano, editors, *Proceedings of the 2nd International Conference on Systems Biology*. Omnipress, Madison, WI, Pasadena, CA, pages 158–167.

20. Bentele, M., I. Lavrik, M. Ulrich, S. Stosser, D. W. Heermann, H. Kalthoff, P. H. Krammer, and R. Eils (2004). Mathematical modeling reveals threshold mechanism in CD95-induced apoptosis. *J. Cell Biol.*, 166(6):839–51.

21. Guo, C. and H. Levine (1999). A thermodynamic model for receptor clustering. *Biophys. J.*, 77(5):2358–65.

22. Lai, R. and T. L. Jackson (2004). A mathematical model of receptor-mediated apoptosis: Dying to know why FASL is a trimer. *Math. Biosci. Eng.*, 1(2):325–338.

23. Bagci, E. Z., Y. Vodovotz, T. R. Billiar, G. B. Ermentrout, and I. Bahar (2006). Bistability in apoptosis: Roles of Bax, Bcl-2 and mitochondrial permeability transition pores. *Biophys. J.*, 90:1546–1559.

24. Legewie, S., N. Blüthgen, and H. Herzel (2006). Mathematical Modeling Identifies Inhibitors of Apoptosis as Mediators of Positive Feedback and Bistability. *PLoS Comput. Biol.*, 2(9):e120.

25. Rehm, M., H. J. Huber, H. Dussmann, and J. H. M. Prehn (2006). Systems analysis of effector caspase activation and its control by X-linked inhibitor of apoptosis protein. *EMBO J.*, 25(18):4338–4349.

26. Stucki, J. W. and H. U. Simon (2005). Mathematical modeling of the regulation of caspase-3 activation and degradation. *J. Theor. Biol.*, 234(1):123–31.

27. Eißing, T., H. Conzelmann, E. D. Gilles, F. Allgöwer, E. Bullinger, and P. Scheurich (2004). Bistability analyses of a caspase activation model for receptor-induced apoptosis. *J. Biol. Chem.*, 279(35):36892–36897.

28. Eißing, T., S. Waldherr, F. Allgöwer, P. Scheurich, and E. Bullinger (2007). Steady state and (bi-) stability evaluation of simple protease signalling networks. *BioSystems*:in press, doi:10.1016/j.biosystems.2007.01.003.
29. Eißing, T., S. Waldherr, E. Bullinger, C. Gondro, O. Sawodny, F. Allgöwer, P. Scheurich, and T. Sauter (2006). Sensitivity analysis of programmed cell death and implications for crosstalk phenomena during Tumor Necrosis Factor stimulation. In *Proceedings of the IEEE International Conference on Control Applications (CCA)*. Munich, Germany, pages 1746–52.
30. Ozbudak, E. M., M. Thattai, H. N. Lim, B. I. Shraiman, and A. V. Oudenaarden (2004). Multistability in the lactose utilization network of Escherichia coli. *Nature*, 427(6976):737–740.
31. Xiong, W. and J. E. Ferrell, Jr. (2003). A positive-feedback-based bistable 'memory module' that governs a cell fate decision. *Nature*, 426(6965):460–5.
32. Angeli, D., J. E. Ferrell, Jr., and E. D. Sontag (2004). Detection of multistability, bifurcations, and hysteresis in a large class of biological positive-feedback systems. *Proc. Natl. Acad. Sci. U. S. A.*, 101(7):1822–1827.
33. Craciun, G., Y. Tang, and M. Feinberg (2006). Understanding bistability in complex enzyme-driven reaction networks. *Proc. Natl. Acad. Sci. U. S. A.*, 103(23):8697–8702.
34. Koshland, D. E. (1998). The era of pathway quantification. *Science*, 280(5365):852–853.
35. Weiss, J. N. (1997). The Hill equation revisited: uses and misuses. *FASEB J.*, 11(11):835–841.
36. Kuznetsov, Y. A. (1995). *Elements of Applied Bifurcation Theory*. Springer-Verlag.
37. Eißing, T., F. Allgöwer, and E. Bullinger (2005). Robustness properties of apoptosis models with respect to parameter variations and stochastic influences. *IEE Syst. Biol.*, 152(4):221–228.
38. McDonald, E. R., 3rd and W. S. El-Deiry (2004). Suppression of caspase-8- and -10-associated RING proteins results in sensitization to death ligands and inhibition of tumor cell growth. *Proc. Natl. Acad. Sci. U. S. A.*, 101(16):6170–5.
39. Rehm, M., H. Dussmann, R. U. Janicke, J. M. Tavare, D. Kogel, and J. H. Prehn (2002). Single-cell fluorescence resonance energy transfer analysis demonstrates that caspase activation during apoptosis is a rapid process. *J. Biol. Chem.*, 277(27):24506–24514.
40. Elowitz, M. B., A. J. Levine, E. D. Siggia, and P. S. Swain (2002). Stochastic gene expression in a single cell. *Science*, 297(5584):1183–6.
41. Barkai, N. and S. Leibler (1997). Robustness in simple biochemical networks. *Nature*, 387(6636):913–7.
42. Stelling, J., U. Sauer, Z. Szallasi, F. J. Dolye, 3rd, and J. Doyle (2004). Robustness of cellular functions. *Cell*, 118(6):675–685.
43. Carlson, J. M. and J. Doyle (2002). Complexity and robustness. *Proc. Natl. Acad. Sci. U. S. A.*, 99 Suppl. 1:2538–45.
44. Carlson, J. M. and J. Doyle (2000). Highly optimized tolerance: robustness and design in complex systems. *Phys. Rev. Lett.*, 84(11):2529–32.
45. Fall, C. P., E. S. Marland, J. M. Wagner, and J. J. Tyson (2002). *Computational Cell Biology*, volume 20 of *Interdisciplinary Applied Mathematics*. Springer-Verlag, New York, NY.
46. Rao, C. V., D. M. Wolf, and A. P. Arkin (2002). Control, exploitation and tolerance of intracellular noise. *Nature*, 420(6912):231–7.
47. Ma, L. and P. A. Iglesias (2002). Quantifying robustness of biochemical network models. *BMC Bioinformatics*, 3:38.

48. Eißing, T., S. Waldherr, F. Allgower, P. Scheurich, and E. Bullinger (2007). Response to Bistability in Apoptosis: Roles of Bax, Bcl-2, and Mitochondrial Permeability Transition Pores. *Biophys. J.*:in press, doi:10.1529/biophysj.106.100362.

49. Schmidt, H. and E. Jacobsen (2004). Identifying feedback mechanisms behind complex cell behavior. *IEEE Cont. Sys. Mag.*, 24(4):91–102.

50. Schmidt, H. and M. Jirstrand (2005). Systems Biology Toolbox for MATLAB: a computational platform for research in Systems Biology. *Bioinformatics*, 22(4):514–5.

51. Klipp, E., R. Herwig, A. Kowald, C. Wierling, and H. Lehrach (2005). *Systems Biology in Practice. Concepts, Implementation and Application.* Wiley-VCH.

52. Schilling, M., T. Maiwald, S. Bohl, M. Kollmann, C. Kreutz, J. Timmer, and U. Klingmüller (2005). Quantitative data generation for systems biology: the impact of randomisation, calibrators and normalisers. *IEE Syst. Biol.*, 152(4):193–200.

53. Stelling, J., E. D. Gilles, and F. J. Doyle, 3rd (2004). Robustness properties of circadian clock architectures. *Proc. Natl. Acad. Sci. U. S. A.*, 101(36):13210–5.

54. Wolf, J., S. Becker-Weimann, and H. R. (2005). Analysing the robustness of cellular rhythms. *IEE Syst. Biol.*, 2:35–41.

55. Aldridge, B. B., G. Haller, P. K. Sorger, and D. A. Lauffenburger (2006). Direct Lyapunov exponent analysis enables parametric study of transient signalling governing cell behaviour. *IEE Syst. Biol.*, 153(6):425–432.

56. Kim, J., I. Postlethwaite, L. Ma, and P. A. Iglesias (2006). Robustness analysis of biochemical network models. *IEE Syst. Biol.*, 153(3):96–104.

57. Hucka, M., A. Finney *et al.* (2003). The systems biology markup language (SBML): a medium for representation and exchange of biochemical network models. *Bioinformatics*, 19(4):524–31.

58. Kollmann, M., L. Løvdok, K. Bartholomé, J. Timmer, and V. Sourjik (2005). Design principles of a bacterial signalling network. *Nature*, 438(7067):504–507.

59. Sontag, E. (2004). Some new directions in control theory inspired by systems biology. *IEE Syst. Biol.*, 1:9–18.

A Petri Net Approach to Persistence Analysis in Chemical Reaction Networks

David Angeli, Patrick De Leenheer, and Eduardo Sontag

[1] Dep. of Systems and Computer Science, University of Florence, Italy
`angeli@dsi.unifi.it`
[2] Dep. of Mathematics, University of Florida, Gainesville, FL
`deleenhe@math.ufl.edu`
[3] Dep. of Mathematics, Rutgers University, Piscataway, NJ
`sontag@math.rutgers.edu`

Summary. A positive dynamical system is said to be persistent if every solution that starts in the interior of the positive orthant does not approach the boundary of this orthant. For chemical reaction networks and other models in biology, persistence represents a non-extinction property: if every species is present at the start of the reaction, then no species will tend to be eliminated in the course of the reaction. This paper provides checkable necessary as well as sufficient conditions for persistence of chemical species in reaction networks, and the applicability of these conditions is illustrated on some examples of relatively high dimension which arise in molecular biology. More specific results are also provided for reactions endowed with mass-action kinetics. Overall, the results exploit concepts and tools from Petri net theory as well as ergodic and recurrence theory.

1 Introduction

Molecular systems biology is a cross-disciplinary and currently very active field of science which aims at the understanding of cell behavior and function at the level of chemical interactions. A central goal of this quest is the characterization of qualitative dynamical features, such as convergence to steady states, periodic orbits, or possible chaotic behavior, and the relationship of this behavior to the structure of the corresponding chemical reaction network. The interest in these questions in the context of systems biology lies on the hope that their understanding might shed new light on the principles underlying the evolution and organization of complex cellular functionalities. Understanding the long-time behavior of solutions is, of course, a classical topic in dynamical systems theory, and is usually formulated in the language of ω-*limit sets*, that is, the study of the set of possible limit points of trajectories of a dynamical system. That is the approach taken in this paper.

Persistency

Persistency is the property that, *if every species is present at the initial time, no species will tend to be eliminated in the course of the reaction*. Mathematically,

I. Queinnec et al. (Eds.): Bio. & Ctrl. Theory: Current Challenges, LNCIS 357, pp. 181–216, 2007.
springerlink.com

we ask that the ω-limit set of any trajectory which starts in the interior of the positive orthant (all concentrations positive) does not intersect the boundary of the positive orthant (more precise definitions are given below). Roughly speaking, persistency can be interpreted as non-extinction: if the concentration of a species would approach zero in the continuous differential equation model, for the corresponding stochastic discrete-event model this could be interpreted by thinking that it would completely disappear in finite time due to its discrete nature.

Thus, one of the most basic questions that one may ask about a chemical reaction is if persistency holds for that network. Also from a purely mathematical perspective persistency is very important, because it may be used in conjunction with other tools in order to guarantee convergence of solutions to equilibria and perform other kinds of Input-Output analysis. For example, if a strictly decreasing Lyapunov function exists on the interior of the positive orthant (see e.g. [26, 27, 15, 16, 17, 40] for classes of networks where this can be guaranteed), persistency allows such a conclusion.

An obvious example of a non-persistent chemical reaction is a simple irreversible conversion $A \to B$ of a species A into a species B; in this example, the chemical A empties out, that is, its time-dependent concentration approaches zero as $t \to \infty$. This is obvious, but for complex networks determining persistency, or lack thereof, is, in general, an extremely difficult mathematical problem. In fact, the study of persistence is a classical one in the (mathematically) related field of population biology (see for example [19, 8] and much other foundational work by Waltman) where species correspond to individuals of different types instead of chemical units; with respect to such studies, chemical networks have one peculiar feature which strongly impacts the invariance property of the boundary and the overall persistence analysis. Lotka-Volterra systems, indeed, are characterized by the property that any extinct species will never make its way back into the ecosystem. As a matter of fact species only interact by influencing the reciprocal death and birth rates but cannot convert into each other, which is instead the typical situation in chemistry.

Petri Nets

Petri nets, also called place/transition nets, were introduced by Carl Adam Petri in 1962 [36], and they constitute a popular mathematical and graphical modeling tool used for concurrent systems modeling [35, 45]. Our modeling of chemical reaction networks using Petri net formalism is a well-establibed idea: there have been many works, at least since [37],which have dealt with biochemical applications of Petri nets, in particular in the context of metabolic pathways, see e.g. [20, 25, 30, 33, 34], and especially the excellent exposition [44]. However, there does not appear to have been previous work using Petri nets for a nontrivial study of dynamics. In this paper, we provide a new set of tools for the robust analysis of persistence in chemical networks modeled by ordinary differential equations endowed both with arbitrary as well as mass-action kinetics (in the latter case we exploit the knowledge of the convergence speed to zero of mass-action reaction rates in approaching the orthant boundary in order to relax some of the assumptions needed in the general case).

Our conclusions are robust in the sense that persistence is inferred regardless of the specific values assumed by kinetic constants and comes as a result of both structural (for instance topology of the network) as well as dynamical features of the system (mass-action rates).

Application to a Common Motif in Systems Biology

In molecular systems biology research, certain "motifs" or subsystems appear repeatedly, and have been the subject of much recent research. One of the most common ones is that in which a substrate S_0 is ultimately converted into a product P, in an "activation" reaction triggered or facilitated by an enzyme E, and, conversely, P is transformed back (or "deactivated") into the original S_0, helped on by the action of a second enzyme F. This type of reaction is sometimes called a "futile cycle" and it takes place in signaling transduction cascades, bacterial two-component systems, and a plethora of other processes. The transformations of S_0 into P and vice versa can take many forms, depending on how many elementary steps (typically phosphorylations, methylations, or additions of other elementary chemical groups) are involved, and in what order they take place. Figure 1 shows two examples, (a) one in which a single step takes place changing S_0 into $P = S_1$, and (b) one in which two sequential steps are needed to transform S_0 into $P = S_2$, with an intermediate transformation into a substance S_1. A chemical reaction model for such a set of transformations incorporates intermediate species, compounds corresponding to the binding of the enzyme and substrate. (In "quasi-steady state" approximations, a singular perturbation approach is used in order to eliminate the intermediates. These approximations are much easier to study, see e.g. [2].) Thus, one model for (a) would be through the following reaction network:

$$E + S_0 \leftrightarrow ES_0 \rightarrow E + S_1$$
$$F + S_1 \leftrightarrow FS_1 \rightarrow F + S_0 \tag{1}$$

(double arrows indicate reversible reactions) and a model for (b) would be:

$$E + S_0 \leftrightarrow ES_0 \rightarrow E + S_1 \leftrightarrow ES_1 \rightarrow E + S_2$$
$$F + S_2 \leftrightarrow FS_2 \rightarrow F + S_1 \leftrightarrow FS_1 \rightarrow F + S_0 \tag{2}$$

where "ES_0" represents the complex consisting of E bound to S_0 and so forth.

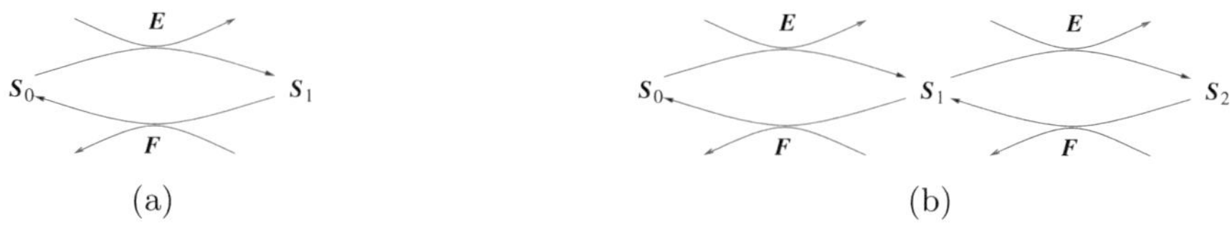

Fig. 1. (a) One-step. (b) Two-step transformations.

As a concrete example, case (b) may represent a reaction in which the enzyme E reversibly adds a phosphate group to a certain specific amino acid in the protein S_0, resulting in a single-phosphorylated form S_1; in turn, E can then bind to S_1 so as to produce a double-phosphorylated form S_2, when a second amino acid site is phosphorylated. A different enzyme reverses the process. (Variants in which the individual phosphorylations can occur in different orders are also possible; we discuss several models below.) This is, in fact, one of the mechanisms believed to underlie signaling by MAPK cascades. *Mitogen-activated protein kinase (MAPK) cascades* constitute a motif that is ubiquitous in signal transduction processes [28, 31, 43] in eukaryotes from yeast to humans, and represents a critical component of pathways involved in cell apoptosis, differentiation, proliferation, and other processes. These pathways involve chains of reactions, activated by extracellular stimuli such as growth factors or hormones, and resulting in gene expression or other cellular responses. In MAPK cascades, several steps as in (b) are arranged in a cascade, with the "active" form S_2 serving as an enzyme for the next stage.

Single-step reactions as in (a) can be shown to have the property that all solutions starting in the interior of the positive orthant globally converge to a unique (subject to stoichiometry constraints) steady state, see [4], and, in fact, can be modeled by monotone systems after elimination of the variables E and F, cf. [1]. The study of (b) is much harder, as multiple equilibria can appear, see e.g. [32, 13]. We will show how our results can be applied to test persistence of this model, as well as several variants.

Organization of Paper

The remainder of paper is organized as follows. Section 2 sets up the basic terminology and definitions regarding chemical networks, as well as the notion of persistence, Section 3 shows how to associate a Petri net to a chemical network, Sections 4 and 5 illustrate, respectively, necessary and sufficient conditions for persistence analysis of broad classes of biochemical networks, regardless of the specific kinetics considered; in Section 6, we show how our results apply to the enzymatic mechanisms described above. Section 7 draws some parallels between liveness analysis for standard and stochastic Petri nets (the so called Commoner's Theorem) and the main result in Section 5. Section 8 further motivates the systematic study of persistence by illustrating a simple toy example for which stochastic and deterministic analysis yield different predictions in terms of qualitative behaviour, while Section 9 presents specific results for attacking such questions in the case of mass-action kinetics. Finally, Section 10 illustrates applicability of the latter analysis results and draws a comparison with the discrete liveness analysis for two simple networks. Conclusions are drawn in Section 11.

2   Chemical Networks

A *chemical reaction network* ("CRN", for short) is a set of chemical reactions $\mathcal{R}_i$, where the index i takes values in $\mathcal{R} := \{1, 2, \ldots, n_r\}$. We next define precisely what one means by reactions, and the differential equation associated to a CRN, using the formalism from chemical networks theory.

Let us consider a set of chemical species S_j, $j \in \{1, 2, \ldots n_s\} := \mathcal{S}$ which are the compounds taking part in the reactions. Chemical reactions are denoted as follows:

$$\mathcal{R}_i : \quad \sum_{j \in \mathcal{S}} \alpha_{ij} S_j \rightarrow \sum_{j \in \mathcal{S}} \beta_{ij} S_j \tag{3}$$

where the α_{ij} and β_{ij} are nonnegative integers called the *stoichiometry coefficients*. The compounds on the left-hand side are usually referred to as the *reactants*, and the ones on the right-hand side are called the *products*, of the reaction. Informally speaking, the forward arrow means that the transformation of reactants into products only happens in the direction of the arrow. If also the converse transformation occurs, then, the reaction is reversible and we need to also list its inverse in the chemical reaction network as a separate reaction.

It is convenient to arrange the stoichiometry coefficients into an $n_s \times n_r$ matrix, called the *stoichiometry matrix Γ*, defined as follows:

$$[\Gamma]_{ji} = \beta_{ij} - \alpha_{ij}, \tag{4}$$

for all $i \in \mathcal{R}$ and all $j \in \mathcal{S}$ (notice the reversal of indices). This will be later used in order to write down the differential equation associated to the chemical reaction network. Notice that we allow Γ to have columns which differ only by their sign; this happens when there are reversible reactions in the network.

We discuss now how the speed of reactions is affected by the concentrations of the different species. Each chemical reaction takes place continuously in time with its own rate which is assumed to be only a function of the concentration of the species taking part in it. In order to make this more precise, we define the vector $S = [S_1, S_2, \ldots S_{n_s}]'$ of species concentrations and, as a function of it, the vector of reaction rates

$$R(S) := [R_1(S), R_2(S), \ldots R_{n_r}(S)]' .$$

Each reaction rate R_i is a real-analytic function defined on an open set which contains the non-negative orthant $\mathcal{O}_+ = \mathbb{R}^{n_s}_{\geq 0}$ of $\mathbb{R}^{n_s}$, and we assume that each R_i depends only on its respective reactants. (Imposing real-analyticity, that is to say, that the function R_i can be locally expanded into a convergent power series around each point in its domain, is a very mild assumption, verified in basically all applications in chemistry, and it allows stronger statements to be made.) Furthermore, we assume that each R_i satisfies the following monotonicity conditions:

$$\frac{\partial R_i(S)}{\partial S_j} = \begin{cases} \geq 0 \text{ if } \alpha_{ij} > 0 \\ = 0 \text{ if } \alpha_{ij} = 0. \end{cases} \tag{5}$$

We also assume that, whenever the concentration of any of the reactants of a given reaction is 0, then, the corresponding reaction does not take place, meaning that the reaction rate is 0. In other words, if $S_{i_1}, \ldots, S_{i_N}$ are the reactants of reaction j, then we ask that

$$R_j(S) = 0 \text{ for all } S \text{ such that } [S_{i_1}, \ldots, S_{i_N}] \in \partial \mathcal{O}_+ ,$$

where $\partial \mathcal{O}_+ = \partial \mathbb{R}^N_{\geq 0}$ is the boundary of $\mathcal{O}_+$ in $\mathbb{R}^N$. Conversely, we assume that reactions take place if reactants are available, that is:

$$R_j(S) > 0 \text{ whenever } S \text{ is such that } [S_{i_1}, \ldots, S_{i_N}] \in \text{int}[\mathbb{R}^N_{\geq 0}] .$$

A special case of reactions is as follows. One says that a chemical reaction network is equipped with *mass-action kinetics* if

$$R_i(S) = k_i \prod_{j=1}^{n_s} S_j^{\alpha_{ij}} \text{ for all } i = 1, \ldots, n_r .$$

This is a commonly used form for the functions $R_i(s)$ and amounts to asking that the reaction rate of each reaction is proportional to the concentration of each of its participating reactants.

With the above notations, the chemical reaction network is described by the following system of differential equations:

$$\dot{S} = \Gamma R(S). \tag{6}$$

with S evolving in $\mathcal{O}_+$ and where Γ is the stoichiometry matrix.

There are several additional notions useful when analyzing CRN's. One of them is the notion of a *complex*. We associate to the network (3) a set of complexes, C_i's, with $i \in \{1, 2, \ldots, n_c\}$. Each complex is an integer combination of species, specifically of the species appearing either as products or reactants of the reactions in (3). We introduce the following matrix $\tilde{\Gamma}$ as follows:

$$\tilde{\Gamma} = \begin{bmatrix} \alpha_{11} & \alpha_{21} & \cdots & \alpha_{n_r 1} & \beta_{11} & \beta_{21} & \cdots & \beta_{n_r 1} \\ \alpha_{12} & \alpha_{22} & \cdots & \alpha_{n_r 2} & \beta_{12} & \beta_{22} & \cdots & \beta_{n_r 2} \\ \vdots & \vdots & & \vdots & \vdots & \vdots & & \vdots \\ \alpha_{1 n_s} & \alpha_{2 n_s} & \cdots & \alpha_{n_r n_s} & \beta_{1 n_s} & \beta_{2 n_s} & \cdots & \beta_{n_r n_s} \end{bmatrix}$$

Then, a matrix representing the complexes as columns can be obtained by deleting from $\tilde{\Gamma}$ repeated columns, leaving just one instance of each; we denote by $\Gamma_c \in \mathbb{R}^{n_s \times n_c}$ the matrix which is thus constructed. Each of the columns of Γ_c is then associated with a complex of the network. We may now associate to each chemical reaction network, a directed graph (which we call the *C-graph*), whose nodes are the complexes and whose edges are associated to the reactions (3). An edge (C_i, C_j) is in the C-graph if and only if $C_i \to C_j$ is a reaction of the network. Note that the C-graph need not be connected (the C-graph is connected if for any pair of distinct nodes in the graph there is an undirected path linking the nodes), and lack of connectivity cannot be avoided in the analysis. (This is in contrast with many other graphs in chemical reaction theory, which can be assumed to be connected without loss of generality.) In general, the C-graph will have several connected components (equivalence classes under the equivalence relation "is linked by an undirected path to", defined on the set of nodes of the graph).

Let $\mathcal{I}$ be the incidence matrix of the C-graph, namely the matrix whose columns are in one-to-one correspondence with the edges (reactions) of the graph and whose rows are in one-to-one correspondence with the nodes (complexes). Each column contains a -1 in the i-th entry and a $+1$ in the j-th entry (and zeroes in all remaining entries) whenever (C_i, C_j) is an edge of the C-graph (equivalently, when $C_i \to C_j$ is a reaction of the network). With this notations, we have the following formula, to be used later:

$$\Gamma = \Gamma_c \mathcal{I}. \tag{7}$$

We denote solutions of (6) as follows: $S(t) = \varphi(t, S_0)$, where $S_0 \in \mathcal{O}_+$ is the initial concentration of chemical species. As usual in the study of the qualitative behavior of dynamical systems, we will make use of ω-limit sets, which capture the long-term behavior of a system and are defined as follows:

$$\omega(S_0) := \{S \in \mathcal{O}_+ : \varphi(t_n, S_0) \to S \text{ for some } t_n \to +\infty\} \tag{8}$$

(implicitly, when talking about $\omega(S_0)$, we assume that $\varphi(t, S_0)$ is defined for all $t \geq 0$ for the initial condition S_0). We will be interested in asking whether or not a chemical reaction network admits solutions in which one or more of the chemical compounds become arbitrarily small. The following definition, borrowed from the ecology literature, captures this intuitive idea.

Definition 1. *A chemical reaction network (6) is* persistent *if* $\omega(S_0) \cap \partial \mathcal{O}_+ = \emptyset$ *for each* $S_0 \in int(\mathcal{O}_+)$.

We will derive conditions for persistence of general chemical reaction networks. Our conditions will be formulated in the language of Petri nets; these are discrete-event systems equipped with an algebraic structure that reflects the list of chemical reactions present in the network being studied, and are defined as follows. In the present chapter we make an effort to be self-contained with respect, however, for a more in depth introduction see for instance one of the many books devoted to this subject, [36, 45].

3 Petri Nets

We associate to a CRN a bipartite directed graph (i.e., a directed graph with two types of nodes) with weighted edges, called the *species-reaction Petri net*, or SR-net for short. Mathematically, this is a quadruple

$$(V_S, V_R, E, W),$$

where V_S is a finite set of nodes each one associated to a species (usually referred to as "places" in Petri Net literature), V_R is a finite set of nodes (disjoint from V_S), each one corresponding to a reaction (usually named the "transitions" of the network), and E is a set of edges as described below. (We often write S or V_S interchangeably, or R instead of V_R, by identifying species or reactions with their respective indices; the context should make the meaning clear.) The set of all nodes is also denoted by $V \doteq V_R \cup V_S$.

The edge set $E \subset V \times V$ is defined as follows. Whenever a certain reaction R_i belongs to the CRN:

$$\sum_{j \in \mathcal{S}} \alpha_{ij} S_j \quad \rightarrow \quad \sum_{j \in \mathcal{S}} \beta_{ij} S_j \,, \tag{9}$$

we draw an edge from $S_j \in V_S$ to $R_i \in V_R$ for all S_j's such that $\alpha_{ij} > 0$. That is, $(S_j, R_i) \in E$ iff $\alpha_{ij} > 0$, and we say in this case that R_i is an *output reaction for S_j*. Similarly, we draw an edge from $R_i \in V_R$ to every $S_j \in V_S$ such that $\beta_{ij} > 0$. That is, $(R_i, S_j) \in E$ whenever $\beta_{ij} > 0$, and we say in this case that R_i is an *input reaction for S_j*.

Accordingly, we also talk about *input and output reactions* for a given set $\Sigma \subset \mathcal{S}$ of species. This is defined in the obvious way, viz. by considering the union of all input (and respectively output) reactions over all species belonging to Σ.

Notice that edges only connect species to reactions and vice versa, but never connect two species or two reactions.

The last element to fully define the Petri net is the function $W : E \rightarrow \mathbb{N}$, which associates to each edge a positive integer according to the rule:

$$W(S_j, R_i) = \alpha_{ij} \quad \text{and} \quad W(R_i, S_j) = \beta_{ij} \,.$$

Several other definitions which are commonly used in the Petri net literature will be of interest in the following. We say that a row or column vector v is non-negative, and we denote it by $v \succeq 0$ if it is so entry-wise. We write $v \succ 0$ if $v \succeq 0$ and $v \neq 0$. A stronger notion is instead $v \gg 0$, which indicates $v_i > 0$ for all i.

Definition 2. *A* P-semiflow *is any row vector $c \succ 0$ such that $c\, \Gamma = 0$. Its support is the set of indices $\{i \in V_S : c_i > 0\}$. A Petri net is said to be* conservative *if there exists a P-semiflow $c \gg 0$.*

Notice that P-semiflows for the system (6) correspond to non-negative linear first integrals, that is, linear functions $S \mapsto cS$ such that $(d/dt)cS(t) \equiv 0$ along all solutions of (6) (assuming that the span of the image of $R(S)$ is $\mathbb{R}^{n_r}$). In particular, a Petri net is conservative if and only if there is a positive linear conserved quantity for the system. (Petri net theory views Petri nets as "token-passing" systems, and, in that context, P-semiflows, also called *place-invariants*, amount to conservation relations for the "place markings" of the network, that show how many tokens there are in each "place," the nodes associated to species in SR-nets. We do not make use of this interpretation in this paper.)

Definition 3. *A* T-semiflow *is any column vector $v \succ 0$ such that $\Gamma v = 0$. A Petri net is said to be* consistent *if there exists a T-semiflow $v \gg 0$.*

The notion of T-semiflow corresponds to the existence of a collection of positive reaction rates which do not produce any variation in the concentrations of the species. In other words, v can be viewed as a set of *fluxes* that is in equilibrium ([44]). (In Petri net theory, the terminology is "T-invariant," and the fluxes are flows of tokens.)

A chemical reaction network is said to be *reversible* if each chemical reaction has an inverse reaction which is also part of the network. Biochemical models are most often non-reversible. For this reason, a far milder notion was introduced [26, 27, 15, 16, 17]: A chemical reaction network is said to be *weakly reversible* if each connected component of the C-graph is strongly connected (meaning that there is a directed path between any pair of nodes in each connected component). In algebraic terms, weak reversibility amounts to existence of $v \gg 0$ such that $\mathcal{I}v = 0$ (see Corollary 4.2 of [18]), so that in particular, using (7), also $\Gamma v = \Gamma_c \mathcal{I} v = 0$. Hence a chemical reaction network that is weakly reversible has a consistent associated Petri net.

A few more definitions are needed in order to state our main results.

Definition 4. *A nonempty set $\Sigma \subset V_S$ is called a siphon if each input reaction associated to Σ is also an output reaction associated to Σ. A siphon is a deadlock if its set of output reactions is all of V_R. A deadlock is* minimal *if it does not contain (strictly) any other deadlocks. A siphon is* minimal *if it does not contain (strictly) any other siphons. Notice that a minimal deadlock need not be a minimal siphon (and viceversa, which is obvious). A pair of distinct deadlocks Σ_1 and Σ_2 is said to be* nested *if either $\Sigma_1 \subset \Sigma_2$ or $\Sigma_2 \subset \Sigma_1$.*

Similarly one defines the notion of *trap*.

Definition 5. *A non-empty set $\mathcal{T} \subset V_S$ is called a trap if each output reaction associated to $\mathcal{T}$ is also an input reaction associated to $\mathcal{T}$.*

For later use we associate a particular set to a siphon Σ as follows:

$$L_\Sigma = \{x \in \mathcal{O}_+ \mid x_i = 0 \iff i \in \Sigma\}.$$

It is also useful to introduce a binary relation "reacts to", which we denote by $\rightarrowtail$, and we define as follows: $S_i \rightarrowtail S_j$ whenever there exists a chemical reaction $\mathcal{R}_k$, so that

$$\sum_{l \in \mathcal{S}} \alpha_{kl} S_l \rightarrow \sum_{l \in \mathcal{S}} \beta_{kl} S_l$$

with $\alpha_{ki} > 0$, $\beta_{kj} > 0$. If the reaction number is important, we also write

$$S_i \overset{k}{\rightarrowtail} S_j$$

(where $k \in \mathcal{R}$). With this notation, the notion of siphon can be rephrased as follows: $Z \subset \mathcal{S}$ is a siphon for a chemical reaction network if for every $S \in Z$ and $k \in \mathcal{R}$ such that $\tilde{S}_k := \{T \in \mathcal{S} : T \overset{k}{\rightarrowtail} S\} \neq \emptyset$, it holds $\tilde{S}_k \cap Z \neq \emptyset$.

4 Necessary Conditions

Our first result will relate persistence of a chemical reaction network to consistency of the associated Petri net.

Theorem 1. *Let (6) be the equation describing the time-evolution of a conservative and persistent chemical reaction network. Then, the associated Petri net is consistent.*

Proof. Let $S_0 \in \text{int}(\mathcal{O}_+)$ be any initial condition. By conservativity, solutions satisfy $cS(t) \equiv cS_0$, and hence remain bounded, and therefore $\omega(S_0)$ is a nonempty compact set. Moreover, by persistence, $\omega(S_0) \cap \partial\mathcal{O}_+ = \emptyset$, so that $R(\tilde{S}_0) \gg 0$, for all $\tilde{S}_0 \in \omega(S_0)$. In particular, by compactness of $\omega(S_0)$ and continuity of R, there exists a positive vector $v \gg 0$, so that

$$R(\tilde{S}_0) \succeq v \quad \text{for all} \ \tilde{S}_0 \in \omega(S_0) \,.$$

Take any $\tilde{S}_0 \in \omega(S_0)$. By invariance of $\omega(S_0)$, we have $R(\varphi(t, \tilde{S}_0)) \succeq v$ for all $t \in \mathbb{R}$. Consequently, taking asymptotic time averages, we obtain:

$$0 = \lim_{T \to +\infty} \frac{\varphi(T, \tilde{S}_0) - \tilde{S}_0}{T} = \lim_{T \to +\infty} \frac{1}{T} \int_0^T \Gamma R(\varphi(t, \tilde{S}_0)) \, dt \qquad (10)$$

(the left-hand limit is zero because $\varphi(T, \tilde{S}_0)$ is bounded). However,

$$\frac{1}{T} \int_0^T R(\varphi(t, \tilde{S}_0)) \, dt \succeq v$$

for all $T > 0$. Therefore, taking any subsequence $T_n \to +\infty$ so that there is a finite limit:

$$\lim_{n \to +\infty} \frac{1}{T_n} \int_0^{T_n} R(\varphi(t, \tilde{S}_0)) \, dt = \bar{v} \succeq v \,.$$

We obtain, by virtue of (10), that $\Gamma \bar{v} = 0$. This completes the proof of consistency, since $\bar{v} \gg 0$.

5 Sufficient Conditions

In this present Section, we derive sufficient conditions for insuring persistence of a chemical reaction network on the basis of Petri net properties.

Theorem 2. *Consider a chemical reaction network satisfying the following assumptions:*

1. *its associated Petri net is conservative;*
2. *each siphon contains the support of a P-semiflow.*

Then, the network is persistent.

We first prove a number of technical results. The following general fact about differential equations will be useful.

For a real number p, let $\operatorname{sign} p := 1, 0, -1$ if $p > 0$, $p = 0$, or $p < 0$ respectively, and, similarly for any real vector $x = (x_1, \ldots, x_n)$, let $\operatorname{sign} x := (\operatorname{sign} x_1, \ldots, \operatorname{sign} x_n)'$. When x belongs to the closed positive orthant $\mathbb{R}^n_+$, $\operatorname{sign} x \in \{0, 1\}^n$.

Lemma 1. *Let f be a real-analytic vector field defined on some open neighborhood of $\mathbb{R}^n_+$, and suppose that $\mathbb{R}^n_+$ is forward invariant for the flow of f. Consider any solution $\bar{x}(t)$ of $\dot{x} = f(x)$, evolving in $\mathbb{R}^n_+$ and defined on some open interval J. Then, $\operatorname{sign} \bar{x}(t)$ is constant on J.*

Proof. Pick such a solution, and define

$$Z := \{i \mid \bar{x}_i(t) = 0 \text{ for all } t \in J\}.$$

Relabeling variables if necessary, we assume without loss of generality that $Z = \{r+1, \ldots, n\}$, with $0 \leq r \leq n$, and we write equations in the following block form:

$$\dot{y} = g(y, z)$$
$$\dot{z} = h(y, z)$$

where $x' = (y', z')'$ and $y(t) \in \mathbb{R}^r$, $z(t) \in \mathbb{R}^{n-r}$. (The extreme cases $r = 0$ and $r = n$ correspond to $x = z$ and $x = y$ respectively.) In particular, we write $\bar{x}' = (\bar{y}', \bar{z}')'$ for the trajectory of interest. By construction, $\bar{z} \equiv 0$, and the sets

$$B_i := \{t \mid \bar{y}_i(t) = 0\}$$

are proper subsets of J, for each $i \in \{1, \ldots, r\}$. Since the vector field is real-analytic, each coordinate function $\bar{y}_i$ is real-analytic (see e.g. [41], Proposition C.3.12), so, by the principle of analytic continuation, each B_i is a discrete set. It follows that

$$G := J \setminus \bigcup_{i=1}^{r} B_i$$

is an (open) dense set, and for each $t \in G$, $\bar{y}(t) \in \operatorname{int} \mathbb{R}^r_+$, the interior of the positive orthant.

We now consider the following system on $\mathbb{R}^r$:

$$\dot{y} = g(y, 0).$$

This is again a real-analytic system, and $\mathbb{R}^r_+$ is forward invariant. To prove this last assertion, note that forward invariance of the closed positive orthant is equivalent to the following property:

for any $y \in \mathbb{R}^r_+$ and any $i \in \{1, \ldots, r\}$ such that $y_i = 0$, $g_i(y, 0) \geq 0$.

Since $\mathbb{R}_+^n$ is forward invariant for the original system, we know, by the same property applied to that system, that for any $(y, z) \in \mathbb{R}_+^n$ and any $i \in \{1, \ldots, r\}$ such that $y_i = 0$, $g_i(y, z) \geq 0$. Thus, the required property holds (case $z = 0$). In particular, $\operatorname{int} \mathbb{R}_+^r$ is also forward invariant (see e.g. [2], Lemma III.6). By construction, $\bar{y}$ is a solution of $\dot{y} = g(y, 0)$, $\bar{y}(t) \in \operatorname{int} \mathbb{R}_+^r$ for each $t \in G$, Since G is dense and $\operatorname{int} \mathbb{R}_+^r$ is forward invariant, it follows that $\bar{y}(t) \in \operatorname{int} \mathbb{R}_+^r$ for all $t \in J$. Therefore,

$$\operatorname{sign} \bar{x}(t) = (1_r, 0_{n-r})' \quad \text{for all } t \in J$$

where 1_r is a vector of r 1's and 0_{n-r} is a vector of $n - r$ 0's.

We then have an immediate corollary:

Lemma 2. *Suppose that $\Omega \subset \mathcal{O}_+$ is a closed set, invariant for (6). Suppose that $\Omega \cap L_Z$ is non-empty, for some $Z \subset \mathcal{S}$. Then, $\Omega \cap L_Z$ is also invariant with respect to (6).*

Proof. Pick any $S_0 \in \Omega \cap L_Z$. By invariance of Ω, the solution $\varphi(t, S_0)$ belongs to Ω for all t in its open domain of definition J, so, in particular (this is the key fact), $\varphi(t, S_0) \in \mathcal{O}_+$ for all t (negative as well as positive). Therefore, it also belongs to L_Z, since its sign is constant by Lemma 1.

In what follows, we will make use of the Bouligand tangent cone $TC_\xi(K)$ of a set $K \subset \mathcal{O}_+$ at a point $\xi \in \mathcal{O}_+$, defined as follows:

$$TC_\xi(K) \;=\; \left\{ v \in \mathbb{R}^n : \exists k_n \in K, k_n \to \xi \text{ and } \lambda_n \searrow 0 : \frac{1}{\lambda_n}(k_n - \xi) \to v \right\}.$$

Bouligand cones provide a simple criterion to check forward invariance of closed sets (see e.g. [5]): a closed set K is forward invariant for (6) if and only if $\Gamma R(\xi) \in TC_\xi(K)$ for all $\xi \in K$. However, below we consider a condition involving tangent cones to the sets L_Z, which are not closed. Note that, for all index sets Z and all points ξ in L_Z,

$$TC_\xi(L_Z) \;=\; \{v \in \mathbb{R}^n : v_i = 0 \; \forall\, i \in Z\}.$$

Lemma 3. *Let $Z \subset \mathcal{S}$ be non-empty and $\xi \in L_Z$ be such that $\Gamma R(\xi) \in TC_\xi(L_Z)$. Then Z is a siphon.*

Proof. By assumption $\Gamma R(\xi) \in TC_\xi(L_Z)$ for some $\xi \in L_Z$. This implies that $[\Gamma R(\xi)]_i = 0$ for all $i \in Z$. Since $\xi_i = 0$ for all $i \in Z$, all reactions in which S_i is involved as a reactant are shut off at ξ; hence, the only possibility for $[\Gamma R(\xi)]_i = 0$ is that all reactions in which S_i is involved as a product are also shut-off. Hence, for all $k \in \mathcal{R}$, and all $l \in \mathcal{S}$ so that $S_l \xrightarrow{k} S_i$, we necessarily have that $R_k(\xi) = 0$.

Hence, for all $k \in \mathcal{R}$ so that $\tilde{S}_k = \{l \in \mathcal{S} : S_l \xrightarrow{k} S_i\}$ is non-empty, there must exist an $l \in \tilde{S}_k$ so that $\xi_l = 0$. But then necessarily, $l \in Z$, showing that Z is indeed a siphon.

The above Lemmas are instrumental to proving the following Proposition:

Proposition 1. *Let $\xi \in \mathcal{O}^+$ be such that $\omega(\xi) \cap L_Z \neq \emptyset$ for some $Z \subset \mathcal{S}$. Then Z is a siphon.*

Proof. Let Ω be the closed and invariant set $\omega(\xi)$. Thus, by Lemma 2, the non-empty set $L_Z \cap \Omega$ is also invariant. Notice that

$$\mathrm{cl}[L_Z] \;=\; \bigcup_{W \supseteq Z} L_W \,.$$

Moreover, $L_W \cap \Omega$ is invariant for all $W \subset \mathcal{S}$ such that $L_W \cap \Omega$ is non-empty. Hence,

$$\mathrm{cl}[L_Z] \cap \Omega \;=\; \bigcup_{W \supseteq Z} [L_W \cap \Omega]$$

is also invariant. By the characterization of invariance for closed sets in terms of Bouligand tangent cones, we know that, for any $\eta \in \mathrm{cl}[L_Z] \cap \Omega$ we have

$$\Gamma R(\eta) \;\in\; TC_\eta(\Omega \cap \mathrm{cl}(L_Z)) \;\subset\; TC_\eta(\mathrm{cl}(L_Z))\,.$$

In particular, for $\eta \in L_Z \cap \Omega$ (which by assumption exists), $\Gamma R(\eta) \in TC_\eta(L_Z)$ so that, by virtue of Lemma 3 we may conclude Z is a siphon.

Although at this point Proposition 1 would be enough to prove Theorem 2, it is useful to clarify the meaning of the concept of a "siphon" here. It hints at the fact, made precise in the Proposition below, that removing all the species of a siphon from the network (or equivalently setting their initial concentrations equal to 0) will prevent those species from being present at all future times. Hence, those species literally "lock" a part of the network and shut off all the reactions that are therein involved. In particular, once emptied a siphon will never be full again. This explains why a siphon is sometimes also called a "locking set" in the Petri net literature. A precise statement of the foregoing remarks is as follows.

Proposition 2. *Let $Z \subset \mathcal{S}$ be non-empty. Then Z is a siphon if and only if $cl(L_Z)$ is forward invariant for (6).*

Proof.
Sufficiency: Pick $\xi \in L_Z \neq \emptyset$. Then forward invariance of $\mathrm{cl}(L_Z)$ implies that $\Gamma R(\xi) \in TC_\xi(\mathrm{cl}(L_Z)) = TC_\xi(L_Z)$, where the last equality holds since $\xi \in L_Z$. It follows from Lemma 3 that Z is a siphon.

Necessity: Pick $\xi \in \mathrm{cl}(L_Z)$. This implies that $\xi_i = 0$ for all $i \in Z \cup Z'$, where $Z' \subset \mathcal{S}$ could be empty. By the characterization of forward invariance of closed sets in terms of tangent Bouligand cones, it suffices to show that $[\Gamma R(\xi)]_i = 0$ for all $i \in Z$, and that $[\Gamma R(\xi)]_i \geq 0$ for all $i \in Z'$ whenever $Z' \neq \emptyset$. Now by (6),

$$[\Gamma R(\xi)]_i \;=\; \sum_k \beta_{ki} R_k(\xi) - \sum_l \alpha_{li} R_l(\xi) \;=\; \sum_k \beta_{ki} R_k(\xi) - 0 \;\geq\; 0, \qquad (11)$$

which already proves the result for $i \in Z'$. Notice that the second sum is zero because if $\alpha_{li} > 0$, then species i is a reactant of reaction l, which implies that

$R_l(\xi) = 0$ since $\xi_i = 0$. So we assume henceforth that $i \in Z$. We claim that the sum on the right side of (11) is zero. This is obvious if the sum is void. If it is non-void, then each term which is such that $\beta_{ki} > 0$ must be zero. Indeed, for each such term we have that $R_k(\xi) = 0$ because Z is a siphon. This concludes the proof of Proposition 2.

Proof of Theorem 2

Let $\xi \in \mathrm{int}(\mathcal{O}_+)$ be arbitrary and let Ω denote the corresponding ω-limit set $\Omega = \omega(\xi)$. We claim that the intersection of Ω and the boundary of $\mathcal{O}_+$ is empty.

Indeed, suppose that the intersection is nonemty. Then, Ω would intersect L_Z, for some $\emptyset \neq Z \subset \mathcal{S}$. In particular, by Proposition 1, Z would be a siphon. Then, by our second assumption, there exists a non-negative first integral cS, whose support is included in Z, so that necessarily $cS(t_n, \xi) \to 0$ at least along a suitable sequence $t_n \to +\infty$. However, $cS(t, \xi) = c\xi > 0$ for all $t \geq 0$, thus giving a contradiction. $\qquad\square$

6 Applications

We now apply our results to obtain persistence results for variants of the reaction (b) shown in Figure 1 as well as for cascades of such reactions.

6.1 Example 1

We first study reaction (2). Note that reversible reactions were denoted by a "$\leftrightarrow$" in order to avoid having to rewrite them twice. The Petri net associated to (2) is shown if Fig. 2. The network comprises nine distinct species, labeled S_0, S_1, S_2, E, F, ES_0, ES_1, FS_2, FS_1. It can be verified that the Petri net in Fig. 2 is indeed consistent (so it satisfies the necessary condition). To see this, order the species and reactions by the obvious order obtained when reading (2) from left to right and from top to bottom (e.g., S_1 is the fourth species and the reaction $E + S_1 \to ES_1$ is the fourth reaction). The construction of the matrix Γ is now clear, and it can be verified that $\Gamma v = 0$ with $v = [2\,1\,1\,2\,1\,1\,2\,1\,1\,2\,1\,1]'$. The network itself, however, is not weakly reversible, since neither of the two connected components of (2) is strongly connected. Computations show that there are three minimal siphons:

$$\{E, ES_0, ES_1\},$$
$$\{F, FS_1, FS_2\},$$

and

$$\{S_0, S_1, S_2, ES_0, ES_1, FS_2, FS_1\}.$$

Each one of them contains the support of a P-semiflow; in fact there are three independent conservation laws:

$$E + ES_0 + ES_1 = \mathrm{const}_1,$$
$$F + FS_2 + FS_1 = \mathrm{const}_2, \text{ and}$$
$$S_0 + S_1 + S_2 + ES_0 + ES_1 + FS_2 + FS_1 = \mathrm{const}_3,$$

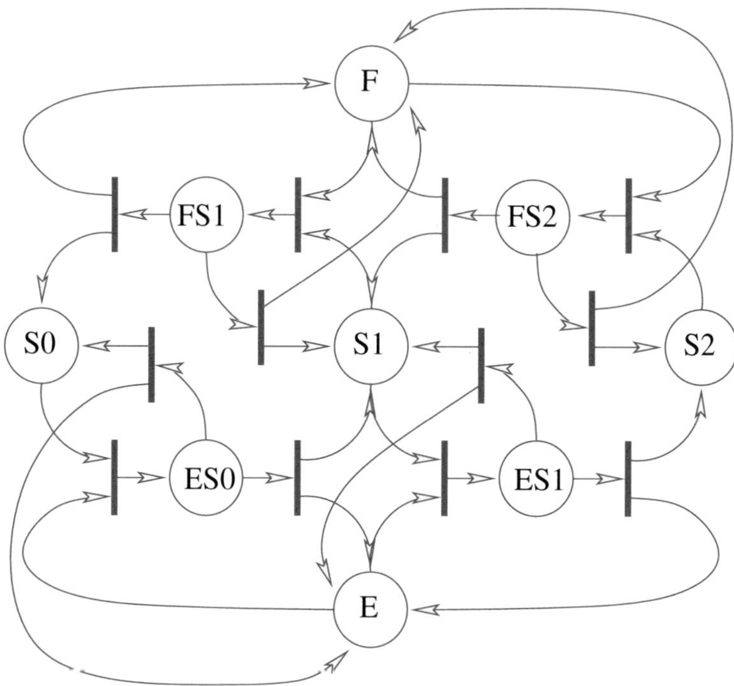

Fig. 2. Petri net associated to reactions (2)

whose supports coincide with the three mentioned siphons. Since the sum of these three conservation laws is also a conservation law, the network is conservative. Therefore, application of Theorem 2 guarantees that the network is indeed persistent.

6.2   Example 2

As remarked earlier, examples as the above one are often parts of cascades in which the product (in MAPK cascades, a doubly-phosphorilated species) S_2 in turn acts as an enzyme for the following stage. One model with two stages is as follows (writing S_2 as $E^\star$ in order to emphasize its role as a kinase for the subsequent stage):

$$
\begin{aligned}
E + S_0 &\leftrightarrow ES_0 \rightarrow E + S_1 \leftrightarrow ES_1 \rightarrow E + E^\star \\
F + E^\star &\leftrightarrow FS_2 \rightarrow F + S_1 \leftrightarrow FS_1 \rightarrow F + S_0 \\
E^\star + S_0^\star &\leftrightarrow ES_0^\star \rightarrow E^\star + S_1^\star \leftrightarrow ES_1^\star \rightarrow E^\star + S_2^\star \\
F^\star + S_2^\star &\leftrightarrow FS_2^\star \rightarrow F^\star + S_1^\star \leftrightarrow FS_1^\star \rightarrow F^\star + S_0^\star.
\end{aligned}
\tag{12}
$$

The overall reaction is shown in Fig. 3. Note – using the labeling of species and reaction as in the previous example – that $\Gamma v = 0$ with $v = [v_1'\, v_1'\, v_1'\, v_1']'$ and

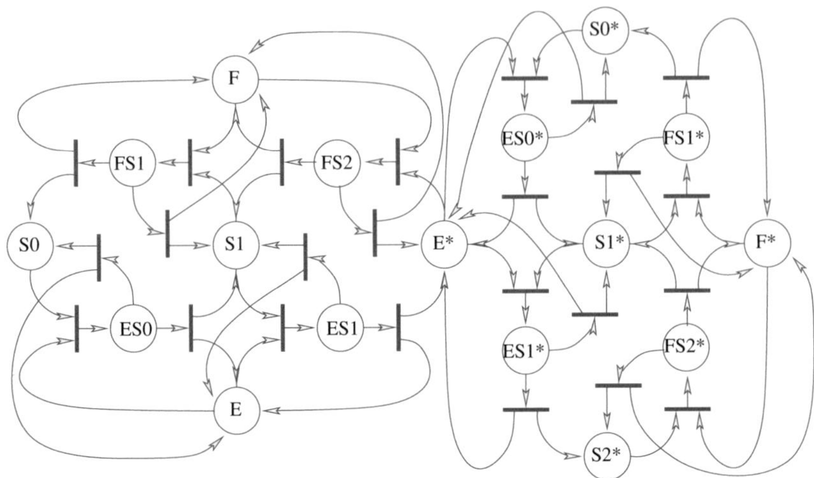

Fig. 3. Petri net associated to reactions (12)

$v_1 = [2\,1\,1\,2\,1\,1]'$, and hence the network is consistent. There are five minimal siphons for this network, namely:

$$\{E, ES_0, ES_1\},$$
$$\{F, FS_2, FS_1\},$$
$$\{F^\star, FS_2^\star, FS_1^\star\},$$
$$\{S_0^\star, S_1^\star, S_2^\star, ES_0^\star, ES_1^\star, FS_2^\star, FS_1^\star\},$$

and

$$\{S_0, S_1, E^\star, ES_0, ES_1, FS_2, FS_1, ES_0^\star, ES_1^\star\}.$$

Each one of them is the support of a P-semiflow, and there are five conservation laws:

$$E + ES_0 + ES_1 = \mathrm{const}_1,$$
$$F + FS_2 + FS_1 = \mathrm{const}_2,$$
$$F^\star + FS_2^\star + FS_1^\star = \mathrm{const}_3,$$
$$S_0^\star + S_1^\star + S_2^\star + ES_0^\star + ES_1^\star + FS_2^\star + FS_1^\star = \mathrm{const}_4,$$

and

$$S_0 + S_1 + E^\star + ES_0 + ES_1 + FS_2 + FS_1 + ES_0^\star + ES_1^\star = \mathrm{const}_5.$$

As in the previous example, the network is conservative since the sum of these conservation laws is also a conservation law. Therefore the overall network is persistent, by virtue of Theorem 2.

6.3   Example 3

An alternative mechanism for dual phosphorilation in MAPK cascades, considered in [32], differs from the previous ones in that it becomes relevant in what order the two phosphorylations occur. (These take place at two different sites,

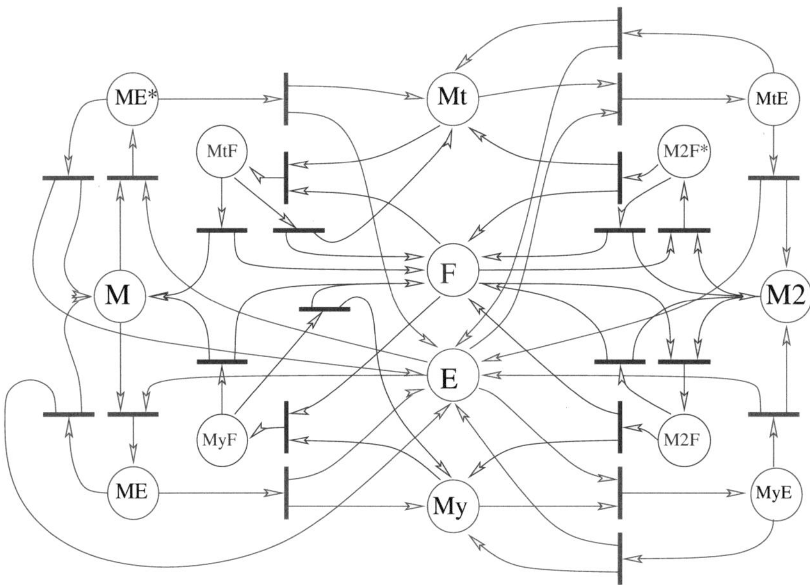

Fig. 4. Petri net associated to the network (13)

a threonine and a tyrosine residue). The corresponding network can be modeled
as follows:

$$
\begin{aligned}
M + E &\leftrightarrow ME \rightarrow M_y + E \leftrightarrow M_y E \rightarrow M_2 + E \\
M + E &\leftrightarrow ME^\star \rightarrow M_t + E \leftrightarrow M_t E \rightarrow M_2 + E \\
M_2 + F &\leftrightarrow M_2 F \rightarrow M_y + F \leftrightarrow M_y F \rightarrow M + F \\
M_2 + F &\leftrightarrow M_2 F^\star \rightarrow M_t + F \leftrightarrow M_t F \rightarrow M + F.
\end{aligned}
\tag{13}
$$

See Fig. 4 for the corresponding Petri net. This network is consistent. Indeed,
$\Gamma v = 0$ for the same v as in the previous example. Moreover it admits three
siphons of minimal support:

$$\{E, ME, ME^\star, M_y E, M_t E\},$$
$$\{F, M_y F, M_t F, M_2 F, M_2 F^\star\},$$
and
$$\{M, ME, ME^\star, M_y, M_t, M_y E, M_t E, M_2, M_2 F, M_2 F^\star, M_t F, M_y F\}.$$

Each of them is also the support of a conservation law, respectively for M,E and
F molecules. The sum of these conservation laws, is also a conservation law and
therefore the network is conservative. Thus the Theorem 2 again applies and the
network is persistent.

7   Discrete vs. Continuous Persistence Results

As a matter of fact, and this was actually the main motivation for the introduction of Petri Nets in [36], each Petri Net (as defined in Section 3) comes

with an associated discrete event system, which governs the evolution of a vector M, usually called the *marking* of the net. The entries of M are non-negative integers, in one-one correspondence with the places of the network, i.e. $M = [m_1, m_2, \ldots, m_{n_s}]' \subset \mathbb{N}^{n_s}$, and the m_is, $i = 1 \ldots n_s$, stand for the number of "tokens" associated to the places $S_1 \ldots S_{n_p}$. In our context, each token may be thought of as a molecule of the corresponding species. Once a certain initial condition $M_0 \subset \mathbb{N}^{n_s}$ has been specified for a given net, we have what is usually called a marked Petri Net, In order to define dynamical behavior, one considers the following *firing rules* for transitions R:

1. a transition R can fire whenever each input place of R is marked with a number of tokens greater or equal than the weight associated to the edge joining such a place to R (in our context a reaction can occur, at a given time instant, only provided that each reagent has a number of molecules greater or equal than the corresponding stoichiometry coefficient); we call such transitions enabled.
2. when a transition R fires, the marking M of the network is updated by subtracting, for each input place, a number of tokens equal to the weight associated to the corresponding edge, while for each output place a number of tokes equal to the weight of the corresponding edge is added.

Together with a rule that specifies the timing of the firings, this specifies a dynamical system describing the evolution of vectors $M \in \mathbb{N}^{n_s}$. There are several ways to specify timings. One may use a deterministic rule in which a specification is made at each time instant of which transition fires (among those enabled). Another possibility is to consider a stochastic model, in which firing events are generated by random processes with exponentially decaying probability distributions, with a specified rate λ. The timing of the next firing of a particular reaction R might depend on R as well as the state vector M. In this way, an execution of the Petri Net is nothing but a realization of a stochastic process (which is Markovian in an appropriate space), whose study is classical not only in Petri Net theory but also in the chemical kinetics literature. In the latter, the equation governing the probability evolution is in fact the so-called *Chemical Master Equation*, which is often simulated by using a method known as "Gillespie's algorithm".

The main results in Sections 4 and 5 are independent of the type of kinetics assumed for the chemical reaction network (for instance mass-action kinetics or Michaelis-Menten kinetics are both valid options at this level of abstraction). This also explains, to a great extent, the similarity between our theorems and their discrete counterparts which arise in the context of liveness's studies for Petri Nets and Stochastic Petri Nets (liveness is indeed the discrete analog of persistence for ODEs, even though its definition is usually given in terms of firing of transitions rather than asymptotic averages of markings, see [45] for a precise definition).

In particular, we focus our attention on the so called *Siphon-Trap Property* which is a sufficient condition for liveness of conservative Petri Nets, and actually a complete characterization of liveness if the net is a "Free Choice Petri Net" (this is known as Commoner's Theorem, [22] and [12]):

Theorem 3. *Consider a conservative Petri Net satisfying the following assumption:*

each (minimal) siphon contains a non-empty trap.

Then, the PN is alive.

Notice the similarity between the assumptions and conclusions in Theorem 2 and in Theorem 3. There are some subtle differences, however. Traps for Petri-Nets enjoy the following invariance property: if a trap is non-empty at time zero (meaning that at least one of its places has tokens), then the trap is non-empty at all future times. In contrast, in a continuous set-up (when tokens are not integer quantities but may take any real value), satisfaction of the siphon-trap property does not prevent (in general) concentrations of species from decaying to zero asymptotically. This is why we needed a strengthened assumption 2., and asked that each siphon contains the support of a P-semiflow (which is always, trivially, also a trap). In other words, in a continuous set-up the notion of a trap looses much of its appeal, since one may conceive situations in which molecules are pumped into the trap at a rate which is lower than the rate at which they are extracted from it, so that, in the limit, the trap can be emptied out even though it was initially full. A similar situation never occurs in a discrete set-up since, whenever a reaction occurs, at least one molecule will be left inside the trap.

8 Networks with Mass-Action Kinetics: A Toy Example

The results presented so far are independent of the type of kinetics assumed for the chemical reaction network. A special case, which is of particular interest in many applications, is that of systems with *mass-action* kinetics, as already mentioned in Section 2. For systems with mass-action kinetics, we will next derive sufficient conditions for persistence that exploit the additional structure in order to relax some of the structural assumptions on the chemical reaction network under consideration. As shown in the proof of Theorem 2, whenever the omega-limit set of an interior solution of a chemical reaction network intersects the boundary of $\mathcal{O}_+$, the zero components of any intersection point correspond to some siphon. There are two ways to rule this out situation. One way is to check whether a siphon contains the support of a P-semiflow, as done in Theorem 2. In this case, we say that the siphon is *structurally non-emptiable*; otherwise, we say that the siphon is *critical*.

The conditions we are seeking will apply to chemical reaction networks whose siphons are allowed to be critical (actually they need not even contain traps).

To further motivate our results, we first of all discuss a toy example which can be easily analyzed both in a deterministic and a stochastic set-up and will illustrate the usefulness of a systematic approach to the problem.

Consider the following simple chemical reaction network:

$$2A + B \to A + 2B \qquad B \to A \qquad\qquad (14)$$

which we assume endowed with mass-action kinetics. The associated Petri Net is shown in Fig. 5 and it has the following properties:

1. it is conservative, with P-semiflow $[1,1]$
2. it is consistent, with T-semiflow $[1,1]'$
3. it admits a unique non-trivial siphon: $\{B\}$, which is also critical

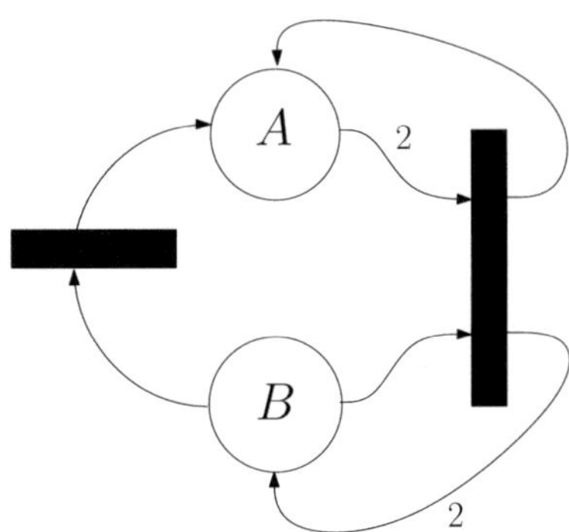

Fig. 5. A persistent chemical network whose associated Petri Net is not alive

The net effect of the first reaction is to transform one molecule of species A into one molecule of species B, and, clearly, the second reaction produces the reverse transformation. Hence, given any positive initial number of molecules for A and B, say n in total, we build the corresponding finite dimensional Markov chain (see [29] for basic definitions), which in this case has the following graphical structure:

$$[1, n-1] \leftrightarrow [2, n-2] \leftrightarrow \ldots \leftrightarrow [n-2, 2] \leftrightarrow [n-1, 1] \to [n, 0],$$

where a pair $[n_a, n_b]$ denotes the number of A and B molecules respectively. Notice that the above graph has a unique absorbing component, corresponding to the node $[n, 0]$. Such a state, when reached, basically shuts off the chemical reaction network, since the reactions do not allow the production of a molecule of A if there are no B molecules. Note that the node $[0, n]$ is never reached from another state, since consumption of an A-molecule requires that at least 2 molecules of A be available beforehand. This is why we do not include it in the diagram.

It turns out that, in the case of Petri Nets, the topology for the associated reachability graph shown in Fig. 6 is not infrequent: namely, there exist one or more absorbing components for the Markov Chain and a central strongly-connected transient component; moreover, as the number of tokens increases,

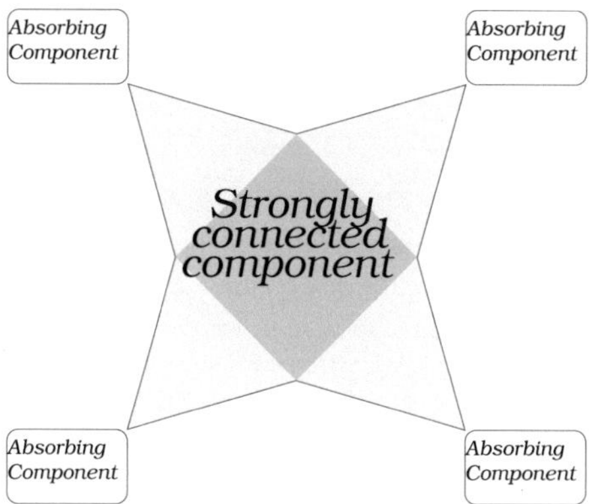

Fig. 6. Reachability graph of a Petri Net with critical siphons

the average-time that it takes to reach the absorbing components (the *time to absorption*) from the central region rapidly grows to infinity. The absorbing components of the graph correspond to the situation in which one or more critical siphons are emptied, while the central region corresponds to situations in which markings are oscillating, yet without reaching simultaneously a zero marking for all places within any given critical siphon.

Asymptotic analysis of Stochastic Petri Nets with an associated reachability graph which has the topology of Fig. 6 may lead to results which are in sharp contrast, to say the least, with what is experienced in practice for any sufficiently large initial number of tokens in the network. In fact, it may be argued that, although in theory the only stationary steady-states are indeed reached when at least one of the critical siphons gets emptied, such evolutions are so unlikely to happen in any finite time (at the scale of what is meaningful to consider for the application at hand), that, though possible in principle, they are however violating some "vague" entropic principle which one expects at the core of chemical kinetics.

Let us analyze our example (14) in further detail. To make our model suitable for computations, we associate to it a homogeneous continuous time Markov chain, assigning to each reaction a positive rate, denoted as k_1 and k_2 for reactions 1 and 2 respectively. If we adopt mass-action kinetics, the matrix corresponding to the associated chemical master equation for a total number of n molecules of A and B is given by:

$$M = \begin{bmatrix} 0 & k_2 & 0 & 0 & \cdots & 0 \\ 0 & \star & 2k_2 & 0 & \cdots & 0 \\ 0 & (n-1)^2 k_1 & \star & 3k_2 & & \vdots \\ 0 & 0 & 2(n-2)^2 k_1 & \star & \ddots & \vdots \\ 0 & \cdots & 0 & & \ddots & \ddots & nk_2 \\ 0 & \cdots & 0 & 0 & (n-1)k_1 & \star \end{bmatrix} \tag{15}$$

where $\star$ is chosen so that the matrix M has each row summing to 0. Hence, the vector $p(t) = [p_{[n,0]}(t), \ldots, p_{[1,n-1]}(t)]'$, evolves according to the following equation:

$$\dot{p}(t) = Mp(t)$$

where $p_{[n_a,n_b]}(t)$ denotes the probability of having n_a molecules of A and n_b molecules of B at time t. One can easily compute the average absorption time for any initial number of molecules of species B. Performing the computation using a symbolic computational package, there results (for k_1 and k_2 equal to 1) the exponential growth rate plotted in Fig. 7. Even with as few as 30 molecules, the average time it takes to have all the Bs transformed into As is so large that no real life experiment nor simulation will ever meet such conditions.

In other words, while a Petri Net graphical analysis leads one to conclude that extinction is theoretically possible, this is an event with vanishingly small probability. On the other hand, as it will be shown next, sometimes such chemical reaction networks can still be proved to be persistent when modeled by means of differential equations for concentrations. Thus, the ODE model (in which no species ever vanishes) provides a more accurate description of the true asymptotic behavior of the physical system in question. Of course, in general it is not clear which modeling framework should be used under what circumstances. Our aim is merely to point out certain discrepancies that may arise between the two kinds of models, in order to further motivate an in depth study of persistence on the basis of ODE techniques.

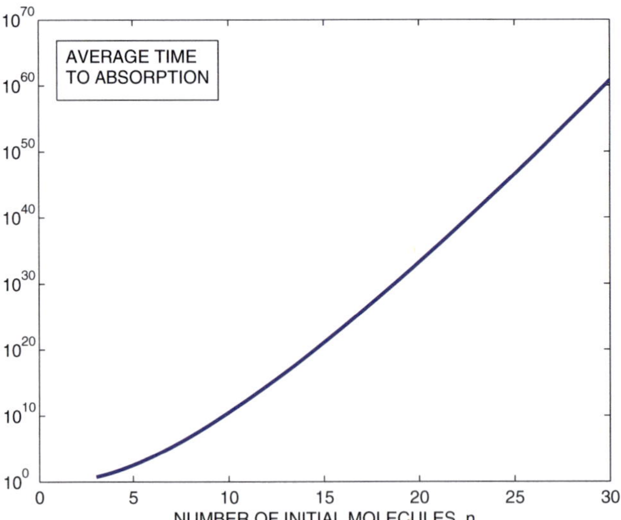

Fig. 7. Exponential increase of average time to absorption

So, let us now perform a simple deterministic analysis of the model. The equations associated to the chemical reaction network are:

$$\dot{a} = -k_1 a^2 b + k_2 b \qquad \dot{b} = -k_2 b + k_1 a^2 b \tag{16}$$

Exploiting the conservation law $a(t) + b(t) = M_{tot}$ we can bring down dimension by 1 and study the simpler system:

$$\dot{a} = (k_2 - k_1 a^2)(M_{tot} - a)$$

with a belonging to $[0, M_{tot}]$. Equilibria of the above equation are located at: $a = M_{tot}$, $a = \pm\sqrt{k_2/k_1}$. Two different scenarios arise, namely:

1. if $\sqrt{k_2/k_1} \geq M_{tot}$ only one equilibrium exists in $[0, M_{tot}]$ and all solutions converge to it; this is a boundary equilibrium and therefore persistence does not hold in this case.
2. if $\sqrt{k_2/k_1} < M_{tot}$, two equilibria exist in $[0, M_{tot}]$ and all solution starting in $[0, M_{tot})$ converge to the interior equilibrium $\sqrt{k_2/k_1}$. In this case persistence holds.

For example, in the above example, we had $k_1 = k_1 = 1$ and $M_{tot} = 30$, so the second case holds.

9   A Notion of Dynamic Non-emptiability

A low-dimensional system such as the example in Section 8 may be easily analyzed by direct computation or phase-plane analysis techniques. However, for higher dimensional examples, it is desirable to have systematic tools that can predict persistence in Petri Nets with critical siphons. We will show next that one can still rule out solutions approaching the set L_Z, for certain kinds of critical siphons Z, by exploiting the additional information that comes from having imposed mass-action kinetics. To this end, we associate to each siphon a hierarchy between its output reactions, as follows.

Let $\Sigma \subset \mathcal{S}$ be a siphon. We say that $R_i \preccurlyeq_\Sigma R_j$ if $\alpha_{ik} \geq \alpha_{jk}$ for all $k \in \Sigma$ and at least one of the inequalities is strict for some $k \in \Sigma$. The meaning of this order relationship becomes clearer thanks to the following Lemma, whose proof is a direct consequence of the definition of mass-action kinetics.

Lemma 4. *Let $\Sigma \subset V_S$ be a siphon and $R_i \preccurlyeq_\Sigma R_j$. Let us consider a network (6) endowed with mass action kinetics. Then, for each $\varepsilon > 0$, and each compact subset $K \subset L_\Sigma$, there exists an open neighborhood U_K of K such that, for all $S \in U_K$, it holds $R_i(S) \leq \varepsilon R_j(S)$.*

Accordingly, for each siphon Σ and each $\varepsilon > 0$, we may define the cone of feasible reaction rates when approaching the boundary region L_Σ, as follows:

$$\mathcal{F}_\varepsilon(\Sigma) := \{v \succeq 0 : v_i \leq \varepsilon v_j, \, \forall\, i, j \in \mathcal{R} : R_i \preccurlyeq_\Sigma R_j\}. \tag{17}$$

The following Lemma is a well-known fact in Petri Net theory and we recall it here for the sake of completeness.

Lemma 5. *Consider a conservative and consistent chemical reaction network and let Σ be an arbitrary subset of V_S. Then either 1. or 2. holds:*

1. there exists $c \succ 0$ such that $c\,\Gamma = 0$ and $c_k = 0$ for all $k \notin \Sigma$.
2. there exists $v \gg 0$ such that $[\Gamma v]_k < 0$ for all $k \in \Sigma$.

Proof. Without loss of generality assume that Σ comprises the first h species of V_S. Accordingly, we may partition Γ as follows:

$$\Gamma = \begin{bmatrix} \Gamma_\Sigma \\ \Gamma_{\bar\Sigma} \end{bmatrix}.$$

Consider the Petri Net associated to Γ_Σ. One of the following conditions holds:

1. the net admits a place which is structurally bounded,
2. the net does not admit structurally bounded places.

By Theorem 15, page 333 of [38], the two conditions are respectively equivalent to:

1. there exists some $c \succ 0$ so that $c\,\Gamma_\Sigma \preceq 0$; in particular then, there exists $\tilde{c} := [c, 0] \succ 0$ such that $\tilde{c}\,\Gamma \preceq 0$; moreover, by consistency of the original net, this is equivalent to $\tilde{c}\,\Gamma = 0$.
2. for each $p \in \Sigma$, there is a $v_p \succeq 0$ such that $\Gamma_\Sigma v_p \succeq e_p$ (p-th canonical basis vector); this, in turn, implies that $[\Gamma \sum_p v_p]_k > 0$ for all $k \in \Sigma$. By consistency, we can find some $w \gg \sum_p v_p$ be such that $\Gamma w = 0$. We pick $v := w - \sum_p v_p \gg 0$. Clearly, $\Gamma v = -\Gamma \sum_p v_p$, which then gives, as desired, that $[\Gamma v]_k < 0$ for all $k \in \Sigma$.

This completes the proof of the Lemma.

In particular, applying the previous Lemma to a siphon Σ, condition 1. is equivalent in our terminology to saying that the siphon is structurally non-emptiable, while condition 2. is therefore a characterization of criticality for a siphon. Notice that condition 2. is equivalent to asking that the following cone has non-empty interior:

$$\mathcal{C}(\Sigma) = \{v \succeq 0 : [\Gamma v]_k \leq 0,\ \forall k \in \Sigma\}$$

Hence, for a critical siphon Σ, we may find suitable positive reaction rates, which overall produce a decrease in the concentration of all of its species. On the other hand, when solutions approach the boundary we know that a certain hierarchy may hold between the output reaction rates of the siphon, due to the mass-action kinetics.

This motivates the following definition which is a key notion needed in the formulation of our main result:

Definition 6. *We say that a critical siphon Σ is dynamically non-emptiable if there is some $\varepsilon > 0$ such that the following condition holds:*

$$\mathcal{C}(\Sigma) \cap \mathcal{F}_\varepsilon(\Sigma) = \{0\}.$$

Intuitively speaking, this definition excludes the possibility of having trajectories which monotonically decrease to the set L_Σ for a given dynamically non-emptiable siphon. Its technical meaning will be clearer in the following developments.

We are now ready to state our main result:

Theorem 4. *Consider a conservative CRN as in (6), endowed with mass-action kinetics. Associate to it a Petri Net and assume that*

1. All of its critical siphons are dynamically non-emptiable.
2. There are no nested distinct critical deadlocks.

Then, the chemical reaction network is persistent.

We start with a result which clarifies the role of dynamic non-emptiability.

Lemma 6. *Consider a chemical reaction network having a dynamically non-emptiable siphon $\Sigma \subset V_S$. Let S_0 be arbitrary in $\mathcal{O}^+ \setminus L_\Sigma$. Then, provided $\omega(S_0)$ is compact, we have that $\omega(S_0) \nsubseteq L_\Sigma$.*

Proof. Assume by contradiction that $\omega(S_0) \subset L_\Sigma$, and pick an increasing sequence $t_n \to +\infty$ such that $S_i(t_n) \geq S_i(t_{n+1})$ for all $i \in \Sigma$ and all $n \in \mathbb{N}$. Let $\varepsilon > 0$ be sufficiently small, as required by the definition of dynamic non-emptiability for Σ. By Lemma 4 (applied with $K = \omega(S_0)$) there exists $T > 0$, so that for all $t \geq T$, it holds:

$$R(S(t)) \in \mathcal{F}_\varepsilon(\Sigma) \tag{18}$$

Taking averages of $\dot{S}(t)$ on intersample intervals of such a sequence yields:

$$\frac{1}{t_{n+1} - t_n} \int_{t_n}^{t_{n+1}} \Gamma R(S(\tau))\, d\tau = \frac{S(t_{n+1}) - S(t_n)}{t_{n+1} - t_n}.$$

Hence, factoring out Γ from the integral above yields

$$\frac{1}{t_{n+1} - t_n} \int_{t_n}^{t_{n+1}} R(S(\tau))\, d\tau \in \mathcal{C}(\Sigma). \tag{19}$$

Now, since $\mathcal{F}_\varepsilon(\Sigma)$ is a closed, convex cone, and exploiting (18), we also have that for all sufficiently large n's, $\frac{1}{t_{n+1}-t_n} \int_{t_n}^{t_{n+1}} R(S(\tau))\, d\tau \in \mathcal{F}_\varepsilon(\Sigma)$. By dynamic non-emptiability of Σ then $\frac{1}{t_{n+1}-t_n} \int_{t_n}^{t_{n+1}} R(S(\tau))\, d\tau = 0$ for all sufficiently large n's; this implies $S(t)$ is an equilibrium for all sufficiently large t's, and therefore, by uniqueness of solutions, S_0 is also an equilibrium. Hence $\{S_0\} = \omega(S_0) \subset L_\Sigma$ which is clearly a contradiction.

The following Lemma is crucial to the proof of Theorem 4.

Lemma 7. *Let $C \subsetneq \omega(S_0)$ be a non-empty closed, invariant set such that there are no other closed invariant sets nearby (i.e., in $[\omega(S_0) \cap U_C] \setminus C$ for some open neighborhood $U_C \supset C$). Then, there exists $\tilde{S}_0 \in [\omega(S_0) \cap U_C] \setminus C$ such that $\omega(\tilde{S}_0) \subset C$.*

Proof. Consider the set $N := [\omega(S_0) \cap V_C] \setminus C$, where V_C is an open neighborhood of C such that $\mathrm{cl}[V_C] \subset U_C$. We claim that N is non-empty. If N were empty, then $\omega(S_0) = [\omega(S_0) \cap V_C] \cup [\omega(S_0) \setminus C]$ would be the union of two non-empty open sets $[\omega(S_0) \cap V_C]$ and $[\omega(S_0) \setminus C]$. Note that their intersection would be N, hence empty, by assumption. This would imply that $\omega(S_0)$ is not connected, a contradiction to connectedness of omega limit sets.

We wish to show that there is some $\tilde{S}_0 \in N$ such that $\omega(\tilde{S}_0) \subset C$. Assume by contradiction that this is not the case, i.e. that $\omega(\tilde{S}_0) \not\subset C$ for all $\tilde{S}_0 \in N$.

Fact 1: All solutions starting in N leave $cl[V_C]$ in forward time.
If not, then there would be some $p \in N$ whose forward orbit is contained in $\mathrm{cl}[V_C]$. But then the definition of omega limit sets implies that $\omega(p) \subset \mathrm{cl}[V_C]$ $(\subset U_C)$ as well. In addition, $p \in \omega(S_0)$ implies that $\omega(p) \subset \omega(S_0)$ (by invariance and closedness of omega limit sets), and thus we have that $\omega(p) \subset \omega(S_0) \cap U_C$. On the other hand, our assumption implies that $\omega(p) \not\subset C$, and therefore the set $\Delta = \omega(p) \setminus C$ is not empty. Moreover, we claim that Δ is invariant. To see that Δ is forward invariant, we argue by contradiction. If not, then there must be some forward solution starting in Δ which must enter C in some finite forward time (since every forward solution starting in Δ certainly remains in $\omega(p)$ by invariance of omega limit sets). But this contradicts backward invariance of C. A similar argument shows that Δ is backward invariant. In conclusion, the set Δ is non-empty, invariant, and contained in $[\omega(S_0) \cap U_C] \setminus C$. This contradicts the hypothesis that there are no invariant sets in $[\omega(S_0) \cap U_C] \setminus C$.

Now we partition N into two subsets: a subset N_1 consistinf of those states whose solutions also leave $\mathrm{cl}[V_C]$ in backward time and a subset N_2 consisting of those states whose backwards solution do not leave $\mathrm{cl}[V_C]$:

$$N_1 := \{\tilde{S}_0 \in N : \exists t < 0 : S(t, \tilde{S}_0) \notin \mathrm{cl}[V_C]\}$$

$$N_2 := \{\tilde{S}_0 \in N : \forall t \leq 0, S(t, \tilde{S}_0) \in \mathrm{cl}[V_C]\}.$$

Fact 2: $N_1 \cap \tilde{U}_C = \emptyset$, for some sufficiently small neighborhood $\tilde{U}_C \supset C$.
If this were not the case, then there would be a sequence of points $S_n \in N_1$ so that $S_n \to S_c$ for some $S_c \in C$. Then we could define $\tilde{S}_n := \varphi(-\tau_n, S_n)$ where $\tau_n > 0$ is the first time τ for which $\varphi(-\tau, S_n)$ belongs to ∂V_C. Let $\tilde{S} := \lim_{n \to +\infty} \tilde{S}_n$ (which without loss of generality always exists after possibly passing to a subsequence). Clearly, $\tilde{S} \in \partial V_C$. We claim that $\omega(\tilde{S}) \subset C$, thus giving rise to a contradiction. The claim can be shown in 3 steps.

1. First we prove that $\tau_n \to +\infty$. If not, then there exists a bounded subsequence admitting a finite limit; without loss of generality, let us relabel this subsequence as τ_n. Let $0 \leq \bar{\tau} = \lim_{n \to +\infty} \tau_n$. By continuity:

$$\varphi(\bar{\tau}, \tilde{S}) = \lim_{n \to +\infty} \varphi(\bar{\tau}, \tilde{S}_n) = \lim_{n \to +\infty} \varphi(\bar{\tau} - \tau_n, S_n) = \varphi(0, S_c) = S_c \in C.$$

   This, however, violates invariance of C.

2. Next, we show that $\varphi(t, \tilde{S}) \in W_C$ for all $t \geq 0$, for some open W_C with $\mathrm{cl}[V_C] \subset W_C$ and $\mathrm{cl}[W_C] \subset U_C$. If not, then there would exist a finite $\bar{t} > 0$ so that $\varphi(\bar{t}, \tilde{S}) \notin \mathrm{cl}[V_C]$. But since $\varphi(\bar{t}, \tilde{S}) = \lim_{n \to \infty} \varphi(\bar{t}, \tilde{S}_n)$, it follows that for all sufficiently large n's, $\varphi(\bar{t}, \tilde{S}_n) \notin \mathrm{cl}[V_C]$. This violates unboundedness of the sequence $\{\tau_n\}$, because by definition of τ_n, there holds that $\varphi(t, \tilde{S}_n) \in \mathrm{cl}[V_C]$ for all $t \in [0, \tau_n]$.
3. Since $\omega(\tilde{S}) \subset \mathrm{cl}[W_C] \subset [\omega(S_0) \cap U_C]$, we are left to conclude that $\omega(\tilde{S}) \subset C$. Indeed, if this were not the case, then it can be proved (using the same arguments used to prove invariance of Δ in the proof of *Fact 1*) that $\omega(\tilde{S}) \backslash C$ is a non-empty invariant set contained in $[\omega(S_0) \cap U_C] \backslash C$. But this contradicts that there are no invariant sets in $[\omega(S_0) \cap U_C] \backslash C$.

Hence we are only left to deal with the smaller set $\tilde{U}_C \cap \omega(S_0)$ where only solutions of type N_2 exist. Notice that, for all $p \in N_2$ we have $\alpha(p) \subseteq C$ (once more, this can be proved by contradiction, by showing that $\alpha(p) \backslash C$ is non-empty and invariant using similar arguments from the proof of invariance of Δ in the proof of *Fact 1*; this in turns yields a contradiction to the fact that there are no invariant sets in $[\omega(S_0) \cap U_C] \backslash C$). On the other hand, by *Fact 1*, the solutions starting in N_2 must leave $\mathrm{cl}[V_C]$ in forward time.

We show next that this situation contradicts chain transitivity of $\omega(S_0)$ (for a proof that this set must be chain transitive, see for instance Lemma 2.1' in [21]), and this will complete the proof of Lemma 7. Let $\varepsilon > 0$ be sufficiently small, and $\tilde{V}_C$ an open neighborhood of C so that $x + z \in \tilde{U}_C$ for all $x \in \tilde{V}_C$ and all z with $|z| \leq \varepsilon$. First notice that by *Fact 1*, for all $S_i \in N_2$, there is some $t_{S_i} > 0$ such that $S(t_{S_i}, S_i) \notin \mathrm{cl}[V_C]$, hence also $S(t_{S_i}, S_i) \notin N_2$. Then by backward invariance of N_2 we obtain the stronger conclusion that $S(t, S_i) \notin N_2$ for all $t \geq t_{S_i}$.

Denote for each $S_i \in N_2$ the infimum of such t_{S_i}'s by t_i (the so-called first crossing time). We claim that $\sup\{t_i | S_i \in [\mathrm{cl}[V_c] \backslash \tilde{V}_C] \cap N_2\} < +\infty$. The proof is based on a standard compactness argument. To see this, fix $S_i \in N_2$, pick some t_{S_i} as above, and consider an open neighborhood V_i of $S(t_{S_i}, S_i)$ which is contained in the complement of $\mathrm{cl}[V_C]$. Then by continuity of the flow, the sets $U_{S_i} := \varphi^{-1}(t_{S_i}, V_i)$ are open neighborhoods of S_i, and are such that for all $x \in U_i \cap N_2$, t_{S_i} is certainly an upper bound of the first crossing-time of the solution starting in x, and in particular -by backward invariance of N_2- there holds that $S(t, x) \notin N_2$ for all $x \in U_i \cap N_2$ and all $t \geq t_{S_i}$. Now since N_2 is compact, and the collection of open sets $\{U_{S_i} \mid S_i \in N_2\}$ is an open cover of N_2, we can extract a finite subcover $\{U_{S_1}, \ldots, U_{S_N}\}$. Let $\tau = \max_{i=1,\ldots N}\{t_{S_i}\}$. Then it follows that $S(t, S_i) \notin N_2$ for all $t \geq \tau$ and all $S_i \in [\mathrm{cl}[V_C] \backslash \tilde{V}_C] \cap N_2$, which proves our claim.

Let $S_1, \ldots, S_N$ be an arbitrary (ε, τ)-chain relative to the flow $\varphi(t, S_0)$ restricted to $\omega(S_0)$ and with $S_1 \in \omega(S_0) \backslash V_C$; hence, there exist $t_1, t_2 \ldots t_{N-1} \geq \tau$ so that $|\varphi(t_j, S_j) - S_{j+1}| \leq \varepsilon$ for all $j \in \{1, 2, \ldots N-1\}$. We claim that $S_j \notin \tilde{V}_C$ for all $j \in \{1 \ldots N\}$. We prove the result by induction. Assume $S_j \notin \tilde{V}_C$ (which is obviously true for $j = 1$); following the flow t_j seconds ahead gives $\varphi(t_j, S_j) \notin N_2$. Indeed, if $S_j \in [\mathrm{cl}[V_C] \backslash \tilde{V}_C] \cap N_2$, this follows from the fact that $t_j \geq \tau$, while if $S_j \notin [\mathrm{cl}[V_C] \setminus \tilde{V}_C] \cap N_2$ (and thus in particular $S_j \notin N_2$), this follows from backward invariance of N_2. Hence, $\varphi(t_j, S_j) \notin \tilde{U}_C$ (since $\omega(S_0) \cap \tilde{U}_C \subset N_2$); therefore, $S_{j+1} \notin \tilde{V}_C$ by our choice of ε. This shows that indeed $\omega(S_0)$ is not chain transitive, since it is not possible to reach C starting outside $\tilde{V}_C$ by means of (ε, τ)-chains (provided that ε and τ are chosen as specified).

Lemma 8. *Assume that all the critical siphons of (6) are dynamically non-emptiable, and let Z be a critical siphon. Suppose that $S_0 \in \mathcal{O}_+$ is such that $\omega(S_0) \cap cl[L_Z] \doteq \Omega$ is non-empty. Assume further that $\omega(S_0) \cap cl[L_Z]$ is separated from $\omega(S_0) \cap cl[L_\Sigma]$ for all deadlocks Σ for which it is not the case that $Z \subseteq \Sigma$. Then, there exists an open neighborhood U of Ω such that $[\omega(S_0) \cap U] \setminus \Omega$ does not contain closed invariant sets.*

Remark. Notice that in the above separation condition, we may assume without loss of generality that the deadlock Σ is critical. Indeed, if it were not critical, and hence structurally non-emptiable, then the arguments in the proof of Theorem 2 show that $\omega(S_0) \cap cl[L_\Sigma] = \emptyset$.

Proof. The lemma is trivial if $\omega(S_0) \subset cl[L_Z]$. Hence, we are only left to deal with the case in which this inclusion does not hold. We recall that $cl[L_Z] = \bigcup_{\Sigma \supseteq Z} L_\Sigma$, so that

$$\omega(S_0) \cap cl[L_Z] = \left\{ \bigcup_{\Sigma \text{ is a siphon: } \Sigma \supseteq Z} L_\Sigma \cap \omega(S_0) \right\} \doteq \Omega,$$

where the restriction to siphons Σ in the union above, follows from Proposition 1. Assume, by contradiction, that every neighborhood U of Ω contains a closed minimal invariant set $C \subset \omega(S_0) \setminus cl[L_Z]$ (every closed invariant set contains a minimal invariant subset, henceforth minimality of C can be assumed without loss of generality). Hence, using the superscript c to denote the complement with respect to $\mathcal{O}^+$:

$$C \subset U \cap \omega(S_0) \cap \left[\bigcup_{\Sigma \text{ is a siphon: } \Sigma \supseteq Z} L_\Sigma \right]^c$$

Now, since

$$\mathcal{O}^+ = \bigcup_{\text{all } \Sigma, \text{ including } \emptyset} L_\Sigma,$$

it follows that

$$C \subset U \cap \omega(S_0) \cap \bigcup_{\text{all } \Sigma, \text{ including } \emptyset: \Sigma \not\supseteq Z} L_\Sigma$$

$$= U \cap \bigcup_{\text{all siphons } \Sigma, \text{ including } \emptyset: \Sigma \not\supseteq Z} L_\Sigma \cap \omega(S_0), \tag{20}$$

where we used Proposition 1 in the last equality. Pick U sufficiently small, so that $\omega(S_0) \cap cl[L_\Sigma] \cap U = \emptyset$ for all deadlocks Σ so that $\Sigma \not\supseteq Z$. As a consequence, we may without loss of generality restrict the union in equation (20) to critical siphons which are not deadlocks. We claim that $R(S) \succ 0$ for all $S \in C$. Suppose the claim is false, Then there is some $S^* \in C \subset L_\Sigma$ with $R(S^*) = 0$ for some critical siphon Σ not being a deadlock. Then for all $i \in \mathcal{R}$, there is some $j \in \Sigma$ (and thus in particular $S_j = 0$), such that $\alpha_{ij} > 0$. This implies that the set of output reactions associated to the siphon Σ consists of all the reactions of the network, and hence Σ is a deadlock. We have a contradiction.

Consider next any $\hat{S} \in C$. By boundedness of solutions, time-averages of reaction rates are also bounded and in particular

$$\lim_{n \to +\infty} \frac{1}{T_n} \int_0^{T_n} R(S(t, \hat{S})) \, dt = v \succeq 0 \tag{21}$$

along some subsequence $T_n \to +\infty$ and for some vector v, possibly depending upon $\hat{S}$. Moreover,

$$0 = \lim_{T_n \to +\infty} \frac{S(T_n, \hat{S}) - \hat{S}}{T_n} = \lim_{n \to +\infty} \frac{1}{T_n} \int_0^{T_n} \Gamma R(S(t, \hat{S})) \, dt = \Gamma v,$$

implying that $v \in \mathrm{Ker}[\Gamma] \subset \mathcal{C}(W)$ (actually for all W).

Next, it is a well known fact in Ergodic Theory, that minimal flows (in our case the flow restricted to C) admit a unique invariant ergodic probability measure. Let $m(\cdot)$ be such a measure; by the Ergodic Theorem (see [7]) for m-almost all $\hat{S} \in C$ it holds:

$$\lim_{T \to +\infty} \frac{1}{T} \int_0^T R(S(t, \hat{S})) \, dt = \int_C R(S) \, dm \tag{22}$$

Hence, (21) and (22) imply that $v = \int_C R(S) \, dm$, and then the above considerations imply that $v \succ 0$. Moreover, compactness of C, the definition (21) of v, and $\mathcal{F}_\varepsilon(\Sigma)$ being a closed convex cone, imply by virtue of Lemma 4, that $v \in \mathcal{F}_\varepsilon(\Sigma)$. So we have found a non-trivial v in $\mathcal{F}_\varepsilon(\Sigma) \cap \mathcal{C}(\Sigma)$, a contradiction to dynamic non-emptiability of the siphon Σ.

Lemma 9. *Consider a chemical reaction network without nested, distinct critical deadlocks. Let Δ_1 and Δ_2 be a critical siphon and deadlock respectively, such that it is not the case that $\Delta_1 \subseteq \Delta_2$. Then, for any $S_0 \in int[\mathcal{O}^+]$, we have $\omega(S_0) \cap cl[L_{\Delta_1}] \cap cl[L_{\Delta_2}] = \emptyset$.*

Proof. Arguing by contradiction, we would have

$$\emptyset \neq \omega(S_0) \cap cl[L_{\Delta_1}] \cap cl[L_{\Delta_2}] = \omega(S_0) \cap cl[L_{\Delta_1 \cup \Delta_2}].$$

As usual, $cl[L_{\Delta_1 \cup \Delta_2}] = \bigcup_{W \supseteq \Delta_1 \cup \Delta_2} L_W$ so that there exists $W \supseteq \Delta_1 \cup \Delta_2$ with $L_W \cap \omega(S_0) \neq \emptyset$. By Proposition 1, W is a critical siphon and therefore, since it contains the deadlock Δ_2, it is also a critical deadlock. Moreover, $W \supsetneq \Delta_2$, but this violates the assumption that critical deadlocks are not nested.

We are now ready to prove an improved version of Lemma 6.

Lemma 10. *Consider a chemical reaction network having a dynamically non-emptiable siphon $\Sigma \subset V_S$ and assume that the network is free of nested critical deadlocks. Let S_0 be arbitrary in $\mathcal{O}^+ \setminus L_\Sigma$. Then, provided $\omega(S_0)$ is compact, we have that $\omega(S_0) \not\subseteq cl[L_\Sigma]$.*

Proof. The proof is carried out by considering two separate cases:

1. $\omega(S_0) \cap L_W = \emptyset$ for all $W \supsetneq \Sigma$; Since $cl[L_\Sigma] = \bigcup_{W \supseteq \Sigma} L_W$, the result follows by Lemma 6, considering that $\omega(S_0) \cap cl[L_\Sigma] = \omega(S_0) \cap L_\Sigma$.
2. Assume that $\exists W \supsetneq \Sigma$ such that $\omega(S_0) \cap L_W \neq \emptyset$ and let W be maximal with this property, so that indeed $\omega(S_0) \cap L_W = \omega(S_0) \cap cl[L_W]$. Clearly, W is a critical siphon (by Proposition 1). Pick any critical deadlock Z (if one exists) so that $\omega(S_0) \cap cl[L_Z] \neq \emptyset$ and it is not the case that $W \subseteq Z$. By Lemma 9, $\omega(S_0) \cap cl[L_Z]$ and $\omega(S_0) \cap cl[L_W]$ are separated, as requested by Lemma 8. Hence, there exists an open neighborhood U_W of $\omega(S_0) \cap cl[L_W]$ so that $\omega(S_0) \cap U_W \setminus cl[L_W]$ does not contain closed invariant sets. Finally, by Lemma 7, there exists $\tilde{S}_0 \in \omega(S_0) \cap U_W \setminus cl[L_W]$ so that $\omega(\tilde{S}_0) \subset cl[L_W] \cap \omega(S_0) = L_W \cap \omega(S_0) \subset L_W$. This however contradicts Lemma 6.

Proof of Theorem 4. The proof will be carried out by contradiction. Assume that the reaction network (6) be not persistent. Then, there exists S_0 in $\mathrm{int}(\mathcal{O}_+)$, so that $\omega(S_0) \cap \partial\mathcal{O}^+ \neq \emptyset$. Let $\mathcal{E} = \{\Sigma \subset V_S : \omega(S_0) \cap L_\Sigma \neq \emptyset\}$; clearly $\mathcal{E}$ is non-empty, and by Proposition 1, its elements are necessarily critical siphons. Pick any pair $\Delta_1, \Delta_2 \in \mathcal{E}$ ($\Delta_1 \neq \Delta_2$) of which Δ_1 is maximal in $\mathcal{E}$ with respect to set inclusion and Δ_2 is a deadlock (if there is not such a pair the next conclusion trivially holds). Of course $\Delta_1 \not\subseteq \Delta_2$ (by maximality of Δ_1) and, as a consequence, by Lemma 9 separation of $\omega(S_0) \cap cl[L_{\Delta_1}]$ and $\omega(S_0) \cap cl[L_{\Delta_2}]$ holds. Let Δ be a maximal element of $\mathcal{E}$, with respect to set inclusion. Two possible cases can be ruled out:

1. $\omega(S_0) \subset cl[L_\Delta]$; this can be ruled out by virtue of Lemma 10 and exploiting dynamical non-emptiability of Δ.
2. $\omega(S_0) \not\subseteq cl[L_\Delta]$; by Lemma 2, $\omega(S_0) \cap L_\Delta$ is invariant. Similarly, for all W such that $\omega(S_0) \cap L_W$ is non-empty, there holds that $\omega(S_0) \cap L_W$ is invariant, and hence $\omega(S_0) \cap cl[L_\Delta]$ is invariant as well since $cl[L_\Delta] = \bigcup_{W \supseteq \Delta} L_W$.

   In this case we may apply Lemma 8 to the siphon Δ (which, as we just proved, satisfies the isolation condition) so that we conclude existence a neighborhood U of $\omega(S_0) \cap cl[L_\Delta] \doteq \Omega$ so that $[U \cap \omega(S_0)] \setminus cl[L_\Delta]$ does not contain closed invariant sets. Application of Lemma 7, then, shows existence of $\tilde{S}_0$ in $\omega(S_0) \setminus \Omega$ such that $\omega(\tilde{S}_0) \subseteq \Omega$. By virtue of Lemma 10, however, this violates dynamical non-emptiability of Δ.

This completes the proof of the Theorem.

10 Examples and Discussion

We illustrate applicability of Theorem 4 through some examples which, despite their apparent simplicity, cannot be treated by the results in Section 5.

Consider the Petri Net displayed in Fig. 8, whose associated CRN is given below:

$$2A + B \to C \to A + 2B \to D \to 2A + B.$$

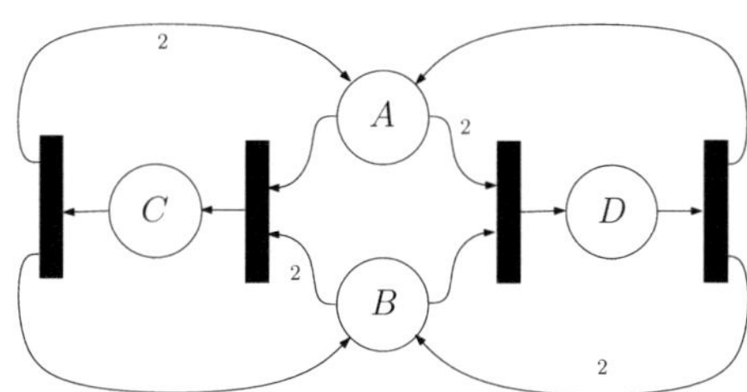

Fig. 8. A live and persistent network with critical siphons

We have that:

$$\Gamma = \begin{bmatrix} -2 & 1 & -1 & 2 \\ -1 & 2 & -2 & 1 \\ 1 & -1 & 0 & 0 \\ 0 & 0 & 1 & -1 \end{bmatrix}.$$

It is easy to verify that the CRN is weakly reversible (and hence its associated Petri Net is consistent, with T-semiflow $[1,1,1,1]'$). Moreover, there is a unique conservation law, $A + B + 3C + 3D$ (associated to the P-semiflow $[1,1,3,3]$), and two non-trivial siphons (in fact, both are deadlocks): $\Sigma_a = \{A, C, D\}$ and $\Sigma_b = \{B, C, D\}$, none of them containing the support of a first integral (both of them are therefore critical). It clearly holds that $R_1 \precnsim_{\Sigma_a} R_3$ and $R_3 \precnsim_{\Sigma_b} R_1$. Notice that both siphons are dynamically non-emptiable. To see this for the critical siphon Σ_a (similar arguments can be used to show it for Σ_b), notice that

$$\mathcal{C}(\Sigma_a) = \{v \succeq 0 : -2v_1 + v_2 - v_3 + 2v_4 \leq 0,\ v_1 - v_2 \leq 0,\ v_3 - v_4 \leq 0\}.$$

This implies in particular that

$$v \in \mathcal{C}(\Sigma_a) \quad \Rightarrow \quad v_3 \leq v_1.$$

Dynamic non-emptiability of Σ_a requires that there is some $\epsilon > 0$ such that the cone $\mathcal{C}$ and the cone

$$\{v \succeq 0 \,|\, v_1 \leq \epsilon v_3\},$$

only intersect in 0. This happens when we choose an ϵ in $(0, 1)$.

Obviously the network does not exhibit nested critical deadlocks since $\Sigma_a \not\subseteq \Sigma_b$ and $\Sigma_b \not\subseteq \Sigma_a$, and therefore Theorem 4 is applicable. We conclude that the network is persistent.

It is worth pointing out that the associated Petri Net does satisfy the assumption of Commoner's theorem; indeed the *traps* of network are the sets $\{A, C, D\}$ and $\{B, C, D\}$ and coincide with the siphons, so that the network is live. In this

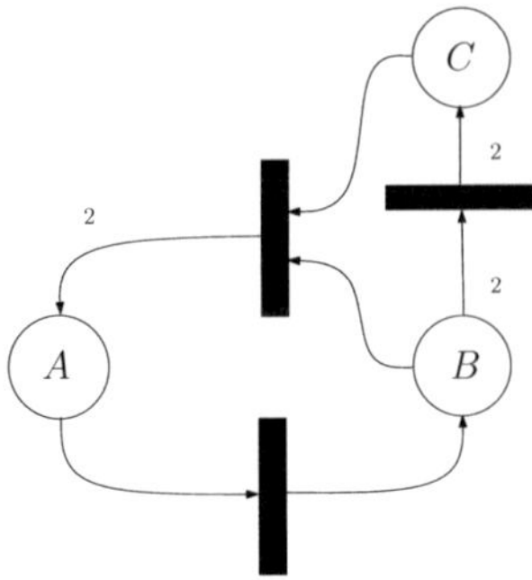

Fig. 9. Non live Petri Net giving rise to persistent CRN

case we expect the stochastic analysis and the deterministic one to give results which are in good agreement with each other.

Consider now the following simple reaction network:

$$A \to B, \quad B + C \to 2A, \quad 2B \to 2C$$

The stoichiometry matrix Γ is given by:

$$\Gamma = \begin{bmatrix} -1 & 2 & 0 \\ 1 & -1 & -2 \\ 0 & -1 & 2 \end{bmatrix}.$$

As before, we associate to it a Petri Net, whose graph is represented in Fig. 9, and compute its invariants. The network is conservative with a unique P-semiflow $[1, 1, 1]$ and consistent with T-semiflow $[4, 2, 1]'$. It exhibits one non-trivial siphon : $\Sigma = \{A, B\}$, which is critical since there cannot contain a support of a P-semiflow. It is worth pointing out that the network does not satisfy the siphon-trap property, in fact there are no non-trivial traps. Indeed, starting with initial marking $[2, 2, 2]$ it is possible to first empty out the A place, by triggering reaction 1 twice, then place B by triggering reaction 2 twice. Once the siphon is emptied, it will be such for all future times and indeed no reaction can take place henceforth. This situation is called a *deadlock* in Petri Net terminology and indeed shows that the net is not live.

However, further analysis of siphon Σ shows that indeed it is a dynamically non-emptiable siphon. In fact,

$$\mathcal{C}(\Sigma) = \{v \geq 0 : -v_1 + 2v_2 \leq 0 \ \text{ and } \ v_1 - v_2 - v_3 \leq 0\}.$$

In combination with the constraint $v_3 \leq \epsilon v_2$ which follows taking into account $R_3 \preccurlyeq_\Sigma R_2$ we get

$$2v_2 \leq v_1 \leq v_2 + v_3 \leq (1 + \epsilon)v_2$$

so that indeed for $\epsilon \in (0, 1)$ we obtain $v_2 = 0$ and consequently v_1 and $v_3 = 0$ as well. We can thus apply Theorem 4 and conclude persistence of the chemical reaction network for all values of the kinetic constants.

11 Conclusions

Persistence is the property that species (for instance in chemical reactions or in ecology) will remain asymptotically non-zero provided that they were present at the initial time. This paper provided necessary as well as sufficient conditions for the analysis of persistence in chemical reaction networks.

The results in the first part of the paper were based only upon structural and topological features of the network. Such results are "robust" with respect to uncertainty in model parameters such as kinetic constants and cooperativity indices, and they are in the same spirit as the work of Clarke [11], Horn and Jackson [26, 27], Feinberg [15, 16, 17], and many others in the context of complex balancing and deficiency theory, as well as the work of Hirsch and Smith [39, 23] and many others (including the present authors [2, 14, 3, 10]) in the context of monotone systems.

On the other hand, the knowledge of the functional dependency of reaction rates upon coefficients of the stoichiometry matrix, as in mass action kinetics, allows one to obtain tighter sufficient conditions for robust persistence of chemical reaction networks, again on the basis of topological information and regardless of the kinetics parameters involved of which only positivity is assumed. The second part of the paper takes advantage of such information. In particular, the conditions given here allow one to isolate certain classes of networks for which stochastic and deterministic analysis provide results which are qualitatively very different; in particular, Theorem 4 may sometimes be useful when one needs to decide that a certain chemical reaction network which is not "live" when considered as a stochastic discrete system, turns out to be persistent in a deterministic context, even regardless of parameter values. Our result may also serve as preliminary steps towards the construction of a systematic Input/Output theory for chemical reaction networks, by allowing systems with inflows and outflows.

Acknowledgments

PDL was supported in part by NSF grant DMS-0614651. EDS was supported in part by NSF grants DMS-0504557 and DMS-0614371. DA was supported in part by project SOSSO2, at INRIA de Rocquencourt, France.

References

1. D. Angeli, P. De Leenheer, E.D. Sontag, "On the structural monotonicity of chemical reaction networks," Proc. IEEE Conf. Decision and Control, San Diego, Dec. 2006, IEEE Publications, (2006), to appear.
2. D. Angeli, E.D. Sontag, "Monotone control systems," *IEEE Trans. Autom. Control* **48** (2003), pages 1684–1698.
3. D. Angeli, J.E. Ferrell, Jr., E.D. Sontag, "Detection of multi-stability, bifurcations, and hysteresis in a large class of biological positive-feedback systems," *Proceedings of the National Academy of Sciences USA* **101** (2004), pages 1822–1827.

4. D. Angeli, E.D. Sontag, "A global convergence result for strongly monotone systems with positive translation invariance," *Nonlinear Analysis Series B: Real World Applications*, to appear.

5. J-P. Aubin, A. Cellina, *Differential Inclusions: Set-Valued Maps and Viability Theory*, Springer-Verlag, 1984.

6. N.P. Bhatia, G.P. Szegö, *Stability Theory of Dynamical Systems*, Springer-Verlag, Berlin, 1970.

7. L.A. Bunimovich, S.G. Dani, R.L. Dobrushin, et al., *Dynamical Systems, Ergodic Theory and Applications*, Springer-Verlag, 2000.

8. G. Butler, P. Waltman, "Persistence in dynamical systems," *J. Differential Equations* **63** (1986), pages 255-263.

9. G. Butler, H.I. Freedman, P. Waltman, "Uniformly persistent systems," *Proc. Am. Math. Soc.* **96** (1986), pages 425-430.

10. M. Chaves, E.D. Sontag, R.J. Dinerstein, "Steady-states of receptor-ligand dynamics: A theoretical framework," *J. Theoretical Biology* **227** (2004), pages 413–428.

11. B.L. Clarke, "Stability of complex reaction networks," *Adv. Chem. Phys.* **43** (1980), pages 1-216.

12. F. Commoner, "Deadlocks in Petri Nets," *Tech. Report, Applied Data Research Inc. Wakefield, Massachussetts* (1972),

13. C. Conradi, J. Saez-Rodriguez, E.-D. Gilles, J. Raisch "Using chemical reaction network theory to discard a kinetic mechanism hypothesis," in *Proc. FOSBE 2005 (Foundations of Systems Biology in Engineering), Santa Barbara, Aug. 2005.* pages 325-328.

14. P. De Leenheer, D. Angeli, E.D. Sontag, "Monotone chemical reaction networks," *J. Mathematical Chemistry* (2006), to appear.

15. M. Feinberg, F.J.M. Horn, "Dynamics of open chemical systems and algebraic structure of underlying reaction network," *Chemical Engineering Science* **29** (1974), pages 775-787.

16. M. Feinberg, "Chemical reaction network structure and the stabiliy of complex isothermal reactors - I. The deficiency zero and deficiency one theorems," Review Article 25, *Chemical Engr. Sci.* **42**(1987), pp. 2229-2268.

17. M. Feinberg, "The existence and uniqueness of steady states for a class of chemical reaction networks," *Archive for Rational Mechanics and Analysis* **132**(1995), pp. 311-370.

18. M. Feinberg, "Lectures on chemical reaction networks," *Lectures at the Mathematics Research Center, University of Wisconsin*, 1979.
    http://www.che.eng.ohio-state.edu/feinberg/

19. T.C. Gard, "Persistence in food webs with general interactions," *Math. Biosci.* **51** (1980), pages 165–174.

20. H. Genrich, R. Küffner, K. Voss, "Executable Petri net models for the analysis of metabolic pathways," *Int. J. on Software Tools for Technology Transfer (STTT)* **3** (2001), pages 394-404.

21. M.W. Hirsch, H.L. Smith, X. Zhao, "Chain transitivity, attractivity, and strong repellors for semidynamical systems," *Journal of Dynamics and Differential Equations* **13** (2001), pages 107-131.

22. M.H.T. Hack, "Analysis of production schemata by Petri-Nets," *Master Thesis, MIT* (1972),

23. M. Hirsch, H.L. Smith, in *Handbook of Differential Equations, Ordinary Differential Equations (second volume)* (A. Canada, P. Drabek, and A. Fonda, eds.), Elsevier, 2005.

24. J. Hofbauer, J.W.-H. So, "Uniform persistence and repellors for maps," *Proceedings of the American Mathematical Society* **107** (1989), pages 1137-1142.

25. R. Hofestädt, "A Petri net application to model metabolic processes," *Syst. Anal. Mod. Simul.* **16** (1994), pages 113-122.

26. F.J.M. Horn, R. Jackson, "General mass action kinetics," *Arch. Rational Mech. Anal.* **49**(1972), pp. 81-116.

27. F.J.M. Horn, "The dynamics of open reaction systems," in *Mathematical aspects of chemical and biochemical problems and quantum chemistry (Proc. SIAM-AMS Sympos. Appl. Math., New York, 1974)*, pp. 125-137. SIAM-AMS Proceedings, Vol. VIII, Amer. Math. Soc., Providence, 1974.

28. C.-Y.F. Huang, Ferrell, J.E., "Ultrasensitivity in the mitogen-activated protein kinase cascade," *Proc. Natl. Acad. Sci. USA* **93** (1996), pages 10078–10083.

29. J. G. Kemeny, A.W. Knapp, J.L. Snell and J.G. Kemeny, *Denumerable Markov Chains*, Graduate Texts in Mathematics, Springer-Verlag, 1976.

30. R. Küffner, R. Zimmer, T. Lengauer, "Pathway analysis in metabolic databases via differential metabolic display (DMD)," *Bioinformatics* **16** (2000), pages 825-836.

31. A.R. Asthagiri and D.A. Lauffenburger, "A computational study of feedback effects on signal dynamics in a mitogen-activated protein kinase (MAPK) pathway model," *Biotechnol. Prog.* **17** (2001), pages 227–239.

32. N.I. Markevich, J.B. Hoek, B.N. Kholodenko, "Signaling switches and bistability arising from multisite phosphorilation in protein kinase cascades" *Journal of Cell Biology*, Vol. 164, N.3, pp. 353-359, 2004

33. J.S. Oliveira, C.G. Bailey, J.B. Jones-Oliveira, Dixon, D.A., Gull, D.W., Chandler, M.L.A., "A computational model for the identification of biochemical pathways in the Krebs cycle," *J. Comput. Biol.* **10** (2003), pages 57-82.

34. M. Peleg, M., I. Yeh, R. Altman, "Modeling biological processes using workflow and Petri net models," *Bioinformatics* **18** (2002), pages 825-837.

35. J.L. Peterson, *Petri Net Theory and the Modeling of Systems* Prentice Hall, Lebanon, Indiana 1981.

36. C.A. Petri, *Kommunikation mit Automaten* Ph.D. Thesis, University of Bonn, 1962.

37. V.N. Reddy, M.L. Mavrovouniotis, M.N. Liebman, "Petri net representations in metabolic pathways.," *Proc. Int. Conf. Intell. Syst. Mol. Biol.* **1** (1993), pages 328-336.

38. G. Rozenberg, W. Reisig, *Lectures on Petri Nets Basic Models: Basic Models, Lecture Notes in Computer Science 1491, Springer-Verlag, 1998.*

39. H.L. Smith, *Monotone dynamical systems: An introduction to the theory of competitive and cooperative systems, Mathematical Surveys and Monographs, vol. 41* (AMS, Providence, RI, 1995).

40. E.D. Sontag, "Structure and stability of certain chemical networks and applications to the kinetic proofreading model of T-cell receptor signal transduction," *IEEE Trans. Autom. Control* **46** (2001), pages 1028–1047. (Errata in *IEEE Trans. Autom. Control* **47**(2002): 705.)

41. E.D. Sontag, *Mathematical Control Theory: Deterministic Finite Dimensional Systems, Second Edition* Springer, New York 1998.

42. H.R. Thieme, "Uniform persistence and permanence for non-autonomous semiflows in population biology," *Math. Biosci.* **166** (2000), pages 173-201.
43. C. Widmann, G. Spencer, M.B. Jarpe, G.L. Johnson, G.L., "Mitogen-activated protein kinase: Conservation of a three-kinase module from yeast to human," *Physiol. Rev.* **79** (1999), pages 143–180.
44. I. Zevedei-Oancea, S. Schuster, "Topological analysis of metabolic networks based on Petri net theory," *In Silico Biol.* **3** (2003), paper 0029.
45. M. Zhou, *Modeling, Simulation, and Control of Flexible Manufacturing Systems: A Petri Net Approach* World Scientific Publishing, Hong Kong, 1999.

Geometric Ideas in the Stability Analysis of Delay Models in Biosciences

Silviu-Iulian Niculescu[1], Constantin-Irinel Morărescu[2], Wim Michiels[3], and Keqin Gu[4]

[1] Laboratoire de Signaux et Systèmes (L2S),
Supélec, 3, rue Joliot Curie, 91190, Gif-sur-Yvette, France
`Silviu.Niculescu@lss.supelec.fr`
[2] Department of Mathematics, University "Politehnica" of Bucharest,
313, Splaiul Independenţei, RO-060032, Bucharest, Romania
`constantin.morarescu@math.pub.ro`
[3] Department of Computer Science, K.U. Leuven,
Celestijnenlaan 200A, B-3001 Heverlee, Belgium
`Wim.Michiels@cs.kuleuven.ac.be`
[4] Department of Mechanical and Industrial Engineering,
Southern Illinois University at Edwarsville, Edwardsville, IL 62026, USA
`kgu@siue.edu`

Summary. This chapter focuses on the stability analysis of various continuous-time delay systems encountered in biosciences. Our main interest is to characterize the effects induced by the presence of delays on the systems' stability. More precisely, we present some intuitive and easy to follow geometric ideas for characterizing the behavior of the characteristic roots (or eigenvalues) of the corresponding linearized models with respect to the imaginary axis for the single and multiple discrete (constant) delays or for some particular classes of distributed delays (gamma-distribution with a gap). Several examples (human respiration, cell-to-cell spread models in well-mixed configurations or immune dynamics in chronic leukemia) complete the presentation.

Keywords: delay, stabilization, robustness.

1 Introduction

The interconnection between two systems, structures and/or compartments is always accompanied by some transfer phenomena at various levels (substance, energy, and/or information) and related transient behaviors in order to define new equilibria. Some of the phenomena that need to be considered in such modeling, and that are not sufficiently understood are represented by *transport* and *propagation*. Without any loss of generality, we can say that the simplified scheme described above is valid in real and abstract processes and mechanisms from engineering to economics. Besides some unifying tendency that naturally exists in all abstraction processes and mechanisms, the scheme above also applies to a large class of models describing dynamics in biosciences: competition between populations, epidemics or respiration control mechanisms.

I. Queinnec et al. (Eds.): Bio. & Ctrl. Theory: Current Challenges, LNCIS 357, pp. 217–259, 2007.
springerlink.com   © Springer-Verlag Berlin Heidelberg 2007

All the systems briefly mentioned above include a particular common element used to describe a reaction chain (distributed character [28]), a transport process (breathing process in the physiological circuite controlling the carbon-dioxide level in the blood [42]), storing nutrients or cell cycles (in the case of controlling the supply of nutrients to a growing population of microorganisms in some chemostat [38, 1]), latency and short intercellular phases (in epidemics, as for example, cell-to-cell spread models in a particular compartment, the bloodstream [8]). Such an element is the *delay*, and it can be discrete (selective memory) or distributed (over finite- or infinite-time intervals), constant or time-varying. If the memory is selective, we have single or multiple constant delays in rational dependence or not.

The list of examples in biosciences including delays in their systems' representations is far to be closed. For instance, we can mention that a density dependent feedback mechanism to respond to changes in population density never takes place *instantaneously* (see, e.g., [15, 24]), and spread (*propagation*) of infections with a family, and epidemics with intermediate classes (that is the presence of "individuals" for a given period, such that they are "exposed, but not (*necessarily*) yet infectious", see, e.g., [21]), or *recurrent diseases*, as suggested in various relapse-recovery models (after a given period, an infective returns to being fully susceptible again, [21]) are always subject to various time-constants and/or time-intervals in order to be "perceptible" at individual or family levels. Further remarks, discussions on delay models in biosciences together with an abundant list of references can be found in [18, 1, 4, 35, 15, 24], the last two addressing mainly the population dynamics models.

Without discussing the way the biological models considered below are derived and without considering the problems related to the model representation (continuous against discrete system representations, deterministic against stochastic approaches, finite- against infinite-dimensional system representations, etc.), this chapter focuses on presenting some simple *geometric ideas* for the stability analysis of the linearizations of some of the delay models mentioned above represented by delay-differential equations. It is well-known that the delay systems are infinite-dimensional [12, 20], and that the characterization of the stability regions in the parameter-space is still open in the general case, as mentioned by [12] (see also some complexity issues in [41]).

Furthermore, and without any loss of generality, the continuity dependence of the roots of the characteristic function of linear DDEs with respect to the delay parameters in the sense defined by Datko [10] (see also [13]) reduces the (stability) analysis to the following problems:

a) first, to detect *crossings* with respect to the *imaginary axis* since such crossing are related to changes in the stability behavior. In other words, we need to compute the *frequency crossing set*, which consists of all positive frequencies corresponding to the existence of at least one characteristic root on the imaginary axis. Such a characteristic root will be called *critical*. As we shall see in the next section, the frequency crossing set is reduced to a finite set of points (single or commensurate delays case) or to a finite collection of intervals (multiple not commensurate delays and some distributed delay cases). The construction of such sets will be given by using geometric arguments.

b) second, to describe the *behavior* of *critical roots* under changes of parameters, and especially with respect to the delays in the corresponding delay-parameter space. More precisely, in the case of a single delay (or multiple commensurate delays) we will detect *switches* and *reversals* corresponding to the situation when the critical characteristic roots crossing the imaginary axis towards instability, and stability, respectively by increasing the delay value. If in the case of multiple constant delays, the definition of the delay-parameter space appears naturally, the gamma-distribution with a gap considered here as distributed delay case will lead to the definition of the delay-parameter space by the mean delay and the corresponding gap. Excepting some explicit computation of the *crossing direction*, we will also briefly discuss the smoothness properties together with some appropriate classification of the *stability crossing boundaries*. The geometric arguments will be also useful in defining our classification.

Each of the problems above received a lot of attention in the literature, and some of existing results and analysis methodologies in frequency-domain are briefly discussed and outlined in [36, 16] (see the references therein). In the next sections, we will mainly discuss some *simple principles* helping in characterizing the behavior of the roots of the corresponding characteristic equation with respect to the imaginary axis together with various interpretations, extensions and applications in biosciences. It is worth to mention that the considered examples have a particular *system structure* that, to the best of the authors' knowledge, was not sufficiently exploited in deriving stability results the literature.

The chapter is organized as follows: some introductory biological models including time delays are briefly presented in Section 2, and the problem formulation ends the section. The main ideas are presented in section 3. More precisely, we consider the stability analysis problem in the delay-parameter space, and we present various geometric ways to characterize the stability crossing surfaces and curves (single and multiple delay cases, as well as some particular class of distributed delays). The presentation is intuitive and follows the lines of [17, 33, 32, 37]. Next, the methodology presented is explicitly applied to the illustrative examples considered in section 2: a second-order system including one transport delay and encountered in human respiration, a second-order system with distributed delay encountered in cell-to-cell spread mechanisms in HIV-1 infections, and finally a second-order model with multiple delays for the immune dynamics in chronic leukemia. Some concluding remarks, notes and references end the chapter.

Throughout the chapter, the following standard notations are considered: $\mathbb{C}$ ($\mathbb{C}^+$, $\mathbb{C}^-$) is the set of complex numbers (with strictly positive, and strictly negative real parts), and $j = \sqrt{-1}$. For $z \in \mathbb{C}$, $\angle(z) \in (-\pi \ \pi]$, $\Re(z)$ and $\Im(z)$ define the argument, the real part and the imaginary part of z. $\mathbb{R}$ ($\mathbb{R}^+$, $\mathbb{R}^-$) denotes the set of real numbers (larger or equal to zero, smaller or equal to zero). Next, $\mathbb{N}$ is the set of natural numbers, including zero, and $\mathbb{Z}$ the set of integers. Finally, RHP denotes the right-half plane of $\mathbb{C}$.

2 Introductory Examples and Problem Formulation

In the sequel, we shall introduce three examples of biological systems encountered in the literature (human respiration, cell-to-cell spread models, and immune dynamics). All the examples describe nonlinear phenomena featuring a particular type of *delays*: single delay in the human respiration case, a distributed delay with some gap in the cell-to-cell spread models in HIV-1 infection, and finally, small (discrete) constant delays in interaction with large constant delays in some immune dynamics model in chronic leukemia.

The section ends with some remarks concerning the common points of the examples under consideration and the problem formulation: (asymptotic) stability analysis of such biological systems. Although the examples are relatively independent each-other, their choice was finally dictated by some interesting facts: a particular system structure easy to exploit correlated with a particular potential interpretation of the delay terms as developed in the next sections.

2.1 Physiology: Delays in Human Respiration

Human respiration is an extremely complicated mechanism, and a large variety of dynamical models describing its behavior exists. In general and as expected, the delays represent the *transport time* between the lung and the peripheral and central chemoreceptors.

In the sequel, we shall consider a two-compartment representation (lungs and tissues) as an interconnection between some "plant" (in which CO_2 exchange takes place) and some "controller" (which regulates the CO_2 partial pressures in the body), as discussed by Vielle and Chauvet [42] (see, e.g. [40] for further discussions on respiratory system modeling). Their relatively simple model writes as a systems of DDEs in two variables:

$$
\begin{cases}
\dot{P}_T(t) = \dfrac{\bar{Q}}{V_T}(P_L(t) - P_T(t)) + \dfrac{\bar{M}}{\alpha V_T} \\[2mm]
\dot{P}_L(t) = \dfrac{\alpha \bar{Q} B}{V_L}(P_T(t) - P_L(t)) - \dfrac{1}{V_L}(P_L(t) - P_1)F(P_L(t - \tau)),
\end{cases}
\tag{1}
$$

where the variables P_L, P_T denote the CO_2 partial pressure, and $F(\cdot)$ is the controller function. Here, the subscript "L" ("T") denotes lungs (tissues). Next, the parameters V_T (volume, tissues), V_L (volume, lungs), B (barometric pressure minus the water vapor pressure), $\bar{Q}$ (blood flow), $\bar{M}$ (CO_2 metabolic production rate), α (CO_2 dissociation curve slope) and P_1 are positive.

The transport delay appears in the equation by the controller action F, which is an appropriate nonlinear function of the CO_2 partial pressure P_L. In order to analyze the asymptotic stability properties of the system (1), let us note that it has a *unique equilibrium point* $(\bar{P}_T, \bar{P}_L)$. By writing $P_T(t) = \bar{P}_T + x_1(t)$

and $P_L(t) = \bar{P}_L + x_2(t)$, and neglecting the non-linear second-order terms, the linearization of the system (1) rewrites as:

$$\begin{cases} \dfrac{\mathrm{d}x_1}{\mathrm{d}t} = -ax_1(t) + ax_2(t), \\[2mm] \dfrac{\mathrm{d}x_1}{\mathrm{d}t} = bx_1(t) - (b+c)x_2(t) - dx_2(t-\tau) \end{cases} \tag{2}$$

where all the parameters are positive:

$$a = \frac{\bar{Q}}{V_T} > 0, \; b = \frac{\alpha \bar{Q} B}{V_L} > 0, \; c = \frac{F(\bar{P}_L)}{V_L} > 0, \; d = \frac{F'(\bar{P}_L)(\bar{P}_L - P_I)}{V_L} > 0.$$

In other words, we arrive to a (state-space representation) system of the form:

$$\dot{x}(t) = A_0 x(t) + A_1 x(t-\tau),$$

under appropriate initial conditions, where $x \in \mathbb{R}^2$ ($x^T = [x_1 \; x_2]$), and the matrices $A_0, A_1 \in \mathbb{R}^{2 \times 2}$, with A_1 of *rank one*.

Some simple computations lead to the following characteristic function Δ : $\mathbb{C} \times \mathbb{R}_+ \mapsto \mathbb{C}$ given by:

$$\Delta(\lambda; \tau) = Q(\lambda) + P(\lambda)\mathrm{e}^{-\lambda\tau}$$

with

$$Q(\lambda) = \lambda^2 + \lambda(a+b+c) + ac, \quad P(\lambda) = \lambda d + ad,$$

where the coefficients a, b, c and d are positive, and defined as above.

2.2 Delays and Immunological Response in Cell-to-Cell Spread Models

In the sequel we will briefly present a two-dimensional model of cell-to-cell spread of HIV-1 in tissue cultures, assuming that infection is spread directly from infected cells to healthy cells and neglecting the effects of free virus [8]. The intracellular incubation period is given by a probability distribution and the model is given by two differential equations with distributed delay. More precisely, we consider the system

$$\begin{cases} \dfrac{\mathrm{d}C}{\mathrm{d}t} = r_C C(t) \left(1 - \dfrac{C(t) + I(t)}{C_m}\right) - k_I C(t)I(t), \\[3mm] \dfrac{\mathrm{d}I}{\mathrm{d}t} = k_I' \displaystyle\int_{-\infty}^{t} C(\theta)I(\theta)g(t-\theta)\mathrm{d}\theta - d_I I(t), \end{cases} \tag{3}$$

where C (I) represents the concentration of healthy (infected) cells, r_C is the reproductive rate of healthy cells, C_m is the effective carrying capacity of the system, k_I represents the infection of healthy cells by the infected cells in a *well-mixed* system, k_I'/k_I is the fraction of cells surviving the incubation period, d_I is the death rate of infected cells, and finally g represents a probability distribution that defines the infectious process history. Explicitly, we assume that the cells which are productively infectious at time t were infected θ time ago, where θ is distributed according to the probability density g.

The initial condition of the system (3) above is given by:

$$C(t) = \phi(t) \geq 0, \quad I(t) = \psi(t) \geq 0, \quad t \in (-\infty, 0]$$

where ϕ and ψ are continuous functions on $(-\infty, 0]$.

The system (3) has three equilibrium points: the trivial equilibrium $(0,0)$, the "healthy" equilibrium $(C_m, 0)$ and the "infected" equilibrium $(\overline{C}, \overline{I})$, where

$$\overline{C} = \frac{d_I}{k_I'}, \quad \overline{I} = \frac{r_C(k_I' C_m - d_I)}{k_I'(r_C + k_I C_m)}$$

if $k_I' > d_I/C_m$. Note that the equilibrium points of the model do *not depend* on the *choice* of the *probability density* g. One of the interesting problems concerns the stability of the infected equilibrium point $(\overline{C}, \overline{I})$.

The probability density g that appears in system (3) is often replaced by some Dirac densities $g(\theta) = \delta(\theta)$ (under the assumption of instantaneous infection), $g(\theta) = \delta(\theta - \tau)$ (assuming selective memory infection) where τ is a constant or by a gamma distributed kernel $g(\theta) = (\alpha^{n+1}\theta^n e^{-\alpha\theta})/n!$ (distribution over the past history $(-\infty, t]$, $t \geq 0$ without latency or "eclipse phase" of the infection), where $\alpha > 0$ and n is a positive integer (see, for instance [8, 30] and the references therein). Further discussions on the use of gamma-distribution in population dynamics modeling and some of the frequency-domain techniques in handling such systems can be found in [28] (see also [9] for an introduction to distributed delays).

Let us briefly discuss some particular cases. Connections with the human respiration linearized model above will be also outlined. First, consider the instantaneous infection assumption, that is $g(\theta) = \delta(\theta)$. Then, the model becomes the following set of ordinary differential equations (ODEs):

$$\begin{cases} \dfrac{dC}{dt} = r_C C(t)\left(1 - \dfrac{C(t) + I(t)}{C_m}\right) - k_I C(t) I(t), \\ \dfrac{dI}{dt} = k_I' C(t) I(t) - d_I I(t), \end{cases} \tag{4}$$

wit the initial conditions given by

$$C(0) = C_0 \geq 0, \quad I(0) = I_0 \geq 0,$$

where C_0 and I_0 are constants.

Under the assumption of selective memory infection, that is $g(\theta) = \delta(\theta - \tau)$ for some constant delay τ describing the selective memory effect, the system (3) becomes the following DDEs with a discrete delay:

$$\begin{cases} \dfrac{dC}{dt} = r_C C(t)\left(1 - \dfrac{C(t) + I(t)}{C_m}\right) - k_I C(t) I(t), \\ \dfrac{dI}{dt} = k_I' C(t-\tau) I(t-\tau) - d_I I(t). \end{cases} \tag{5}$$

In this case, the initial conditions are

$$C(t) = \phi(t) \geq 0, \quad I(t) = \psi(t) \geq 0, \quad t \in [-\tau, 0]$$

where ϕ and ψ are continuous real functions on $[-\tau, 0]$. It is easy to see that the linearization (5) has a similar form to the human respiration linearized model. Indeed, some simple computations allow rewriting the linearization of (5) as $(C(t) = \overline{C} + x_1(t),\ I(t) = \overline{I} + x_2(t))$:

$$\dot{x}(t) = A_0 x(t) + A_1 x(t - \tau),$$

where $x \in \mathbb{R}^2$, and A_0, $A_1 \in \mathbb{R}^{2 \times 2}$, with A_1 of *rank one*. In other words, the analysis which we are proposing to the human respiration model holds also for the HIV-1 linearized model considered above.

Finally, when the assumption of a gamma-distribution over $(-\infty, t]$, the distribution kernel g has the form $g(\theta;\ \alpha, n) := (\alpha^{n+1}\theta^n \mathrm{e}^{-\alpha\theta})/n!$, and the system becomes a set of DDEs with distributed delay. According to MacDonald [28], n is called the *order* of the delay kernel and the *mean delay* is defined by

$$T := \int_0^\infty \theta g(\theta)\mathrm{d}\theta = \frac{n+1}{\alpha}.$$

Further discussions on the use of gamma-distribution in modeling population dynamics can be found in [28, 24]. Such a particular delay kernel was also considered by [8] (see also the references therein).

Upon infection with HIV-1, there is a short intracellular "eclipse phase" (often referred as "latency"). During this period the cell is infected but has not yet begun producing virus. Spouge *et al* [39] did not included in their model a latent period after cells have been infected, but they pointed out that the latency can be modeled either by a delay or by an explicit class of latently infected cells. Therefore, in order to get a more realistic model we will consider in the following that the probability density g is given by a gamma-distribution with a gap:

$$g(\theta;\ \alpha, n, \tau) := \begin{cases} 0, & \theta < \tau \\ \dfrac{\alpha^{n+1}}{n!}(\theta - \tau)^n e^{-a(\theta - \tau)}, & \theta \geq \tau, \end{cases} \tag{6}$$

where τ is a positive constant, known also as the corresponding *gap* (an appropriate time-shift of the gamma-distribution).

The linearization around the equilibrium $(\overline{C}, \overline{I})$ leads to the following characteristic function $\Delta : \mathbb{C} \times \mathbb{R}_+ \mapsto \mathbb{C}$, given by:

$$\Delta(\lambda;\ T) := \lambda^2 + p\lambda + r + (s\lambda + q)G(\lambda;\ T), \tag{7}$$

where

$$p = \frac{d_I(k_I' C_m + r_C)}{k_I' C_m}, \quad q = \frac{r_C d_I(k_I' C_m - 2d_I)}{k_I' C_m},$$

$$r = \frac{r_C d_I^2}{k_I' C_m}, \quad s = -d_I,$$

and $G(\lambda;\ T)$ is the Laplace transform of $g(t;\ \alpha, n)$. Next, replacing g with the gamma distribution with a gap $g(t;\ \alpha, n, \tau)$ defined by (6) we get the following particular characteristic function $\Delta : \mathbb{C} \times \mathbb{R}^2_+ \mapsto \mathbb{C}$, given by:

$$\Delta(\lambda;\ T, \tau) = (\lambda^2 + p\lambda + r)(1 + \lambda T)^n + (s\lambda + q)e^{-\lambda\tau}, \tag{8}$$

with the coefficients not necessarily positive.

2.3   Delays and Immune Dynamics Models in Leukemia

Consider now the following nonlinear model proposed by [11, 22] (see also [23] for further discussions on the modeling) to describe the post-transplantation dynamics of the immune response to chronic myelogenous leukemia:

$$\begin{cases} \dfrac{\mathrm{d}T(t)}{\mathrm{d}t} = -d_T T(t) - kC(t)T(t) + p_2 kC(t-\sigma)T(t-\sigma) \\ \qquad + 2^N p_1 q_1 kC(t-\rho-N\tau)T(t-\rho-N\tau) \\ \qquad + p_1 q_2 kC(t-\rho-v)T(t-\rho-v) \\ \dfrac{\mathrm{d}C(t)}{\mathrm{d}t} = rC(t)\left(1 - \dfrac{C(t)}{K}\right) - \tilde{p}_1 kC(t-\rho)T(t-\rho), \end{cases} \tag{9}$$

where the variable T refers to the anti-cancer T cell population, and C refers to the cancer cell population, both functions of time t. All the other variables are constant and non-negative. The constants p_i, q_i, and $\tilde{p}_1$ are probabilities between 0 and 1. Furthermore, $p_1 + p_2 = 1$, and $0 \leq q_1 + q_2 \leq 1$.

In the system (9) above, there are *four distinct delays*, namely σ, $\rho + N\tau$, $\rho + v$, and ρ. The relevant values considered by [11] in their system's analysis are as follows: $\sigma = 1$ min, $\rho = 5$ min, $\tau = 1$ day, and $v = 1$ day. The parameter N is between 1 and 8 and probably close to 3. These constants respectively represent the time for unreactive interactions between T cells and cancer cells (σ), the time for reactive interactions (ρ), the time for one round of cell division (τ), the T cell recovery time after killing a cancer cell (v), and the average number of T cell divisions after stimulation (N). Due to the scale difference, we can define without any loss of generality ρ and σ as *small delays*, and the remaining delays as *large*. It is well-known that the presence of multiple delays is always accompanied by unexpected stability properties (see, for instance, [26, 27, 36]).

For convenience, let

$$\begin{aligned} b_1 &= d_T, & b_5 &= p_1 q_2 k, & \tilde{\tau} &= \rho + N\tau, \\ b_2 &= k, & c_1 &= r, & \tilde{v} &= \rho + v, \\ b_3 &= p_2 k, & c_2 &= r/K, \\ b_4 &= 2^N p_1 q_1 k, & c_3 &= \tilde{p}_1 k, \end{aligned} \tag{10}$$

and rewrite (9) as

$$\begin{cases} \dfrac{\mathrm{d}T(t)}{\mathrm{d}t} = -b_1 T(t) - b_2 C(t)T(t) + b_3 C(t-\sigma)T(t-\sigma) \\ \qquad + b_4 C(t-\tilde{\tau})T(t-\tilde{\tau}) + b_5 C(t-\tilde{v})T(t-\tilde{v}), \\ \dfrac{\mathrm{d}C(t)}{\mathrm{d}t} = c_1 C(t) - c_2 C(t)^2 - c_3 C(t-\rho)T(t-\rho). \end{cases} \tag{11}$$

Note also that all parameters in (10) are positive. Let $b = -b_2 + b_3 + b_4 + b_5$. Further discussions on the model above can be also found in [37].

The characteristic function of the corresponding linearized model can be rewritten as:

$$p(\lambda;\ \sigma,\rho,\tilde{\tau},\tilde{v}) := \det\left(\lambda I_2 - A_0 - b_{01}c_{0\rho}^T e^{-\lambda\rho} - b_{02}c_{0\sigma}^T e^{-\lambda\sigma} b_{01}c_{0\rho}^T e^{-\lambda\rho}\right.$$
$$\left. -b_{02}c_{0\tilde{v}}^T e^{-\lambda\tilde{v}} - b_{02}c_{0\tilde{\tau}}^T e^{-\lambda\tilde{\tau}}\right), \tag{12}$$

where $b_{01} = [0\ 1]^T$, $b_{02} = [1\ 0]^T$, and $c_{0\rho}$, $c_{0\sigma}$, $c_{0\tilde{v}}$, $c_{0\tilde{\tau}}$ are appropriately defined.

As in the previous examples considered in this section, the "entry" matrices for the delays ρ, $\tilde{\tau}$, $\tilde{v}$, and σ are *all* of *rank one*, a fact which simplifies the characteristic function to $p : \mathbb{C} \times \mathbb{R}_+^4 \mapsto \mathbb{C}$ given by:

$$p(\lambda;\ \sigma,\rho,\tilde{\tau},\tilde{v}) := p_0(\lambda) + p_1(\lambda)e^{-\rho\lambda} + p_2(\lambda)e^{-\sigma\lambda} + p_3(\lambda)e^{-\tilde{\tau}\lambda} + p_4(\lambda)e^{-\tilde{v}\lambda},$$

where the polynomials p_i, $i = 0, 1, 2, 3, 4$ have the following form:

$$\begin{aligned}
p_0(\lambda) &= -(\tilde{b}_1 + \lambda)(\tilde{c}_1 - \lambda), & p_1(\lambda) &= c_3 T_0(b_1 + \lambda), \\
p_2(\lambda) &= b_3 C_0(\tilde{c}_1 - \lambda), & p_3(\lambda) &= b_4 C_0(\tilde{c}_1 - \lambda), \\
p_4(\lambda) &= b_5 C_0(\tilde{c}_1 - \lambda).
\end{aligned}$$

For future reference, we note that the polynomial $p_0(\lambda)$ has no roots on the imaginary axis.

2.4   Problem Formulation

The common points of all these examples can be summarized as follows:

a) all the models include *delays* in their systems representation: *single* (human respiration model) and *multiple* (immune dynamics model in leukemia) *discrete delays* (selective memory) or *distributed delays* (immunological response in cell-to-cell spread models).

b) all the models have a particular *system structure* (matrices corresponding to the delays in the system linearization have rank one), fact leading to some particular forms of characteristic functions.

The problem considered in the sequel is to *characterize the effects induced by the delays on the (asymptotic) stability*, and more precisely *on the stability crossing boundaries*. The arguments to be considered are geometric, simply to understand, and extremely easy to apply. We believe that the analysis proposed below represents an interesting starting point in exploiting such ideas in the stability analysis of biological systems, and represent an extremely simple and appealing approach.

3   Delay-Parameter Space and Geometric Approaches

Without any loss of generality, we shall start with the stability crossing characterization in the single delay. Next, the particular class of a gamma-distribution with a gap is considered. Finally, the two delays case is presented, and its treatment opens some perspectives in handling more general cases.

3.1 Single Delay: Frequency-Sweeping and Crossing Characterizations

Consider the following linear system:

$$\dot{x}(t) = A_0 x(t) + bc^T x(t - h), \tag{13}$$

under appropriate initial conditions defined on the interval $[-\tau, 0]$, with $x \in \mathbb{R}^n$, $A_0 \in \mathbb{R}^n$ and $b, c \in \mathbb{R}^n$, and $h > 0$. For the sake of brevity, assume that the matrix A_0 has no eigenvalues on the imaginary axis. However, the arguments below can be adapted to handle also the remaining case.

In the sequel, we shall derive a simple way to characterize the behavior of the characteristic roots (zeros of the characteristic function associated with the delay system (13) or eigenvalues of the corresponding infinitesimal operator) as a function of the delay h, when the characteristic roots are crossing the imaginary axis. As derived in the first paragraph, the particular structure of the delayed matrix bc^T (rank one) allows to rewrite the characteristic equation in a particular appealing way, way that will be exploited in deriving the main results in this case study.

Frequency Crossing Set and Geometric Interpretations

Consider now the following meromorphic function $q : \mathbb{C} \times \mathbb{R}_+ \mapsto \mathbb{C}$, given by:

$$q(\lambda; \tau) = 1 - a(\lambda)e^{-\lambda h}, \tag{14}$$

where $a(s) = c^T(sI_n - A_0)^{-1}b$ represents the transfer function of a single-input single-output system having the representation (A, B, c^T).

Then, we have the following stability result:

Proposition 1. *The characteristic function associated to (13) and the meromorphic function $q(\lambda; \tau)$ have the same solutions in a neighborhood $\mathcal{V}_\delta$ of the imaginary axis, where:*

$$\mathcal{V}_\delta = \{\lambda \in \mathbb{C} : \quad \delta \geq \Re(\lambda) > -\delta\},$$

for some $\delta > 0$.

In other words, all the characteristic roots on the imaginary axis are captured by using a simple *frequency-sweeping test*:

$$\mid a(j\omega) \mid = 1.$$

The geometric interpretation is quite simple to understand: critical characteristic roots on the imaginary axis are obtained at the intersection of the plot $a(j\omega)$ with the unit circle. It is easy to see that one needs to consider only positive values for the frequency ω. Next, it easy to see that:

$$\lim_{\omega \to +\infty} \mid a(j\omega) \mid = 0.$$

Thus, since $a(j\omega)$ is a smooth curve, then intersection with the unit circle always exists (in other words, crossing always exists) if $\mid a(0) \mid > 1$. Thus, the frequency sweeping test above allows computing the *crossing frequency set* Ω which naturally consists of a finite collection of (isolated) points.

Switches, Reversals, and Crossing Direction

Next, Proposition 1 gives a simple way to analyze the switches (roots crossing the imaginary axis from stability to instability), and reversals (roots crossing the imaginary axis from instability to stability) in the light of the method considered by Cooke and van den Driessche in [7], and corrected by Boese in [5] (see also Walton and Marshall [43] for a slightly different proof argument). Some extensions to the case when the coefficients depend also on the delay parameter can be found in [3]. Thus, we have the following:

Proposition 2. *Consider the system (13). Assume that A_0 has no eigenvalues on the imaginary axis, and that $1 + a(0) \neq 0$, where $a(s) = c^T(sI_n - A)^{-1}b$. Define now the real function $f : \mathbb{R} - \{0\} \mapsto \mathbb{R}$, $f(\omega) = 1 - \mid a(j\omega) \mid$.*

Under these conditions the following statements are true:

(i) Suppose that the equation $f(\omega) = 0$ has no strictly positive roots $\omega > 0$. Then if (13) is stable at $h = 0$, it remains stable for all $h \geq 0$, whereas if it is unstable at $h = 0$ it remains unstable for all $h \geq 0$, and there does not exist any root crossing the imaginary axis when the delay h is increased in $\mathbb{R}_+$.

(ii) Suppose that the equation $f(\omega) = 0$ has at least one positive root, and that each root is simple. As h increases, stability switches may occur. There exists a positive number h^*, such that the system (13) is unstable for all $h > h^*$. As h varies from 0 to h^*, at most a finite number of stability switches may occur.

Furthermore, if for $\omega = \omega_0$ one pair of roots lies on the imaginary axis, the roots will cross the imaginary axis from left to right (or from right to left) if and only if:

$$f'(\omega_0) > 0 \quad (< 0). \tag{15}$$

Remark 1. Condition $f'(\omega_0) > 0$ in (15) is equivalent to:

$$\Re\left[\frac{a(j\omega_0)}{j\omega_0 c^T b}\right] < 0 \qquad (> 0). \tag{16}$$

Based on the result above, it is easy to see that the crossing frequency set Ω is given by:

$$\Omega := \{\omega \in \mathbb{R}_+ : \quad f(\omega) = 0\}.$$

Next, for each *critical characteristic root* $j\omega_0$, with $\omega \in \Omega$ the set of corresponding (critical) delay values is given by:

$$\mathcal{T}_{\omega_0} := \left\{\frac{1}{\omega_0}\left[\angle(a(j\omega_0)) + 2\pi\ell\right] \geq 0, \quad \ell \in \mathbb{Z}\right\}.$$

In conclusion, the delay crossing set is defined as follows:

$$\mathcal{T} := \bigcup_{\omega \in \Omega} \mathcal{T}_\omega.$$

Now, if the system free of delay is asymptotically stable and $\Omega \neq \emptyset$, the first crossing will be always *towards instability*, and the asymptotic stability will be preserved for all delays $\tau \in [0, \tau_m)$, where:

$$\tau_m := \inf \mathcal{T}.$$

The bound τ_m is also called *delay margin*.

Finally, let us comment the results in Proposition 2:

- The condition *(i)* simply says that the stability/instability property is of *delay-independent* type, and in the second case, we will say that the system is (delay-independent) *hyperbolic* (see also [19] for some definitions, and related characterizations). Further comments, and related results on such notions can be found in [36, 16].
- Similarly, the condition *(ii)* will define a *delay-dependent* type property. More precisely, we will obtain a finite set of delay intervals guaranteeing stability. All these intervals are open if the system free of delays is unstable or marginally stable.
- Finally, the stability characterization is complete when the number of *unstable* characteristic roots will be computed explicitly for *each* delay interval by exploiting the crossing direction test. It is important to note that if $\mathrm{card}(\Omega) = 1$, then the crossing direction is always towards instability, and there exists only one delay interval guaranteeing stability if the system free of delays is stable. Note that we can have a *delay-independent* unstable system, and characteristic roots crossing the imaginary axis if for all the corresponding delay intervals, the number of unstable characteristic roots is *strictly* positive.

Consider now the case when the roots on the imaginary axis are *not simple*. Using an operator perturbation approach, Fu *et al.* [14] arrived to the following characterization of the critical roots behavior:

Proposition 3. *Let $j\omega_0$ be a repeated zero of $q(\lambda;\ \tau)$ with multiplicity m with $\omega_0 \in \Omega$ and for some delay value $\tau = \tau_0 \in \mathcal{T}_{\omega_0}$.*

Then for any τ sufficiently close to τ_0 but $\tau > \tau_0$, the zeros corresponding to $j\omega_0$ can be expanded by the Puiseux series

$$j\omega_0 + \left| m! \frac{\frac{d\ q(j\omega_0;\ \tau)}{d\tau}\Big|_{\tau=\tau_0}}{\frac{d^m\ q(\lambda;\ \tau_0)}{d\lambda^m}\Big|_{\lambda=j\omega_0}} \right|^{\frac{1}{m}} e^{j\frac{2k\pi+\pi+\theta}{m}} (\tau - \tau_0)^{\frac{1}{m}} + \cdots, \quad k = 0,\ 1, \cdots,\ m-1,$$

where $\theta \in [0,\ 2\pi]$ is the phase angle of

$$\frac{\frac{d\ q(j\omega_0;\ \tau)}{d\tau}\Big|_{\tau=\tau_0}}{\frac{d^m\ q(\lambda;\ \tau_0)}{d\lambda^m}\Big|_{\lambda=j\omega_0}}.$$

Hence, for τ sufficiently close to τ_0 but $\tau > \tau_0$, the number of critical zeros entering the right-half plane (or vice versa) can be determined by the condition

$$\cos\left(\frac{2k\pi + \pi + \theta}{m}\right) > 0 \quad (< 0), \quad k = 0, 1, \cdots, m - 1.$$

If

$$m! \frac{\frac{d\ q(j\omega_0;\ \tau)}{d\tau}\big|_{\tau=\tau_0}}{\frac{d^m\ q(\lambda;\ \tau_0)}{d\lambda^m}\big|_{\lambda=j\omega_0}} = 0,$$

then for any τ sufficiently close to τ_0 but $\tau > \tau_0$, the zeros corresponding to $j\omega_0$ can be expanded by the Puiseux series

$$j\omega_0 + \left| m! \frac{\frac{d^2\ a(j\omega_0;\ \tau)}{d\tau^2}\big|_{\tau=\tau_0}}{\frac{d^m\ q(\lambda;\ \tau_0)}{d\lambda^m}\big|_{\lambda=j\omega_0}} \right|^{\frac{1}{m}} e^{j\frac{2k\pi+\pi+\theta}{m}} (\tau - \tau_0)^{\frac{1}{m}} + \cdots, \quad k = 0, 1, \cdots, m - 1,$$

where $\theta \in [0,\ 2\pi]$ is the phase angle of

$$\frac{\frac{d^2\ q(j\omega_0;\ \tau)}{d\tau^2}\big|_{\tau=\tau_0}}{\frac{d^m\ q(\lambda;\ \tau_0)}{d\lambda^m}\big|_{\lambda=j\omega_0}}.$$

Hence, for τ sufficiently close to τ_0 but $\tau > \tau_0$, the number of critical zeros entering the right-half plane (or vice versa) can be determined by the condition

$$\cos\left(\frac{2k\pi + \pi + \theta}{m}\right) > 0 \quad (< 0), \quad k = 0, 1, \cdots, m - 1.$$

In the sequel, we shall use the following convention: if the function f has *simple roots*, we shall say that the system (13) satisfies the assumption of a simple crossing. For the sake of brevity, we will focus only on such a case in the remaining paragraphs.

Consider now a partition of the delay crossing set $\mathcal{T}$ such that:

$$\mathcal{T} = \mathcal{T}_+ \bigcup \mathcal{T}_-,$$

where $\mathcal{T}_+$ and $\mathcal{T}_-$ are *ordered* disjoint sets and corresponds to a partition of $\mathcal{T}$ in delay switches ($\mathcal{T}_+$: critical roots crossing imaginary axis towards instability) and delay reversals ($\mathcal{T}_-$: critical roots crossing imaginary axis towards stability), respectively.

Then we have the following result:

Proposition 4. *Consider the system (13), under the assumption of simple crossings. Then for any positive $\widetilde{h}$, the number of unstable roots of the corresponding characteristic equation is given by:*

$$n(\widetilde{h}) = n_+(\widetilde{h}) - n_-(\widetilde{h}) + n_0, \tag{17}$$

where n_0 denotes the number of unstable roots for the system free of delays, and n_+, and n_- are defined as follows:

$$n_+(\widetilde{h}) = 2\mathrm{card}\left\{[0,\widetilde{h}) \cap \mathcal{T}_+\right\}, \tag{18}$$

$$n_-(\widetilde{h}) = 2\mathrm{card}\left\{[0,\widetilde{h}) \cap \mathcal{T}_-\right\} \tag{19}$$

Furthermore, if $\widetilde{h} \in \mathcal{T}$, there exists at least one pair of complex conjugate characteristic roots on the imaginary axis.

3.2   Extension to Distributed Delay Kernels

Consider now the following parameter-dependent characteristic function Δ : $\mathbb{C} \times \mathbb{R}_+^2 \mapsto \mathbb{C}$ described by the following quasipolynomial:

$$\Delta(\lambda;\, T, \tau) = P(\lambda)(1 + \lambda T)^n + Q(\lambda)\mathrm{e}^{-\lambda\tau}. \tag{20}$$

where P and Q are (real) polynomials such that $\deg(Q) < \deg(P)$, T, and τ denote the mean delay and the gap, respectively. Such a characteristic function represents a natural generalization of the one associated to the HIV-1 model considered in the previous section. Assume that P, and Q satisfy the following complementary assumptions (see, e.g. [31] for a complete theory on the subject, and also a full discussions on the assumptions):

Assumption I.a. $P(0) + Q(0) \neq 0$;
Assumption II.a. $P(\lambda)$ and $Q(\lambda)$ do not have common zeros;
Assumption III.a. If $P(\lambda) = \alpha$, $Q(\lambda) = \beta$, where α, β are constant real, then
   $|\alpha| \neq |\beta|$;
Assumption IV.a. $P(0) \neq 0$, $|P(0)| \neq |Q(0)|$;
Assumption V.a. $P'(j\omega) \neq 0$ whenever $P(j\omega) = 0$.

Using the methodology proposed by [31] we describe the *stability crossing curves*, which is the set of (T,τ) such that the characteristic equation (20) has solutions on the imaginary axis. We will denote the stability crossing curves as $\mathcal{T}$. As the parameters (T,τ) cross the stability crossing curves, some characteristic roots cross the imaginary axis from stability (instability) to instability (stability). Therefore, the number of roots on the right half complex plane are different on the two sides of the crossing curves. The analysis will be completed by deriving the crossing direction for each curve. Further discussions can be found in [33] (see also [34] for an algebraic characterization).

Frequency Crossing Set Computation

Similarly to the single delay case, we need to compute first the *frequency crossing set* Ω, $\Omega \subset \mathbb{R}_+$, which is defined as the collection of all frequencies $\omega > 0$ such that there exists a delay-parameter pair (T,τ) such that $\Delta(j\omega;\, T,\tau) = 0$. In other words, as the parameters T and τ vary, the characteristic roots may cross the imaginary axis at $j\omega$ if and only if $\omega \in \Omega$. We have the following:

Proposition 5. *Given any $\omega > 0$, $\omega \in \Omega$ if and only if it satisfies*

$$0 < |P(j\omega)| \leq |Q(j\omega)|. \tag{21}$$

Consider now in more detail the inequality (21). First at all, it is easy to see that Ω is *bounded*, since $\deg(P) > \deg(Q)$. Next, Ω consists of a *finite set* of *intervals of finite length*. Indeed, it is easy to see that there are only a finite number of solutions to each of the following two equations:

$$P(j\omega) = 0, \tag{22}$$

and

$$|P(j\omega)| = |Q(j\omega)|, \tag{23}$$

because P and Q are both co-prime polynomials with $P(0)+Q(0) \neq 0$. Therefore, Ω, which is the collection of ω satisfying (21), consists of a finite number of intervals. Denote these intervals as Ω_1, Ω_2, ..., Ω_N. Then

$$\Omega := \bigcup_{k=1}^{N} \Omega_k.$$

Without any loss of generality, we may order these intervals from left to right, i.e., for any $\omega_1 \in \Omega_{k_1}$, $\omega_2 \in \Omega_{k_2}$, $k_1 < k_2$, we have $\omega_1 < \omega_2$.

Stability Crossing Curves Characterization

Using the results and the notations above, we arrive to the following characterization:

Proposition 6. *The stability crossing curves* $(T, \tau) \in \mathcal{T}$ *are given by*

$$T = \frac{1}{\omega} \left(\left| \frac{Q(j\omega)}{P(j\omega)} \right|^{2/n} - 1 \right)^{1/2}, \tag{24}$$

$$\tau = \tau_m = \frac{1}{\omega} \left(\angle Q(j\omega) - \angle P(j\omega) - n \arctan(\omega T) + \pi + 2m\pi \right), \tag{25}$$
$$m = 0, \pm 1, \pm 2, ...,$$

where $\omega \in \Omega$ *represents a crossing frequency.*

We will not restrict $\angle Q(j\omega)$ and $\angle P(j\omega)$ to a 2π range. Rather, we allow them to vary continuously within each interval Ω_k. Thus, for each fixed m, (24) and (25) represent a continuous curve. We denote such a curve as $\mathcal{T}_m^k$. Therefore, corresponding to a given interval Ω_k, we have an infinite number of continuous stability crossing curves $\mathcal{T}_m^k$, $m = 0, \pm 1, \pm 2, ...$ It should be noted that, for some m, part or the entire curve may be outside of the range $\mathbb{R}_+^2$, and therefore, may not be physically meaningful. The collection of all the points in $\mathcal{T}$ corresponding to Ω_k may be expressed as

$$\mathcal{T}^k = \bigcup_{m=-\infty}^{+\infty} \left(\mathcal{T}_m^k \bigcap \mathbb{R}_+^2 \right).$$

Obviously,

$$\mathcal{T} = \bigcup_{k=1}^{N} \mathcal{T}^k.$$

Classification of Stability Crossing Curves

Let the left and right end points of interval Ω_k be denoted as ω_k^ℓ and ω_k^r, respectively. Due to Assumptions IV.a and V.a, it is not difficult to see that each end point ω_k^ℓ or ω_k^r must belong to one, and only one, of the following three types.

Type 1. It satisfies the equation (23).
Type 2. It satisfies the equation (22).
Type 3. It equals 0.

Denote an end point as ω_0, which may be either a left end or a right end of an interval Ω_k. Then the corresponding points in $\mathcal{T}_m^k$ may be described as follows.

If ω_0 is of type 1, then $T = 0$. In other words, $\mathcal{T}_m^k$ intersects the τ-axis at $\omega = \omega_0$.

If ω_0 is of type 2, then as $\omega \to \omega_0$, $T \to \infty$ and $\tau \to$

$$\frac{1}{\omega_0}\left(\angle Q(j\omega_0) - \lim_{\omega \to \omega_0} \angle P(j\omega) - \frac{n\pi}{2} + \pi + m2\pi\right). \tag{26}$$

Obviously,

$$\lim_{\omega \to \omega_0} \angle P(j\omega) = \angle \left[\frac{d}{d\omega}P(j\omega)\right]_{\omega \to \omega_0} \tag{27}$$

if ω_0 is the left end point ω_k^ℓ of Ω_k, and

$$\lim_{\omega \to \omega_0} \angle P(j\omega) = \angle \left[\frac{d}{d\omega}P(j\omega)\right]_{\omega \to \omega_0} + \pi \tag{28}$$

if ω_0 is the right end point ω_k^r of Ω_k. In other words, $\mathcal{T}_m^k$ approaches a horizontal line. Obviously, only ω_1^ℓ may be of type 3. Due to non-singularity assumptions, if $\omega_1^\ell = 0$, we must have $0 < |P(0)| < |Q(0)|$. In this case, as $\omega \to 0$, both T and τ approach ∞. In fact, (T, τ) approaches a straight line with slope

$$\tau/T \to \frac{(\angle Q(0) - \angle P(0) - n\arctan\alpha + \pi + m2\pi)}{\alpha}, \tag{29}$$

where

$$\alpha = \left(\left|\frac{Q(0)}{P(0)}\right|^{2/n} - 1\right)^{1/2}.$$

We say an interval Ω_k is of type ℓr if its left end is of type ℓ and its right end is of type r. We may accordingly divide these intervals into the following 6 types.

Type 11. In this case, $\mathcal{T}_m^k$ starts at a point on the τ-axis, and ends at another point on the τ-axis.
Type 12. In this case, $\mathcal{T}_m^k$ starts at a point on the τ-axis, and the other end approaches ∞ along a horizontal line.
Type 21. This is the reverse of type 12. $\mathcal{T}_m^k$ starts at ∞ along a horizontal line, and ends at the τ-axis.

Type 22. In this case, both ends of $\mathcal{T}_m^k$ approaches horizontal lines.

Type 31. In this case, $\mathcal{T}_m^k$ begins at ∞ with an asymptote of slope expressed in (29). The other end is at the τ-axis.

Type 32. In this case, $\mathcal{T}_m^k$ again begins at ∞ with an asymptote of slope expressed in (29). The other end approaches ∞ along a horizontal line.

Several examples covering the cases mentioned above can be found in [31] (see also [33]. For the sake of brevity, we will not present them here. We will only discuss our cell-to-cell HIV-1 linearized model in this perspective.

Crossing Direction Characterization

Introduce now the following notations:

$$R_0 = \Re\left(\frac{j}{\lambda}\frac{\partial\Delta(\lambda;\,T,\tau)}{\partial\lambda}\right)_{\lambda=j\omega}, \quad I_0 = \Im\left(\frac{j}{\lambda}\frac{\partial\Delta(\lambda;\,T,\tau)}{\partial\lambda}\right)_{\lambda=j\omega},$$

$$R_1 = \Re\left(\frac{1}{\lambda}\frac{\partial\Delta(\lambda;\,T,\tau)}{\partial T}\right)_{\lambda=j\omega}, \quad I_1 = \Im\left(\frac{1}{\lambda}\frac{\partial\Delta(\lambda;\,T,\tau)}{\partial T}\right)_{\lambda=j\omega},$$

$$R_2 = \Re\left(\frac{1}{\lambda}\frac{\partial\Delta(\lambda;\,T,\tau)}{\partial\tau}\right)_{\lambda=j\omega}, \quad I_2 = \Im\left(\frac{1}{\lambda}\frac{\partial\Delta(\lambda;\,T,\tau)}{\partial\tau}\right)_{\lambda=j\omega}.$$

Then, the crossing direction is given by the following:

Proposition 7. *Let $\omega \in \Omega$ and $(T,\tau) \in \mathcal{T}$ such that $j\omega$ is a simple solution of (20) and*

$$\Delta(j\omega';\,T,\tau) \neq 0,\ \forall\omega' > 0,\ \omega' \neq \omega$$

(i.e. (T,τ) is not an intersection point of two curves or different sections of a single curve of T). Then a pair of solutions of (20) cross the imaginary axis to the right, through $\lambda = \pm j\omega$ if

$$R_2 I_1 - R_1 I_2 > 0.$$

The crossing is to the left if the inequality is reversed.

3.3 Two Delays and Stability Crossing Curves

Consider now some natural extension of the previous case study (gamma-distributed delay with some gap) when the exponent $n \to +\infty$. In such a case, we will arrive to a particular time-delay systems including two delays, but with a particular form (without any term free of delays). In the sequel, we shall present a more general characteristic function, including two independent delays $p : \mathbb{C} \times \mathbb{R}_+^2 \mapsto \mathbb{C}$ of the form:

$$p(\lambda;\,\tau_1,\tau_2) := p_0(\lambda) + p_1(\lambda)e^{-\lambda\tau_1} + p_2(\lambda)e^{-\lambda\tau_2}, \tag{30}$$

where

$$p_l(\lambda) := \sum_{i=1}^{n_l} p_{li}\lambda^i, \qquad l = 0, 1, 2,$$

and we will focus on the characterization of the stability regions in the parameter space defined by the delays τ_1 and τ_2. Furthermore, we will consider only the case $n_0 > \min\{n_1, n_2\}$. This condition implies that the corresponding system is of *retarded* type. The analysis of the neutral case is omitted, but it can be treated in a similar way under some appropriate assumptions on the difference operator (see, for instance, [17]).

Similarly to the previous analysis, our first objective is to identify the regions of (τ_1, τ_2) in $\mathbb{R}_+^2$ such that $p(\lambda; \tau_1, \tau_2)$ has zeros on the imaginary axis. We first exclude some simple trivial cases and restrict the analysis to the cases when $p(\lambda; \tau_1, \tau_2)$ satisfies the following conditions (avoiding invariant root at the origin, avoiding common factors, etc.):

Assumption I.b. Zero frequency

$$p_0(0) + p_1(0) + p_2(0) \neq 0; \tag{31}$$

Assumption II.b. The polynomials $p_0(\lambda)$, $p_1(\lambda)$ and $p_2(\lambda)$ do not have any common zeros.

In the sequel, we will need also the following definition [17]:

Definition 1. *Let $\mathcal{C}_k : [a, b] \to \mathbb{R}^2$, $k = 1, 2, \ldots$ be a series of curves satisfying*

$$\mathcal{C}_k(b) - \mathcal{C}_k(a) = A, \ k = 1, 2, \ldots$$

where $A \in \mathbb{R}^2$ is a constant 2-dimensional vector independent of k, and

$$\mathcal{C}_{k+1}(a) = \mathcal{C}_k(b).$$

Then, the curve $\mathcal{C}$ formed by connecting all the curves $\mathcal{C}_k$, $k = 1, 2, \ldots$

$$\mathcal{C} = \bigcup_{k=1}^{\infty} \mathcal{C}_k$$

is known as a spiral-like curve, *and A is known as its* axis. *If in addition,*

$$\mathcal{C}_{k+1}(\xi) = \mathcal{C}_k(\xi) + A \ for \ all \ \xi \in [a, b],$$

then $\mathcal{C}$ is known as a spiral.

The definition above simply says that a spiral is formed by connecting identical curves head to tail. It is important to point out that the composite curves in a spiral-like curve do not have to be identical. In the sequel, in the spiral-like curves case, $\mathcal{C}_{k+1}$ can often be viewed as formed from $\mathcal{C}_k$ with a small deformation, which justifies the term "spiral-like curve" .

Identification of Crossing Points

Let us adapt the notations from gamma-distributed delay with a gap to the delay-parameter space defined by τ_1 and τ_2. Thus, $\mathcal{T}$ denote the set of all the points of (τ_1, τ_2) in $\mathbb{R}_+^2$ such that $p(\lambda;\ \tau_1, \tau_2)$ has at least one zero on the imaginary axis. Any $(\tau_1, \tau_2) \in \mathcal{T}$ is known as a *crossing point*. The set $\mathcal{T}$, which is the collection of all the crossing points, is called the *stability crossing curves*.

Define now $a_l(\lambda) = p_l(\lambda)/p_0(\lambda)$, $l = 1, 2$, and

$$a(\lambda;\ \tau_1, \tau_2) := 1 + a_1(\lambda)e^{-\lambda\tau_1} + a_2(\lambda)e^{-\lambda\tau_2}.$$

Using a similar argument to the one proposed in the rank one delayed matrix case, for any given τ_1 and τ_2, as long as $p_0(\lambda)$ does not have characteristic roots on the imaginary axis, $p(\lambda;\ \tau_1, \tau_2)$ and $a(\lambda;\ \tau_1, \tau_2)$ *share* all the zeros in some neighborhood of the imaginary axis. Therefore, in general, we may obtain all the crossing points and directions of crossing from:

$$a(\lambda;\ \tau_1, \tau_2) = 0 \tag{32}$$

instead of $p(\lambda;\ \tau_1, \tau_2) = 0$.

For each given $\lambda = j\omega$, $\omega > 0$, we may consider the three terms in $a(j\omega;\ \tau_1, \tau_2)$ as three vectors in the complex plane, with the magnitudes 1, $|a_1(j\omega)|$, and $|a_2(j\omega)|$, respectively. Furthermore, if we adjust the values of τ_1 and τ_2, we may arbitrarily adjust the directions of the vectors represented by the second and third terms. Equation (32) means that if we put these vectors head to tail, they form a triangle as illustrated in Figure 1. This allows us to conclude the following proposition (see, for instance, [17] for a complete proof):

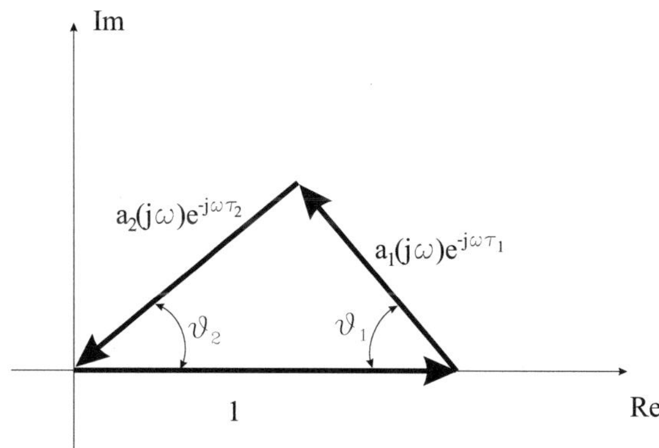

Fig. 1. Triangle formed by 1, $|a_1(j\omega)|$ and $|a_2(j\omega)|$

Proposition 8. *For each ω, $\omega \neq 0$, $p_0(j\omega) \neq 0$, $\lambda = j\omega$ can be a solution of $p(\lambda;\ \tau_1, \tau_2) = 0$ for some $(\tau_1, \tau_2) \in \mathbb{R}_+^2$ if and only if*

$$|a_1(j\omega)| + |a_2(j\omega)| \geq 1, \tag{33}$$

$$-1 \leq |a_1(j\omega)| - |a_2(j\omega)| \leq 1. \tag{34}$$

For $\omega \neq 0$ satisfying $p_0(j\omega) = 0$, $\lambda = j\omega$ can be a zero of $p(\lambda;\ \tau_1, \tau_2)$ for some $(\tau_1, \tau_2) \in \mathbb{R}_+^2$ if and only if

$$|p_1(j\omega)| = |p_2(j\omega)|. \tag{35}$$

Let Ω be the set of all $\omega > 0$ which satisfy (33) and (34) if $p_0(j\omega) \neq 0$ and (35) if $p_0(j\omega) = 0$. Similar to the commensurate and single delay cases, we will refer to Ω as the *crossing set*. It contains all the ω such that some zero of $p(\lambda;\ \tau_1, \tau_2)$ may cross the imaginary axis at $j\omega$. Then, for any given $\omega \in \Omega$, $p_l(j\omega) \neq 0$, $l = 0, 1, 2$, one may easily find all the pairs of (τ_1, τ_2) satisfying (32) as follows.

$$\tau_1 = \tau_1^{u\pm}(\omega) = \frac{\angle a_1(j\omega) + (2u-1)\pi \pm \theta_1}{\omega} \geq 0,\ u = u_0^{\pm}, u_0^{\pm}+1, u_0^{\pm}+2, ..., \tag{36}$$

$$\tau_2 = \tau_2^{v\pm}(\omega) = \frac{\angle a_2(j\omega) + (2v-1)\pi \mp \theta_2}{\omega} \geq 0,\ v = v_0^{\pm}, v_0^{\pm}+1, v_0^{\pm}+2, ..., \tag{37}$$

where $\theta_1, \theta_2 \in [0, \pi]$ are the internal angles of the triangle in Figure 1, and can be calculated by the law of cosine as

$$\theta_1 = \cos^{-1}\left(\frac{1 + |a_1(j\omega)|^2 - |a_2(j\omega)|^2}{2|a_1(j\omega)|} \right), \tag{38}$$

$$\theta_2 = \cos^{-1}\left(\frac{1 + |a_2(j\omega)|^2 - |a_1(j\omega)|^2}{2|a_2(j\omega)|} \right), \tag{39}$$

and u_0^+, u_0^-, v_0^+, v_0^- are the smallest possible integers (may be negative and may depend on ω) such that the corresponding $\tau_1^{u_0^+ +}$, $\tau_1^{u_0^- -}$, $\tau_2^{v_0^+ +}$, $\tau_2^{v_0^- -}$ calculated are nonnegative. Notice, $u_0^+ \leq u_0^-$, $v_0^+ \geq v_0^-$. The position in Figure 1 corresponds to $(\tau_1^{u+}, \tau_2^{v+})$. The position corresponding to $(\tau_1^{u-}, \tau_2^{v-})$ is its mirror image about the real axis. Next, let $\mathcal{T}_{\omega,u,v}^+$ and $\mathcal{T}_{\omega,u,v}^-$ be the singletons defined by

$$\mathcal{T}_{\omega,u,v}^{\pm} = \{(\tau_1^{u\pm}(\omega), \tau_2^{v\pm}(\omega))\},$$

and define

$$\mathcal{T}_\omega = \left(\bigcup_{u \geq u_0^+, v \geq v_0^+} \mathcal{T}_{\omega,u,v}^+ \right) \bigcup \left(\bigcup_{u \geq u_0^-, v \geq v_0^-} \mathcal{T}_{\omega,u,v}^- \right),$$

which generalizes the delay crossing set notion encountered in the commensurate or single delay cases. Here, $\mathcal{T}_\omega$ represents the set of all (τ_1, τ_2) such that $p(\lambda;\ \tau_1, \tau_2)$ has a zero at $\lambda = j\omega$. In the following remark, we will discuss the degenerate cases of $p_i(j\omega) = 0$ for at least one $i\ (= 1, 2)$.

Remark 2. If $p_0(j\omega) = 0$, $\omega \in \Omega$. Then $p(j\omega) = 0$ and assumption II imply $|p_1(j\omega)| = |p_2(j\omega)| \neq 0$. In this case, $\mathcal{T}_\omega$ consists of the solutions of

$$\angle p_1(j\omega) - \omega\tau_1 + 2\pi u = \angle p_2(j\omega) - \omega\tau_2 + 2\pi v + \pi$$

in $\mathbb{R}^2_+$ for integers u, v. Instead of isolated points, $\mathcal{T}_\omega$ now consists of an infinite number of straight lines of slope 1 of equal distance.

On the other hand, if $p_0(j\omega) \neq 0$, $\omega \in \Omega$, and $p_1(j\omega) = 0$, then $a_1(j\omega) = 0$ and $|a_2(j\omega)| = 1$, we have $\theta_2 = 0$, and θ_1 can assume all the values in $[0, \pi]$, and $\mathcal{T}^{\pm}_{\omega,u,v}$ contains all the points calculated by (36) and (37) with $\theta_1 \in [0, \pi]$, $\theta_2 = 0$. The corresponding $\mathcal{T}_\omega$ is a series of horizontal lines. Similarly, for $\omega \in \Omega$ satisfying $p_0(j\omega) \neq 0$, $p_2(j\omega) = 0$, the corresponding $\mathcal{T}^{\pm}_{\omega,u,v}$ contains all the points calculated by (36) and (37) with $\theta_1 = 0$, $\theta_2 \in [0, \pi]$, and $\mathcal{T}_\omega$ is a series of vertical lines.

Obviously,

$$\mathcal{T} = \{\mathcal{T}_\omega \mid \omega \in \Omega\}.$$

Since the behavior of the degenerate cases discussed in the above remark is easily understood, for the brevity, we will exclude these degenerate situations in the sequel, by imposing the following last assumption:

Assumption III.b. Non-degeneracy

$$p_l(j\omega) \neq 0 \text{ for all } \omega \in \Omega \text{ and } l = 0, 1, 2. \tag{40}$$

Stability Crossing Curves

We give now the complete characterization of the crossing set Ω and of the stability crossing curves $\mathcal{T}$. The presentation follows closely [17]. We have the following result:

Proposition 9. *The crossing set Ω consists of a finite number of intervals of finite length, including the cases which may violate (40).*

From the Propositions 8 and 9, it follows that the end intervals frequency points ω of the crossing set Ω should satisfy one of the following conditions:

$$|a_1(j\omega)| + |a_2(j\omega)| = 1, \tag{41}$$

or

$$|a_1(j\omega)| - |a_2(j\omega)| = 1, \tag{42}$$

or

$$|a_2(j\omega)| - |a_1(j\omega)| = 1. \tag{43}$$

Let these intervals be Ω_k, $k = 1, 2, ..., N$, arranged in such an order that the left end point of Ω_k increases with increasing k. Then

$$\Omega := \bigcup_{k=1}^{N} \Omega_k.$$

It is worth clarifying that $0 \notin \Omega$ by definition even if $\omega = 0$ satisfies (33) and (34). Indeed, if (33) and (34) are satisfied for $\omega = 0$ and sufficiently small positive values of ω, then, $\Omega_1 = (0, \omega_1^r]$, and we will let $\omega_1^l = 0$ in this case. Otherwise, $\Omega_1 = [\omega_1^l, \omega_1^r]$, $\omega_1^l \neq 0$. For $k \geq 2$, $\Omega_k = [\omega_k^l, \omega_k^r]$. We will subdivide the intervals if necessary so that for any $\omega \in (\omega_k^l, \omega_k^r)$, none of the three equations (41), (42) and (43) is satisfied.

Let

$$\mathcal{T}_{u,v}^{\pm k} = \bigcup_{\omega \in \Omega_k} \mathcal{T}_{\omega,u,v}^{\pm} = \{(\tau_1^{u\pm}(\omega), \tau_2^{v\pm}(\omega)) \mid \omega \in \Omega_k\},$$

and

$$\mathcal{T}^k = \bigcup_{u=-\infty}^{\infty} \bigcup_{v=-\infty}^{\infty} (\mathcal{T}_{u,v}^{+k} \cup \mathcal{T}_{u,v}^{-k}) \cap \mathbb{R}_+^2$$

$$= \bigcup_{\omega \in \Omega_k} \mathcal{T}_\omega. \tag{44}$$

Then,

$$\mathcal{T} = \bigcup_{k=1}^{N} \mathcal{T}^k.$$

Note that we allow part of $\mathcal{T}_{u,v}^{+k}$ or $\mathcal{T}_{u,v}^{-k}$ to be outside of $\mathbb{R}_+^2$ in some cases for the convenience of discussions. We should, however, keep in mind that the part of $\mathcal{T}_{u,v}^{+k}$ or $\mathcal{T}_{u,v}^{-k}$ outside of $\mathbb{R}_+^2$ no longer represents the boundary of a meaningful change of the number of RHP zeros of $p(\lambda; \tau_1, \tau_2)$. We will not restrict $\angle a_l(j\omega)$ to be within a range of 2π but make it a continuous function of ω within each Ω_k. This is always possible due to the way Ω_k is defined. As a result, for a fixed pair of integers (u, v), each $\mathcal{T}_{u,v}^{+k}$ or $\mathcal{T}_{u,v}^{-k}$ is a continuous curve.

To study how each $\mathcal{T}_{u,v}^{+k}$ or $\mathcal{T}_{u,v}^{-k}$ is connected in $\mathcal{T}^k$ at the ends of Ω_k, we make the following observation - under our standing non-degenerate assumption (40), the end points of the intervals, ω_k^l, $k = 2, 3, ...$ and ω_k^r, $k = 1, 2, ...$ must satisfy one and only one of the three equations (41), (42) and (43). Accordingly, we can classify these end points into three types according to which equation $\omega = \omega_k^l$ or $\omega = \omega_k^r$ satisfies. The left end of Ω_1 may have an additional type if $\omega_1^l = 0$. A careful examination of the equations (36) and (37) allows us to arrive at the following list [17]:

Type 1. (42) is satisfied. In this case, $\theta_1 = 0$, $\theta_2 = \pi$, and $\mathcal{T}_{u,v}^{+k}$ is connected with $\mathcal{T}_{u,v-1}^{-k}$ at this end.

Type 2. (43) is satisfied. In this case, $\theta_1 = \pi$, $\theta_2 = 0$, and $\mathcal{T}_{u,v}^{+k}$ is connected with $\mathcal{T}_{u+1,v}^{-k}$ at this end.

Type 3. (41) is satisfied. In this case, $\theta_1 = \theta_2 = 0$, and $\mathcal{T}_{u,v}^{+k}$ is connected with $\mathcal{T}_{u,v}^{-k}$ at this end.

Type 0. $\omega_k^l = 0$. This requires that $\omega = 0$ satisfy (33) and (34). In this case, as $\omega \to 0$, $\mathcal{T}_{u,v}^{+k}$ and $\mathcal{T}_{u,v}^{-k}$ approach ∞ with asymptotes passing through the points $(\hat{a}_1 \pm \hat{\theta}_1, \hat{a}_2 \mp \hat{\theta}_2)$ with slopes of

$$\frac{\tau_2^{v\pm}}{\tau_1^{u\pm}} \to \kappa_{u,v}^{\pm} = \frac{\angle a_2(0) + (2v-1)\pi \mp \theta_2(0)}{\angle a_1(0) + (2u-1)\pi \pm \theta_1(0)}, \tag{45}$$

where $\theta_1(0)$ and $\theta_2(0)$ are evaluated by (38) and (39) using $a_1(0)$ and $a_2(0)$, respectively, and

$$\hat{a}_l = \frac{d}{d\omega}[\angle a_l(j\omega)]_{\omega=0}, \tag{46}$$

$$\hat{\theta}_l = \frac{d}{d\omega}\theta_l(j\omega)|_{\omega=0}. \tag{47}$$

Correspondingly, we say an interval Ω_k is of type lr if the left end of Ω_k is of type l and its right end is of type r. There are a total of $4 \times 3 = 12$ possible types of such intervals. According to the types of Ω_k, $\mathcal{T}^k$ may have different shapes, as specified in the following proposition.

Proposition 10. *Under the standing assumption III.b (40), the stability cross-ing curves $\mathcal{T}^k$ corresponding to Ω_k must be an intersection of $\mathbb{R}^2_+$ with a series of curves belonging to one of the following categories:*

A. *A series of closed curves;*
B. *A series of spiral-like curves with axes oriented either horizontally, vertically, or diagonally.*
C. *A series of open ended curves with both ends approaching ∞.*

The validity of the above proposition can be done by a detailed list of scenarios (see, for instance, [17] for more details and illustrative examples corresponding to each situation). For the sake of brevity, we will focus only on *closed curves*. As an illustration of this situation, examine first $\mathcal{T}^k$ corresponding to Ω_k of type 11. In this case, for given u and v, $\mathcal{T}^{+k}_{u,v}$ and $\mathcal{T}^{-k}_{u,v-1}$ are connected on both ends to form a closed curve. As u and v vary, a series of deformed versions of such closed curves are generated along the horizontal and vertical directions. $\mathcal{T}^k$ is the intersection of $\mathbb{R}^2_+$ with this series of closed curves. Similarly, it is easily shown that a $\mathcal{T}^k$ corresponding to Ω_k of type 22 or type 33 also form a similar series of closed curves. In the case of type 22, a closed curved is formed by connecting both ends of $\mathcal{T}^{+k}_{u,v}$ and $\mathcal{T}^{-k}_{u+1,v}$. For type 33, a closed curve is formed by connecting both ends of $\mathcal{T}^{+k}_{u,v}$ and $\mathcal{T}^{-k}_{u,v}$.

Tangent, Smoothness and Crossing Direction

We discuss in detail the smoothness of the stability crossing curves, characterize their tangents and derive expressions for the corresponding direction of crossing.

Tangents and Smoothness

For a given k, we will discuss the smoothness of the curves in $\mathcal{T}^k$ and thus of $\mathcal{T}$. For this purpose, we consider τ_1 and τ_2 as implicit functions of $\lambda = j\omega$ defined by (32). As λ moves along $j\mathbb{R}$, $(\tau_1,\tau_2) = (\tau_1^{u\pm}(\omega), \tau_2^{v\pm}(\omega))$ moves along $\mathcal{T}^k$. For a given $\omega \in \Omega_k$, let

$$R_0 = \Re \left(\frac{j}{\lambda} \frac{\partial a(\lambda;\ \tau_1, \tau_2)}{\partial \lambda} \right)_{\lambda = j\omega}$$

$$= \frac{1}{\omega} \Re \left([a_1'(j\omega) - \tau_1 a_1(j\omega)] e^{-j\tau_1 \omega} + [a_2'(j\omega) - \tau_2 a_2(j\omega)] e^{-j\tau_2 \omega} \right), \quad (48)$$

$$I_0 = \Im \left(\frac{j}{\lambda} \frac{\partial a(\lambda;\ \tau_1, \tau_2)}{\partial \lambda} \right)_{\lambda = j\omega}$$

$$= \frac{1}{\omega} \Im \left([a_1'(j\omega) - \tau_1 a_1(j\omega)] e^{-j\tau_1 \omega} + [a_2'(j\omega) - \tau_2 a_2(j\omega)] e^{-j\tau_2 \omega} \right), \quad (49)$$

and

$$R_l = -\Re \left(\frac{1}{\lambda} \frac{\partial a(\lambda;\ \tau_1, \tau_2)}{\partial \tau_k} \right)_{\lambda = j\omega} = \Re \left(a_k(j\omega) e^{-j\tau_k \omega} \right), \quad (50)$$

$$I_l = -\Im \left(\frac{1}{\lambda} \frac{\partial a(\lambda;\ \tau_1, \tau_2)}{\partial \tau_k} \right)_{\lambda = j\omega} = \Im \left(a_k(j\omega) e^{-j\tau_k \omega} \right), \quad (51)$$

for $l = 1, 2$. Then, since $a(\lambda;\ \tau_1, \tau_2)$ is an analytic function of λ, τ_1 and τ_2, the implicit function theorem indicates that the tangent of $\mathcal{T}^k$ can be expressed as

$$\begin{pmatrix} \frac{d\tau_1}{d\omega} \\ \frac{d\tau_2}{d\omega} \end{pmatrix} = \begin{pmatrix} R_1 & R_2 \\ I_1 & I_2 \end{pmatrix}^{-1} \begin{pmatrix} R_0 \\ I_0 \end{pmatrix} = \frac{1}{R_1 I_2 - R_2 I_1} \begin{pmatrix} R_0 I_2 - I_0 R_2 \\ I_0 R_1 - R_0 I_1 \end{pmatrix}, \quad (52)$$

provided that

$$R_1 I_2 - R_2 I_1 \neq 0. \quad (53)$$

Next, the curve $\mathcal{T}^k$ is smooth everywhere except possibly at the points where either (53) is not satisfied, or when

$$\frac{d\tau_1}{d\omega} = \frac{d\tau_2}{d\omega} = 0. \quad (54)$$

A careful examination of these cases allows us concluding with the following (see also [17]):

Proposition 11. *Under the standing assumptions including (40), the curves in $\mathcal{T}^k$ are smooth everywhere except possibly at the degenerate points corresponding to ω in any one of the following three cases:*

(1)$\lambda = j\omega$ is a multiple solution of $a(\lambda;\ \tau_1, \tau_2) = 0$.
(2)ω is a type 3 end point of Ω_k, and

$$\frac{d}{d\omega} \left(|a_1(j\omega)| + |a_2(j\omega)| \right) = 0.$$

(3)ω is a type 1 or type 2 end point of Ω_k, and

$$\frac{d}{d\omega} \left(|a_1(j\omega)| - |a_2(j\omega)| \right) = 0.$$

Furthermore, if the point is not among the three cases, then the tangents of the curves in $\mathcal{T}^k$ can be expressed as

$$\frac{d\tau_2}{d\tau_1} = \begin{cases} \frac{1/\tan\varphi_0 - 1/\tan\varphi_1}{1/\tan\varphi_0 - 1/\tan\varphi_2}, & \omega \in (\omega_k^l, \omega_k^r), \\ -\frac{|a_1(j\omega)|}{|a_2(j\omega)|}, & \omega \text{ is a type 3 end point of } \Omega_k, \\ \frac{|a_1(j\omega)|}{|a_2(j\omega)|}, & \omega \text{ is a type 1 or 2 end point of } \Omega_k, \end{cases} \tag{55}$$

where

$$\varphi_0 = \angle\left([a_1'(j\omega) - \tau_1 a_1(j\omega)]\, e^{-j\tau_1\omega} + [a_2'(j\omega) - \tau_2 a_2(j\omega)] e^{-j\tau_2\omega} \right),$$

$$\varphi_k = \angle\left(a_k(j\omega) e^{-j\tau_k\omega} \right), \; k = 1, 2.$$

Direction of Crossing

Next, we will discuss the direction in which the solutions of (32) cross the imaginary axis as (τ_1, τ_2) deviates from a curve in $\mathcal{T}^k$. We will call the direction of the curve that corresponds to increasing ω the *positive direction*. Notice, as the curve passes through the points corresponding to the end points of Ω_k, the positive direction is reversed. We will also call the region on the left hand side as we head in the positive direction of the curve *the region on the left*. Again, due to the possible reversion of parametrization, the same region may be considered on the left with respect to one point of the curve, and be considered as on the right on another point of the curve.

For the purpose of discussing the direction of crossing, we need to consider τ_1 and τ_2 as functions of

$$\lambda = \sigma + j\omega,$$

i.e., functions of two real variables σ and ω, and partial derivative notation needs to be adopted instead. Since the tangent of $\mathcal{T}^k$ along the positive direction is $(\partial\tau_1/\partial\omega, \partial\tau_2/\partial\omega)$, the normal to $\mathcal{T}^k$ pointing to the left hand side of the positive direction is $(-\partial\tau_2/\partial\omega, \partial\tau_1/\partial\omega)$.

Also, as a pair of complex conjugate solutions of (32) cross the imaginary axis to the RHP, (τ_1, τ_2) moves along the direction $(\partial\tau_1/\partial\sigma, \partial\tau_2/\partial\sigma)$. We can therefore conclude that if the inner product of these two vectors are positive, i.e.,

$$\left[\frac{\partial\tau_1}{\partial\omega}\frac{\partial\tau_2}{\partial\sigma} - \frac{\partial\tau_2}{\partial\omega}\frac{\partial\tau_1}{\partial\sigma} \right]_{s=j\omega} > 0, \tag{56}$$

the region on the left of $\mathcal{T}^k$ at ω has two more solutions on the RHP. On the other hand, if the inequality in (56) is reversed, then the region on the left of $\mathcal{T}^k$ has two fewer solutions on the right hand side of the complex plane. We can very easily express, parallel to (52), that,

$$\begin{pmatrix} \frac{\partial\tau_1}{\partial\sigma} \\ \frac{\partial\tau_2}{\partial\sigma} \end{pmatrix}_{\lambda=j\omega} = \begin{pmatrix} R_1 & R_2 \\ I_1 & I_2 \end{pmatrix}^{-1} \begin{pmatrix} I_0 \\ -R_0 \end{pmatrix} = \frac{1}{R_1 I_2 - R_2 I_1} \begin{pmatrix} R_0 R_2 + I_0 I_2 \\ -R_0 R_1 - I_0 I_1 \end{pmatrix}, \tag{57}$$

where R_l and I_l, $l = 0, 1, 2$, are defined in (48) to (51). This allows concluding with the following:

242   S.-I. Niculescu et al.

Proposition 12. *Let $\omega \in (\omega_k^l, \omega_k^r)$ and $(\tau_1, \tau_2) \in \mathcal{T}^k$ such that $j\omega$ is a simple solution of $a(\lambda;\ \tau_1, \tau_2) = 0$, and*

$$a(j\omega';\ \tau_1, \tau_2) \neq 0, \text{ for any } \omega' > 0,\ \omega' \neq \omega. \tag{58}$$

Then as (τ_1, τ_2) moves from the region on the right to the region on the left of the corresponding curve in $\mathcal{T}^k$, a pair of solutions of (32) cross the imaginary axis to the right if

$$\Im(a_1(j\omega)a_2(-j\omega)e^{j\omega(\tau_2-\tau_1)}) = R_2 I_1 - R_1 I_2 > 0. \tag{59}$$

The crossing is in the opposite direction if the inequality is reversed.

The condition (58) simply means that (τ_1, τ_2) is not an intersection point of two curves or different sections of a single curve in $\mathcal{T}$.

Finally, any given direction, $d = (d_1, d_2)$, with $\|d\|_2 = 1$, is to the left hand side of the curve if its inner product with the left hand side normal $(-\partial\tau_2/\partial\omega, \partial\tau_1/\partial\omega)$ is positive, i.e.,

$$-d_1\partial\tau_2/\partial\omega + d_2\partial\tau_1/\partial\omega > 0, \tag{60}$$

from which we have the following:

Corollary 1. *Let ω, τ_1 and τ_2 satisfy the same condition as Proposition 12. Then as (τ_1, τ_2) crosses the curve along the direction (d_1, d_2), a pair of solutions of (32) cross the imaginary axis to the right if*

$$d_1(R_0 I_1 - I_0 R_1) + d_2(R_0 I_2 - I_0 R_2) > 0. \tag{61}$$

The crossing is in the opposite direction if the inequality is reversed.

4   Illustrative Examples

In the sequel, we will reconsider the examples introduced in Section 2 in the light of the characterizations proposed in Section 3.

4.1   On Some Human Respiration Model in Physiology

The characteristic function of the linearized system of the human respiration model writes as follows:

$$p(\lambda;\ \tau) := Q(\lambda) + P(\lambda)e^{-\lambda\tau} = Q(\lambda)\left(1 + h(\lambda)e^{-\lambda\tau}\right),$$

where $h(\lambda) = P(\lambda)/Q(\lambda)$, and the polynomials P and Q are given by:

$$Q(\lambda) = \lambda^2 + \lambda(a + b + c) + ac, \qquad P(\lambda) = \lambda d + ad.$$

Since a, b, c and d are strictly positive, it is clear that $p(\lambda;\ 0)$ is Hurwitz stable, that is all its roots have strictly negative real parts. Furthermore, the polynomial Q is also Hurwitz, and thus the quasipolynomial $p(\lambda;\ \tau)$ shares the same characteristic roots on the imaginary axis with the meromorphic function:

$$q(\lambda;\ \tau) := 1 + h(\lambda)e^{-\lambda\tau}.$$

In the sequel we will analyze the stability of the model (2) by taking into account the particular form and properties of its characteristic function $p(\lambda;\ \tau)$.

4.2 Stability Analysis and Delay Intervals

In Section 2, we proposed a procedure for computing the frequency crossing set Ω together with the delay intervals guaranteeing asymptotic stability by using the meromorphic function $q(\lambda; \tau)$ instead of the original quasipolynomial $p(\lambda; \tau)$.

In our case, since the strictly proper transfer a is stable (the denominator Q is Hurwitz), it follows that it is bounded on the imaginary axis. In this context, it is easy to see that the crossing set Ω, that is the set of all crossing roots $\omega \in \mathbb{R}_+$ w.r.t. the imaginary axis $\mid h(j\omega) \mid = 1$, is given by the solution of the polynomial equation:

$$\mid P(j\omega) \mid = \mid Q(j\omega) \mid .$$

As expected, only two situations can occur:

(i) $\mid P(j\omega) \mid < \mid Q(j\omega) \mid$, for all $\omega \in \mathbb{R}_+^*$, or
(ii) there exists at least one frequency $\omega^* > 0$, such that $\mid P(j\omega^*) \mid = \mid Q(j\omega^*) \mid$

Let us analyze each case separately.

Delay-Independent Stability

The condition (i) simply says that the system (2) is hyperbolic, that is there are no characteristic roots crossing the imaginary axis if the delay parameter is increased from 0 to ∞. In such a case, the stability of the system free of delays will be preserved for all positive delays. In other words, the system is *delay-independent* asymptotically stable. Under the assumption of stability of the strictly proper transfer function h, the *frequency-sweeping* test above $\mid h(j\omega) \mid < 1$, for all $\omega \in \mathbb{R}_+$ is nothing else than the Tsypkin criterion encountered in control engineering area (see, for instance, [36, 16]).

Geometrically speaking, we will have delay-independent asymptotic stability if and only if the graph of $-h(j\omega)$, for all $\omega \in \mathbb{R}_+^*$ will stay inside the unit circle of the complex plane, with some eventual tangency at the point $(-1, 0)$ for the frequency $\omega = 0$ under the assumption of a delay-free stable system.

A *necessary condition* for no roots crossing the imaginary axis is:

$$\mid Q(0) \mid \geq \mid P(0) \mid,$$

that is $c \geq d$. It is important to point out that we can have the equality above since 0 is not a root of the characteristic function $p(\lambda; \tau)$ for any positive delay $\tau \in \mathbb{R}_+$. Geometrically speaking, the equality simply describes the tangency property mentioned above, since $P(0) = ad$ will be equal to $Q(0) = ac$ if $c = d$.

Now let us check if the condition $c \geq d$ is also *sufficient* for getting delay-independent stability. Simple computations prove that there does not exist crossing roots if and only if the following second-order equation:

$$x^2 + \left[(a+b)^2 + 2bc + c^2 - d^2\right] x + a^2(c^2 - d^2) = 0 \tag{62}$$

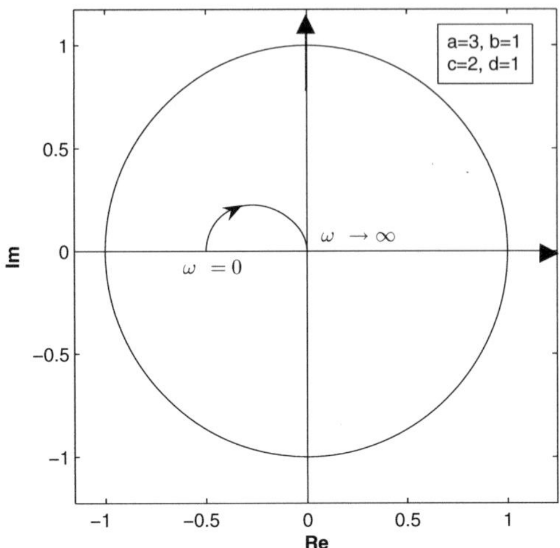

Fig. 2. Delay-independent stability since the intersection between the ratio curve $-h(j\omega) = -P(j\omega)/Q(j\omega)$ and the unit circle is empty and the system free of delay is asymptotically stable

has *no strictly positive* roots. This last condition holds if and only if the following inequalities are satisfied simultaneously:

$$\begin{cases} a^2(c^2 - d^2) \geq 0, \\ (a + b)^2 + 2bc + c^2 - d^2 \geq 0, \end{cases}$$

under the constraint of positive parameters a, b, c and d. In conclusion, we obtain the following simple *delay-independent stability* condition:

$$c \geq d. \tag{63}$$

The Figure 2 depicts such a delay-independent stability condition for some positive c, and d, such that $c > d$. As explained in Section 2, the plot of $-P(j\omega)/Q(j\omega)$ and its intersection with the unit circle are not sufficient to conclude on asymptotic stability of the corresponding system. We can only expect to detect the presence or not of crossing roots. Finally, notice that the same condition was obtained in [42], but using a different argument.

Delay-Dependent Stability

As expected, the characteristic function $p(\lambda; \tau)$ will have zeros crossing the imaginary axis if and only if the second-order equation (62) has *at least one strictly positive root*. Based on the particular form of this equation, it follows that such a situation appears if and only if:

$$d > c. \tag{64}$$

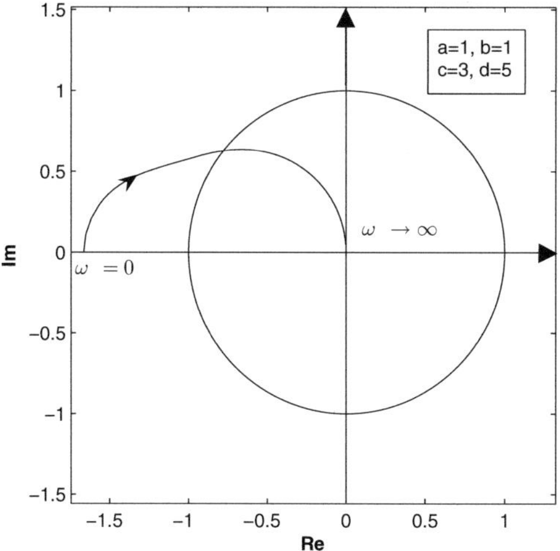

Fig. 3. Delay-dependent stability since the intersection between the ratio curve $-h(j\omega) = -P(j\omega)/Q(j\omega)$ and the unit circle consists of only one point, and there exists points outside and inside the unit circle

More precisely, if $d > c$, the equation (62) will always have two real roots of opposite sign which implies that *only one root* x_+ will be *positive*. The *frequency crossing set* Ω is given by:

$$\Omega = \{\omega_+\},$$

where $\omega_+ = \sqrt{x_+}$, with x_+ the only positive root of (62). In conclusion, since only one crossing root exists, the corresponding crossing direction is always *towards instability*. Thus, the system will be asymptotically stable for *all delays* $\tau \in [0, \tau_m)$, where the delay margin τ_m is given by the formula:

$$\tau_m = \frac{\angle\left(-\frac{P(j\omega_+)}{Q(j\omega_+)}\right)}{\omega_+}, \qquad \omega_+ \in \Omega. \tag{65}$$

The corresponding crossing direction is towards instability, and it is independent of the parameter values. Therefore, we can conclude that the corresponding linearization is *unstable* for all $\tau \geq \tau_m$. Finally, notice that the same stability conditions were obtained in [42], but using a slightly different argument. The particular form of the characteristic function proves that the essential parameters in defining the delay-induced instabilities are c, and d given by:

$$c = \frac{F(\bar{P}_L)}{V_L}, \qquad d = \frac{F'(\bar{P}_L)(\bar{P}_L - P_I)}{V_L},$$

that is the evaluation of the continuous controller function F and of its derivative F' relating the air flow in lungs to the delayed partial pressure in arterial blood, under the "standard" assumption that the partial pressure in arterial blood is greater than "outside". Please note that the partial pressures in arterial blood and in lungs are identical (equilibrium) [42]. In conclusion, the delay-induced stability is completely characterized by the value F'/F evaluated at $\bar{P}_L$, and compared to $1/(\bar{P}_L - P_I)$, which are directly defined by the corresponding equilibrium, and by the controller law. Some discussions on the linear and Hill controllers, and appropriate (physiological) assumptions on F can be found in [42].

Remark 3. The arguments and the analysis procedure above also apply to the second-order system with a single and constant considered by [2], system modeling arterial partial pressures of O_2, and CO_2 and a peripheral controller.

4.3 Distributed Delays in Some Immunological Response in Cell-to-Cell Spread Models

We derive now the stability regions of (8) in delay parameter space (T, τ). First, it is easy to see that in our case study, i.e. characteristic equation $\Delta(\lambda; T, \tau)$ given by (8), the Assumptions I.a and V.a are automatically satisfied, and the remaining reduces to the comparison of $P(0) = p$ with $Q(0) = q$.

Consider now $C_m = 2 \times 10^6/mL$, $k_I = 2 \times 10^{-6}/mL/day$, $k'_I = 1.5 \times 10^{-6}$, $d_I = 0.3/day$, $r_C = 0.68/day$ (see [25, 8, 39]). Then the characteristic equation of the HIV-1 model becomes:

$$(\lambda^2 + 0.368\lambda + 0.0204)(1 + \lambda T)^n + (-0.3\lambda + 0.1632)e^{-\lambda\tau} = 0. \tag{66}$$

In this case, Ω consists of one interval $(0, 3.995]$ and, for $n = 1$, the first three $(m = 0, 1, 2)$ crossing curves can be seen in Figure 4 (Up). Note that the system free of delay is stable and all the crossings are towards instability. Therefore, the system has *only one stability region* plotted in Figure 4 (Down).

A simple analysis show us that increasing the order of the delay kernel (n), although the shape of the crossing curves is the same, the slopes of the tangents to the crossing curves become bigger and the stability region becomes smaller. The behavior of the slopes can be seen also comparing Figures 4 (Up) and 5. The fact that the stability region gets smaller is illustrated in Figure 6.

4.4 Crossing Boundaries of Immune Response Dynamics in Chronic Myelogenous Leukemia

Recall the corresponding characteristic function of the linearized model:

$$p(\lambda; \sigma, \rho, \tilde{\tau}, \tilde{v}) := p_0(\lambda) + p_1(\lambda)e^{-\rho\lambda} + p_2(\lambda)e^{-\sigma\lambda} + p_3(\lambda)e^{-\tilde{\tau}\lambda} + p_4(\lambda)e^{-\tilde{v}\lambda}.$$

As mentioned in Section 2, $p_0(\lambda)$ has no purely imaginary roots. Then, we write

$$p(\lambda, \sigma, \rho, \tilde{\tau}, \tilde{v}) = p_0(\lambda)(1 + a_1(\lambda)e^{-\rho\lambda} + a_2(\lambda)e^{-\sigma\lambda} + a_3(\lambda)e^{-\tilde{\tau}\lambda} + a_4(\lambda)e^{-\tilde{v}\lambda}), \tag{67}$$

where $a_i(\lambda) = p_i(\lambda)/p_0(\lambda)$, for all $i = 1, \ldots, 4$. In conclusion, $p(\lambda; \sigma, \rho, \tilde{\tau}, \tilde{v})$ and the meromorphic function $a : \mathbb{C} \times \mathbb{R}_+^4 \mapsto \mathbb{C}$ given by:

$$a(\lambda; \sigma, \rho, \tilde{\tau}, \tilde{v}) := 1 + a_1(\lambda)e^{-\rho\lambda} + a_2(\lambda)e^{-\sigma\lambda} + a_3(\lambda)e^{-\tilde{\tau}\lambda} + a_4(\lambda)e^{-\tilde{v}\lambda}$$

share the same characteristic roots on the imaginary axis.

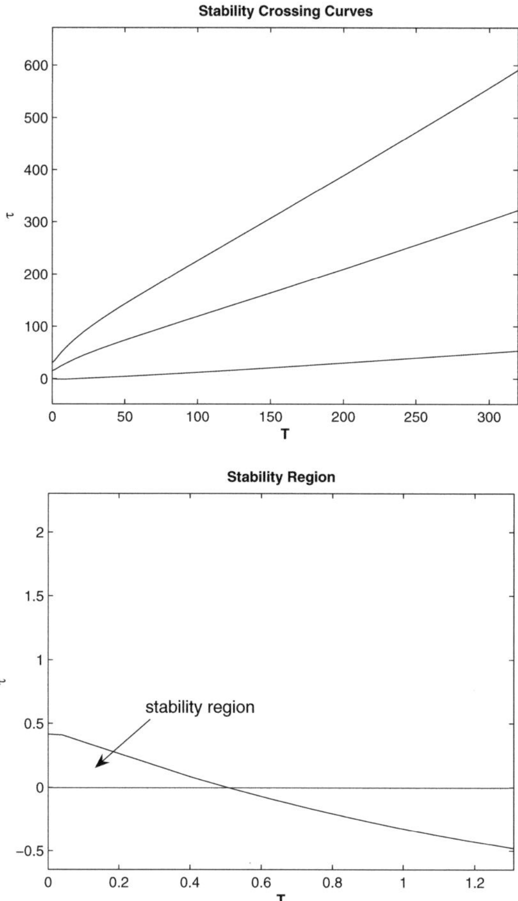

Fig. 4. (Up) Stability crossing curves in delay parameter space T, τ for the system given by (66); (Down) Stability region bounded by the first crossing curve for the system expressed by (66)

Stability Crossing Curves in the Delay-Parameter Space

In the sequel, we apply the methodology proposed in Section 3 (two delays case) for characterizing the stability crossing curves. One way to visualize the crossing surface of such a system is to fix two delays and determine the crossing curves for the other two delays. The particular form of the characteristic equation and of the delays scales suggests considering a (natural) delays *partition* in *small*, and *large* delays.

Associate now an *auxiliary system* with the characteristic function $p_{\rho,\sigma} : \mathbb{C} \times \mathbb{R}^2_+ \mapsto \mathbb{C}$ given by:

$$p_{\rho,\sigma}(\lambda;\ \rho,\sigma) := p_0(\lambda) + p_1(\lambda)e^{-\rho\lambda} + p_2(s)e^{-\sigma\lambda} = 0. \tag{68}$$

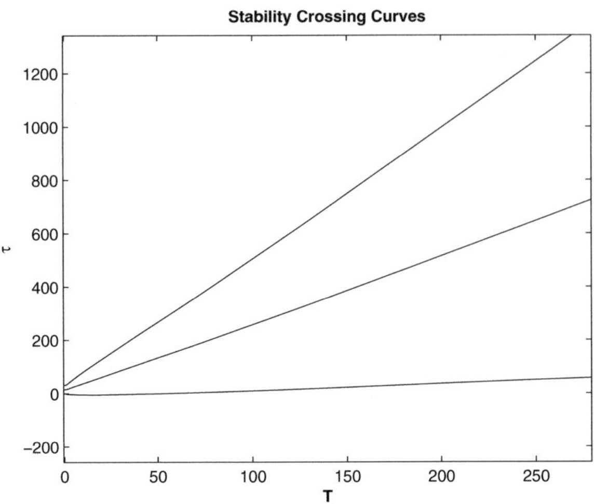

Fig. 5. Stability crossing curves in delay parameter space T, τ for the system given by (66) for $n = 2$

Using the *geometric idea* detailed in Section 3 (two delays case), we can easily characterize the *stability crossing curves* of $p_{\rho,\sigma}$ given by (68) in the delay-parameter space defined by the small delays ρ, and σ. Based on such a characterization, and using a standard continuity argument with respect to the delay parameters, we make the following assumption:

Assumption I.c. Let $\mathcal{I}_\rho \subset \mathbb{R}_+$, and $\mathcal{I}_\sigma \subset \mathbb{R}_+$ be some real intervals for which there exists some $\delta > 0$, such that

$$p_{\rho,\sigma}(\lambda;\ \rho, \sigma) \neq 0, \qquad \forall(\sigma, \rho) \in \mathcal{I}_\sigma \times \mathcal{I}_\rho$$

for all $\lambda \in \mathcal{V}_\delta$, where $\mathcal{V}_\delta$ is defined by:

$$\mathcal{V}_\delta = \{\lambda \in \mathbb{C}:\quad -\delta < \Re(\lambda) < \delta\}. \tag{69}$$

Such an assumption is not restrictive, and it simply describes some *regularity* condition for the original linearized model. Further comments and explanations can be found in [37]. All these results below are developed in analogy with the characterization for the two delays case.

Delay-Independent Type Results, and Weak T/C Interactions

Define $a_{\tilde{\tau},\tilde{v}}$ by:

$$a_{\tilde{\tau},\tilde{v}}(\lambda, \tilde{\tau}, \tilde{v}) = 1 + a_{\tilde{\tau}}(\lambda)e^{-\lambda\tilde{\tau}} + a_{\tilde{v}}(\lambda)e^{-\lambda\tilde{v}}, \tag{70}$$

where:

$$a_{\tilde{\tau}}(\lambda) = \frac{p_3(\lambda)}{p_{\rho,\sigma}(\lambda, \rho, \sigma)}, \quad a_{\tilde{v}}(\lambda) = \frac{p_4(\lambda)}{p_{\rho,\sigma}(\lambda, \rho, \sigma)},$$

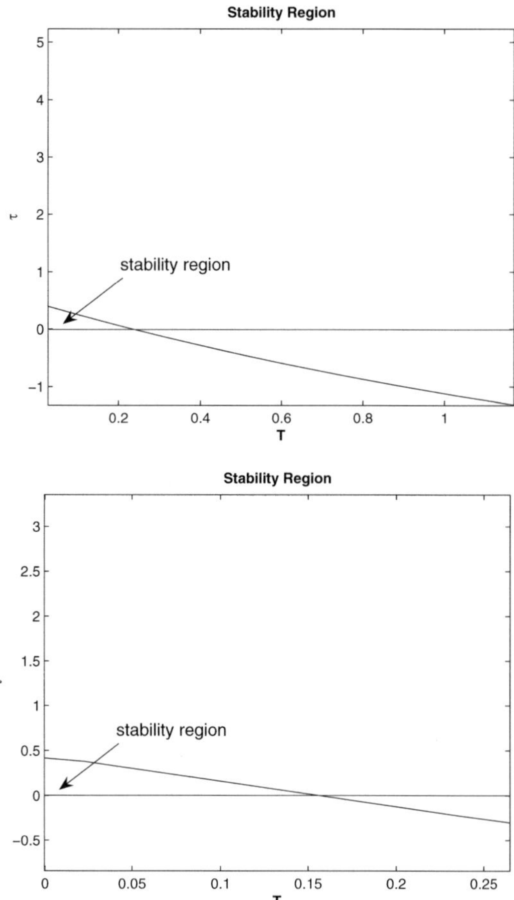

Fig. 6. Stability region bounded by the first crossing curve for the system expressed by (66) for Up:$n = 2$, Down: $n = 3$

for all $(\sigma, \rho) \in \mathcal{I}_\sigma \times \mathcal{I}_\rho$. With the notations, and the results above, we have the following result (see [37] for more details):

Proposition 13. *Assume that the auxiliary system given by the characteristic equation (68) satisfies Assumption I.c, and that $a_{\tilde{\tau}, \tilde{v}}(0;\, 0, 0) \neq 0$, where $a_{\tilde{\tau}, \tilde{v}}$ is defined by (70).*

Then the following statements are equivalent:

(a) If the auxiliary system (68) is stable for some pair $(\rho_0, \sigma_0) \in \mathcal{I}_\rho \times \mathcal{I}_\sigma$, and if the linearized system free of delays ($\sigma = \rho = \tilde{\tau} = \tilde{v} \equiv 0$) is stable, then the system (11) is stable for all pairs $(\tilde{\tau}, \tilde{v}) \in \mathbb{R}_+ \times \mathbb{R}_+$, and there does not exist any root crossing the imaginary axis when the delays $\tilde{\tau}$, and $\tilde{v}$ are increased in $\mathbb{R}_+$.

(b) The following frequency-sweeping test holds:

$$\frac{|C_0|\sqrt{\tilde{c}_1^2 + \omega^2}}{|p_{\rho,\sigma}(j\omega;\ \rho_0,\sigma_0)|} < \frac{1}{(2^N q_1 + q_2)p_1 k}, \quad \forall \omega > 0, \quad \forall(\rho_0,\sigma_0) \in \mathcal{I}_\rho \times \mathcal{I}_\sigma. \tag{71}$$

The same equivalence holds if the stability property is replaced by the instability of the system with a prescribed number of unstable roots.

Such a result simply describes the situation where the stability or instability of the linearized second-order model is *independent* of the delays $\tilde{v}$ and $\tilde{\tau}$. Since such delays describe the *T/C cell interactions*, such a situation can be simply called *weak T/C cell interaction*, and it defines a reduced probability of reactive interactions between anti-cancer T cells and cancer cells.

The frequency-sweeping test (71) can be used in defining a *measure* for characterizing the T/C interaction type in the following sense:

- the T/C interaction will be called *weak* if the probabilities (q_1, q_2), and the average number of cell division N verify the condition:

$$(2^N q_1 + q_2)p_1 k \left(\sup_{\omega\in\mathbb{R},(\rho_0,\sigma_0)\in\mathcal{I}_\rho\times\mathcal{I}_\sigma} \frac{|C_0|\sqrt{\tilde{c}_1^2 + \omega^2}}{|p_{\rho,\sigma}(j\omega;\ \rho_0,\sigma_0)|} \right) < 1. \tag{72}$$

The left-hand-side of (72) gives the corresponding *T/C interaction measure*.

In this context, it appears naturally that the average number N of cell division plays a central role in defining the T/C interaction character, since the quantity $2^N q_1 + q_2$ is an increasing function of N.

Strong T/C Interactions, and Identification of the Crossing Points

It is easy to see that the existence of crossing sets in the delay-parameter space $(\tilde{\tau}, \tilde{v})$ is related to the fact that the inequality (71) is not satisfied for all $\omega > 0$, or in other words that (q_1, q_2, N) do not satisfy the measure condition (72) for the T/C weak interaction. Such a situation is called a *strong T/C cells interaction*.

Based on the remarks above, it follows that we have a relatively simple condition for checking the T/C strong interaction character[1]:

Proposition 14. *The T/C interaction is strong if the equilibrium $(T_0, C_0) = \neq (0,0)$ satisfies the following inequality:*

$$\left|(d_T + k(1 - p_2)C_0)\,\mathrm{sign}(2C_0 - K) + \frac{\tilde{p}_1 k K d_T}{r}\frac{T_0}{|2C_0 - K|}\right| <$$

$$< |C_0|(2^N q_1 + q_2)p_1 k. \tag{73}$$

---

[1] Such a case simply parallelizes the situation $c < d$ and $a(0) > 1$ in the respiration second-order model in the previous section, for which the stability is always of *delay-dependent* type.

Let us characterize now the strong T/C interactions. Based on the discussions in Section 3, the condition that $a_{\tilde{\tau},\tilde{v}}$ defined by (70) has at least one root $j\omega_0$ on the imaginary axis is reduced geometrically to the condition that the "lengths" 1, $|\,a_v(j\omega_0)\,|$, and $|\,a_\mu(j\omega_0)\,|$ define a triangle. Thus, some simple computations lead to the following criterion for the *identification of the crossing points*:

Proposition 15. *Assume that the auxiliary system given by the characteristic equation (68) satisfies the Assumption I.c. Then each $\omega \in \mathbb{R}_+$ can be a solution of the characteristic function associated to the original linearized system for some $(\tilde{\tau}, \tilde{v}) \in \mathbb{R}_+^2$ if and only if:*

$$\frac{1}{(2^N q_1 + q_2)p_1 k} \leq \frac{|\,C_0\,|\,\sqrt{\tilde{c}_1^2 + \omega^2}}{|\,p_{\rho,\sigma}(j\omega;\ \rho,\sigma)\,|} \leq \frac{1}{|\,2^N q_1 - q_2\,|\,p_1 k}. \tag{74}$$

Then, the *crossing set* Ω will be defined by all $\omega \in \mathbb{R}_+$, for which the frequency condition (74) holds. Define now $\mathcal{T}_\omega$ as the set of all $(\tilde{\tau}, \tilde{v})$ such that $a_{\tilde{\tau},\tilde{v}}$ has one zero on the imaginary axis at $\lambda = j\omega$. It is easy to see that Ω is *bounded*.

In the stability analysis performed in the delay-parameter space in Section 3, the characterization of the stability crossing curves for a general system including *two delays* was based on an important property of the corresponding *stability crossing set*, namely the fact that it consisted of a finite number of intervals of finite length. In our case study, and in order to completely characterize the crossing curves in the parameter space defined by the large delays, it will be interesting to have a similar property. Introduce now the following assumption (see, for instance, [37]):

Assumption II.c. The following condition holds:

$$\frac{d}{d\omega}\left(\frac{|\,C_0\,|\,\sqrt{\tilde{c}_1^2 + \omega^2}}{|\,p_{\rho,\sigma}(j\omega;\ \rho,\sigma)\,|}\right) \neq 0,$$

whenever

$$\frac{|\,C_0\,|\,\sqrt{\tilde{c}_1^2 + \omega^2}}{|\,p_{\rho,\sigma}(j\omega;\ \rho,\sigma)\,|} = \frac{1}{|\,2^N q_1 \pm q_2\,|\,p_1 k}$$

for some $\omega \in \mathbb{R}_+$.

With the remarks and the assumptions above, we have:

Proposition 16. *Under Assumptions I.c and II.c, the crossing set Ω consists of a finite number of intervals of finite length.*

The proof can be found in [37], and is by contradiction, by assuming that inside a given frequency-interval $[0, \omega_M]$, such that $\Omega \subset [0, \omega_M]$, an appropriate real function cannot have an infinite number of roots.

Characterization of the Stability Crossing Curves

The next step is to characterize the crossing curves of the system (11), or, equivalently, all the crossing curves satisfying $a_{\tilde{\tau},\tilde{v}}(\lambda;\ \tilde{\tau},\tilde{v}) = 0$ for $\lambda = j\omega$, $\omega \in \Omega$. Using a similar classification of the crossing points as in Section 3, define now by $\Omega_k \subset \Omega$ some interval of crossing set Ω, and let $\mathcal{T}^k \subset \mathcal{T}$ be the corresponding stability crossing curves for some positive integer k. We have the following result:

Proposition 17. *Under the standard Assumption I.c, the stability crossing curves $\mathcal{T}^k$ corresponding to Ω_k must be an intersection of $\mathbb{R}^2_+$ with a series of curves belonging to one of the following categories:*

A. A series of closed curves;
B. A series of spiral-like curves with axes oriented either horizontally, vertically, or diagonally.
C. A series of open ended curves with both ends approaching ∞.

Tangent, Smoothness and Crossing Direction

All these properties follow straightforwardly from the stability crossing curves analysis proposed in the previous section.

Tangent, smoothness
More precisely, we have the following result:

Proposition 18. *Under standard assumptions including Assumption I.c, the curves in $\mathcal{T}^k$ are smooth everywhere except possibly at degenerate points corresponding to a root $\lambda = j\omega$ in any one of the following two cases:*

1. $\lambda = j\omega$ is a multiple solution of $a_{\tilde{\tau},\tilde{v}}(j\omega) = 0$.
2. ω is an end point, and

$$\frac{d}{d\omega}\left(\frac{|C_0|\sqrt{\tilde{c}_1^2+\omega^2}}{|p_{\rho,\sigma}(j\omega)|}\right) = 0.$$

Direction of crossing
Next, for a given $\omega \in \Omega_k$, introduce:

$$R_l = -\Re\left(\frac{1}{\lambda}\frac{\partial a_{\tilde{\tau},\tilde{v}}(\lambda;\ \tilde{\tau},\tilde{v}))}{\partial \tau_k}\right)_{\lambda=j\omega}, \tag{75}$$

$$I_l = -\Im\left(\frac{1}{\lambda}\frac{\partial a_{\tilde{\tau},\tilde{v}}(\lambda;\ \tilde{\tau},\tilde{v})}{\partial \tau_k}\right)_{\lambda=j\omega}, \tag{76}$$

for $l = 1,2$, and τ_1, τ_2 correspond to $\tilde{\tau}$, and $\tilde{v}$, respectively. This allows concluding:

Proposition 19. *Let $\omega \in \Omega_k$, but an end point, and $(\tilde{\tau}_0,\tilde{v}_0) \in \mathcal{T}^k$ such that $\lambda = j\omega$ is a simple solution of $a_{\tilde{\tau},\tilde{v}}(\lambda;\ \tilde{\tau}_0,\tilde{v}_0)) = 0$, and*

$$a_{\tilde{\tau},\tilde{v}}(j\omega';\ \tilde{\tau}_0,\tilde{v}_0) \neq 0,\ \text{for any}\ \omega' > 0,\ \omega' \neq \omega. \tag{77}$$

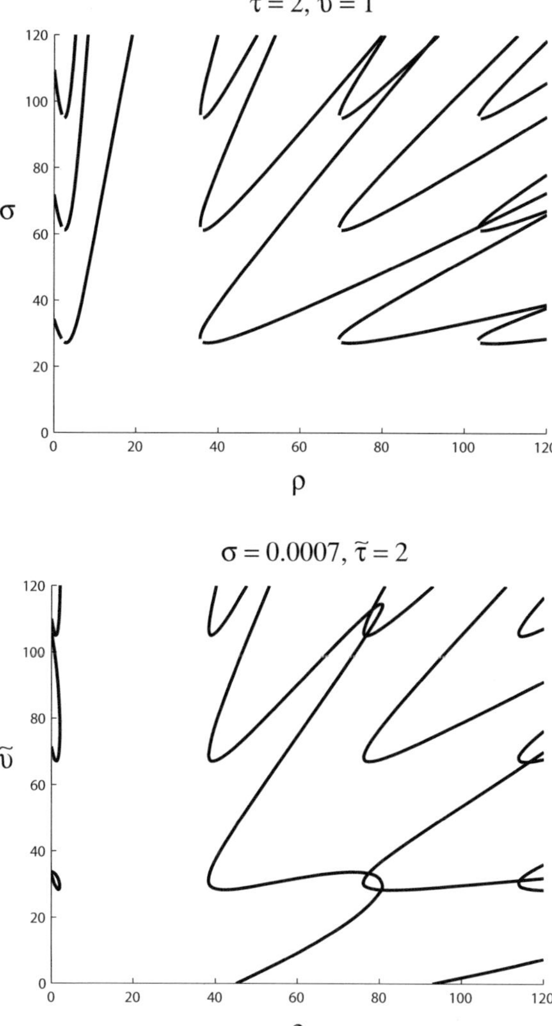

Fig. 7. Crossing curves for σ vs. ρ and $\tilde{\upsilon}$ vs. ρ for given values of $(\tilde{\tau}, \tilde{\upsilon})$ and $(\sigma, \tilde{\tau})$ respectively

Then as $(\tilde{\tau}, \tilde{\upsilon})$ moves from the region on the right to the region on the left of the corresponding curve in $\mathcal{T}^k$, a pair of solutions of $a_{\tilde{\tau},\tilde{\upsilon}}(\lambda; \; \tilde{\tau}, \tilde{\upsilon}) = 0$ crosses the imaginary axis to the right if

$$R_2 I_1 - R_1 I_2 > 0. \tag{78}$$

The crossing is in the opposite direction if the inequality is reversed.

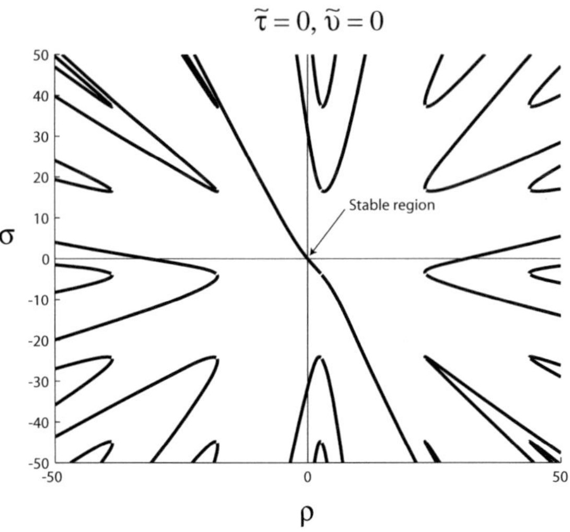

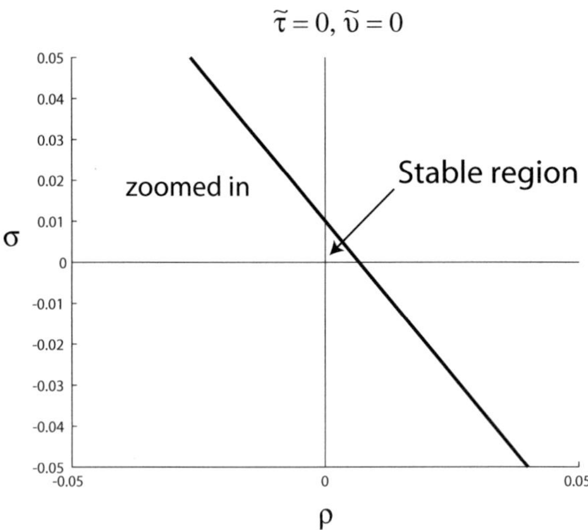

Fig. 8. Crossing curves for σ vs. ρ with $(\tau, \upsilon) = (0,0)$ for the linearization of (11) around fixed point III, $\left(\frac{c_1 - c_2 b_1/b}{c_3}, \frac{b_1}{b} \right)$

Illustrative Example

For our application in [11, 37], we estimated values of the parameters to be approximately

$$
\begin{aligned}
d_T &= 0.2, & p_1 &= 0.5, & \rho &= 0.0035, \\
r &= 0.2, & p_2 &= 0.5, & \sigma &= 0.0007, \\
k &= 1, & q_1 &= 0.5, & \tilde{\tau} &= 2.0035, \\
N &= 2, & q_2 &= 0.5, & \tilde{\upsilon} &= 1.0035. \\
K &= 200, & \tilde{p}_1 &= 0.5,
\end{aligned}
\tag{79}
$$

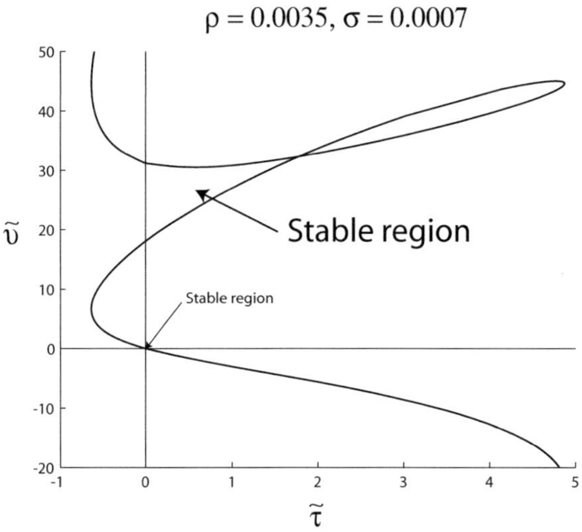

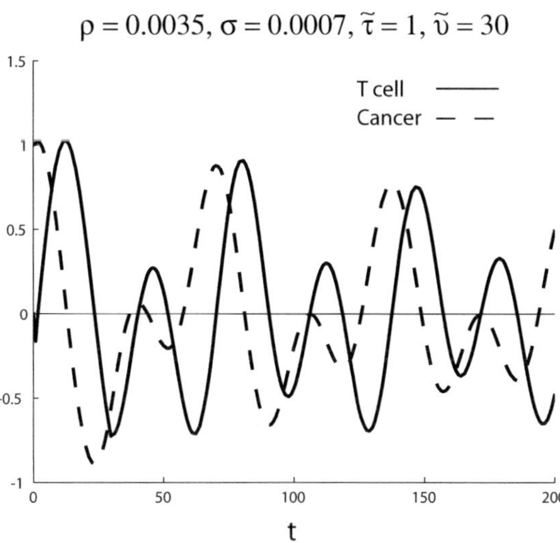

Fig. 9. (above) Crossing curves for v vs. τ with $(\rho, \sigma) = (0.0035, 0.0007)$. (below) Solution of the corresponding linearized system around the fixed point III.

Hence, b_1, c_1, and b are of order 1 or 0.1, while $c_1/c_2 = K$, the carrying capacity of the cancer population, is around 200. As already mentioned in the previous section, we have three *fixed points*. Consider first the case *free of delays*. In such a case, these fixed points have the following properties:

(a) The fixed-point I $(T_0, C_0) = (0, 0)$ is a saddle point.
(b) The fixed point II $(T_0, C_0) = (0, c_1/c_2)$ is unstable.
(c) The fixed point III $(T_0, C_0) = \left(\frac{c_1 - c_2 b_1/b}{c_3}, \frac{b_1}{b} \right)$ is stable.

Similarly to the previous examples, we focus on the third fixed point. As previously discussed, one way to visualize the crossing surface of the characteristic function p is to fix two delays and determine the crossing curves for the other two delays. This procedure is demonstrated in Figure 7.

In this case, any pairwise combination of ρ, σ, or τ gives open curves such as the ones shown on the upper-side of Figure 7. Any pair containing v leads to spiral-like curves such as the ones shown on the down-side of Figure 7. For the choice of parameters in (79), the fixed point III is stable in the un-delayed case, so there exists a stability region for sufficiently small delays. However, this region is very small and disappears quickly as the delays are increased. Figure 8 shows the crossing curves for σ vs. ρ, when $(\tau, v) = (0,0)$ for the linearization of (11) around the fixed point III, $\left(\frac{c_1 - c_2 b_1 / b}{c_3}, \frac{b_1}{b} \right)$.

In particular, the delays for T cell division, $\tilde{\tau}$, and recovery from a cytotoxic process, $\tilde{v}$, are about 2 and 1 days, respectively, so fixed point III is unstable. However, for low values of ρ and σ, we find another stable region in $(\tilde{\tau}, \tilde{v})$-space away from the origin. For $\rho = 0.0035$ and $\sigma = 0.0007$, the crossing curves for $\tilde{\tau}$ and $\tilde{v}$ are shown in Figure 9(up). A stable solution is shown in Figure 9(down).

In this region, the values for ρ and σ are 5 and 1 minutes as estimated in (79). The delay $\tilde{\tau}$, corresponding to $N = 2$ cell divisions, is about 1 day, which is a little fast, but still reasonable. On the other hand, the delay $\tilde{v}$, corresponding to the turn around time for T cell recovery after cytotoxic responses, is around 20 to 30 days, which is far longer than the expected 1 day turn around time. Discussion and further interpretations on the stability regions by changing delay parameters but also some non-delay parameters have been considered by [37]. For the sake of brevity, we will not present such arguments here. However, we point out that for most parameters, the stability region around the origin is very small. The parameters that influence the size of the stability region the most are the kinetic coefficient k and the T cell death rate d_T.

5 Concluding Remarks, Notes and References

This chapter addressed the stability analysis of various models and systems encountered in biosciences. We proposed several simple approaches (geometric ideas combined with appropriate frequency-sweeping tests) for characterizing the effects induced by the delay presence on the (asymptotic) stability of the corresponding systems. We considered constant (single and multiple, not necessarily rationally-dependent) as well as distributed delays (gamma-distribution with a gap). Several models have been used as illustrative examples: human respiration, HIV-1 cell-to-cell spread models, and immune dynamics in chronic myelogenous leukemia. The methodology considered in this chapter can be found in extent in [36, 16] (frequency-sweeping tests, and related characterizations), [31, 33] (distributed delays and related applications in engineering area), [17] (two delays case). The adaptation of the geometric approach in [17] to the four delays stability analysis of immune dynamics in chronic myelogenous leukemia can be found in [37]. Finally, a more general framework for treating all the sensitivity problems mentioned here can be found in [29].

Acknowledgements

The work of C.-I. MORĂRESCU was (partially) supported through a European Community Marie Curie Fellowship and in the framework of the CTS, contract number: HPMT-CT-2001-00278. The work of S.-I. NICULESCU and K. GU was partially funded by the CNRS-US Grant: *"Delays in interconnected dynamical systems: Analysis, and applications"* (2005-2007). The work of W. MICHIELS was supported by the Fund for Scientific Research - Flanders (Belgium), the Belgian Programme on Interuniversity Poles of Attraction, initiated by the Belgian State, Prime Minister's Office for Science, Technology and Culture (IAP P6), project DYSCO: "Dynamical Systems, Control and Optimization" and the Centre of Excellence on Optimization in Engineering of KU Leuven. Finally, the work of W. MICHIELS and S.-I. NICULESCU was partially funded by the H. CURIEN French-Belgium Bilateral Cooperation Program: *"Distributed delays in dynamical systems: analysis and applications"* (2006-2007).

References

1. Baker, C.T.H., Bocharov, G.A., and Rihan, F.A.: *A report on the use of delay differential equations in numerical modelling in the biosciences*. Numerical Analysis Report No. 343, Manchester Centre for Computational Mathematics, Manchester, UK (1999).
2. Batzel, J.J. and Tran, H.T.: Stability of the human respiratory control system I. Analysis of a two-dimensional delay state-space model. *J. Math. Biol.*, **41** (2000) 45-79.
3. Beretta, E. and Kuang, Y.: Geometric stability switch criteria in delay differential systems with delay dependent parameters. *SIAM J. Math. Anal.* **33** (2002) 1144-1165.
4. Bocharov, G.A. and Rihan, F.A.: Numerical modelling in biosciences using delay differential equations. *J. Comput. Applied Math.* **125** (2000) 183-199.
5. Boese, F.G.: Stability with respect to the delay: On the paper of K.L. Cooke and P. van den Driessche. *J. Math. Anal.Appl.*, **228** (1998) 293-321.
6. Chao, D. L., Forrest, S., Davenport, M. P., Perelson, A. S.: Stochastic stage-structured modeling of the adaptive immune system. *Proc. of the Computational Systems Bioinformatics*, 2003.
7. Cooke, K. L. and van den Driessche, P.: On zeroes of some transcendental equations. in *Funkcialaj Ekvacioj* **29** (1986) 77-90.
8. Culshaw, R.V., Ruan, S. and Webb, G.: A mathematical model of cell-to-cell spread of HIV-1 that includes a time delay. *J. Math. Biol.* **46** (2003) 425-444.
9. Cushing, J.M.: Volterra integrodifferential equations in population dynamics. in *Mathematics of Biology* (M. IANNALLI, EDS. Ligouri Editore: Naples) (1981) 81-148.
10. Datko, R.: A procedure for determination of the exponential stability of certain differential-difference equations. *Quart. Appl. Math.* **36** (1978) 279-292.

11. DeConde, R., Kim, P. S., Levy, D., Lee, P. P.: Post-transplantation dynamics of the immune response to chronic myelogenous leukemia. *J. Theor. Biol.* **236** (2005) 39-59.

12. Diekmann, O., van Gils, S. A., Verduyn-Lunel, S. M. and Walther, H. -O.: *Delay equations, Functional-, Complex and Nonlinear Analysis* (Appl. Math. Sciences Series, **110**, Springer-Verlag, New York, 1995).

13. El'sgol'ts, L. E. and Norkin, S. B.: *Introduction to the theory and applications of differential equations with deviating arguments* (Mathematics in Science and Eng., **105**, Academic Press, New York, 1973).

14. Fu, P., Chen, J. and Niculescu, S.-I.: High-order analysis of critical stability properties of linear time-delay systems. in *Proc. 2007 American Contr. Conf.* (July 2007) New York, NY, USA (to be presented).

15. Gopalsamy, K.: *Stability and oscillations in delay differential equations of population dynamics* (Kluwer Academic Publishers, Math. Its Appl. Series, **74**, 1992).

16. Gu, K., Kharitonov, V.L. and Chen, J.: *Stability of time-delay systems* (Birkhauser: Boston, 2003).

17. Gu, K., Niculescu, S.-I., and Chen, J.: On stability of crossing curves for general systems with two delays. in *J. Math. Anal. Appl.*, **311** (2005) 231-253.

18. Hadeler, K.P.: Delay equations in biology. in *Functional differential equations and approximation of fixed points* (H. -O. PEITGEN and H. -O. WALTHER, Eds.), Lecture Notes Math., Springer-Verlag, Berlin **730** (1979) 136-157.

19. Hale, J. K., Infante, E. F. and Tsen, F. S. -P.: *Stability in linear delay equations.* J. Math. Anal. Appl., **105**. 533-555. (1985).

20. Hale, J. K. and Verduyn Lunel, S. M.: *Introduction to Functional Differential Equations* (Applied Math. Sciences, **99**, Springer-Verlag, New York, 1993).

21. Hoppensteadt, F.C. and Peskin, C.S.: *Mathematics in Medicine and Life Sciences* (Springer-Verlag: New York, TAM, vol. 10, 1992).

22. Kim, P.S., DeConde, R., Levy, D., and Lee, P.: Post-transplantation dynamics of the immune response to chronic myelogenous leukemia. in presented at *Int. Conf. on Diff. Eqs. and Appli. in Math. Biology*, Nanaimo, BC, Canada, July 2004.

23. Kim, P.S., Lee, P.P. and Levy, D.: Mini-transplants for chronic myelogenous leukemia: A modeling perspective. (this volume).

24. Kuang, Y.: *Delay differential equations with applications in population dynamics* (Academic Press, Boston, 1993).

25. Leonard, R., Zagury, D., Desports, I., Bernard, J., Zagury, J.-F., Gallo, R.C.: Cytopathic effect of human immunodeficiency virus in T4 cells is linked to the last stage of virus infection. *Proc. Natl. Acad. Sci. USA*, **85**, 3570-3574 (1988).

26. MacDonald, N.: Two delays may not destabilize although either delay can. *Math Biosciences*, **82** (1986) 127-140.

27. MacDonald, N.: An interference effect of independent delays. in *IEE Proc. Contr. Theory & Appl.*, Pt. **D 134** (1987) 38-42.

28. MacDonald, N.: *Biological delay systems: linear stability theory.* (Cambridge University Press, Cambridge, 1989).

29. Michiels, W. and Niculescu, S.-I.: *Stability stabilization of time-delay systems. An eigenvalue-based approach.* (SIAM: Philadelphia, 2007, to be published).

30. Mittler, J.E., Sulzer, B., Neumann, A.U. and Perelson, A.S.: Influence of delayed viral production on viral dynamics in HIV-1 infected patients. in *Math. Biosci.* **152** (1998) 143-163.

31. Morărescu, C.I.: *Qualitative analysis of distributed delay systems: Methodology and algorithms* (Ph.D. thesis, University of Bucharest/Université de Technologie de Compiègne, September 2006).

32. Morărescu, C.-I., Niculescu, S.-I. and Gu, K.: Remarks on the stability crossing curves of linear systems with distributed delay, *Conference on Differential & Difference Equations and Applications*, Melbourne, FL, USA, August 2005.
33. Morărescu, C.-I., Niculescu, S.-I. and Gu, K.: Stability crossing curves of shifted gamma-distributed delay systems. *Internal Note HeuDiaSyC'06* (submitted).
34. Morărescu, C.-I., Niculescu, S.-I. and Michiels, W.: Asymptotic stability of some distributed delay systems: An algebraic approach, *Int. J. Tomography & Statistics* (Special Issue on Control Applications of Optimization: Optimization methods, differential games, time-delay control games, economics, and management), **7** (2007) 128-134.
35. Murray, J.D.: *Mathematical Biology* (BioMath. **18**, Springer: Berlin, 2nd Edition, 1993).
36. Niculescu, S. -I.: *Delay effects on stability: A robust control approach* (Springer-Verlag: Heidelberg, Germany, LNCIS, vol. **269**, May 2001).
37. Niculescu, S.-I., Kim, P.S., Gu, K. and Levy, D.: Stability crossing boundaries of delay systems modeling immune dynamics in leukemia. *Internal Note HeuDiaSyC'06* (submitted).
38. Smith, H.L. and Waltman, P.: *The theory of the chemostat* (Cambridge University Press: Cambridge, 1994).
39. Spouge, J.L., Shrager, R.I. and Dimitrov, D.S.: HIV-1 infection kinetics in tissue cultures. i n *Math. Biosci.*, **138** (1996) 1-22.
40. Timischl; S.: *A global model for the cardiovascular and respiratory system* (Ph.D. Thesis, University of Graz, Austria, August 1998).
41. Toker, O. and Özbay, H.: Complexity issues in robust stability of linear delay-differential systems. *Math., Contr., Signals, Syst.*, **9** (1996) 386-400.
42. Vielle, B. and Chauvet, G.: Delay equation analysis of human respiratory stability. *Math. Biosciences* **152** (1998) 105-122.
43. Walton, K. and Marshall, J. E.: Direct method for TDS stability analysis. *IEE Proc.* **134** part **D** (1987) 101-107.

Part III

Analysis and Control Aspects

Modeling and Control of Anesthetic Pharmacodynamics

Carolyn Beck[1], Hui-Hung Lin[2], and Marc Bloom[3]

[1] University of Illinois at Urbana-Champaign
[2] National Cheng-Kung University
[3] New York University School of Medicine

1 Introduction

Engineering and control technology have played and continue to play a major role in medicine over the past half-century, from the invention of the pacemaker in 1950 and ventricular assist devices in the 1980's, to more recent advances incorporating robotic and image-guided surgery. Modeling and control of drug dosing in clinical pharmacology is one area of medicine in which mathematical modeling is used extensively, and hence is well-suited for applications of control design and analysis techniques. Obvious examples of potential application areas for control in clinical pharmacology include insulin delivery for control of diabetes, chemotherapy dosing and timing for treatment of HIV, and anesthetic dosing during surgery to optimally control sedation level and manage hemodynamic functions. The increasing use of computers in the operating room combined with the recent and ongoing development of non-invasive yet effective means of measuring a number of the goals of anesthesia, such as the *bispectral index* or BIS measure of sedation, real-time measurements of exhaled gas concentrations by spectroscopic methods, and the use of electromyographic methods to measure lack of movement, promise to make the incorporation of control techniques into the anesthetic delivery process imminent.

In this chapter, we focus on the development of multivariable modeling and control methods aimed at closed loop (or semi-closed loop) control of anesthesia delivery during surgery. This is not only a viable control goal by itself, but also serves as a platform for other clinical pharmacology control studies.

1.1 Anesthetic Pharmacodynamics and Control

During surgery, the attending anesthesiologist continuously adjusts the delivery of anesthetic agents given to the patient in an effort to maintain a consistent and adequate level of *anesthetic depth*, that is, adequate levels of *hypnosis*, or lack of consciousness; *analgesia*, or lack of pain perception and the resulting autonomous system effects (e.g., increased heart rate and blood pressure); and muscle relaxation or lack of movement. Simultaneously, the anesthesiologist

I. Queinnec et al. (Eds.): Bio. & Ctrl. Theory: Current Challenges, LNCIS 357, pp. 263–289, 2007.
springerlink.com   © Springer-Verlag Berlin Heidelberg 2007

maintains the ventilation parameters for the patient and monitors cardiovascular and respiratory functions such as heart rate (HR), blood pressure (BP), oxygen saturation and end-tidal (exhaled) carbon dioxide (CO_2) levels. Invasive montitoring is sometimes used by attending anesthesiologists to directly measure not only arterial blood pressure, but right-heart filling pressures (CVP), left-heart filling pressures (PCWP), and pulmonary arterial pressures. Cardiac output (CO) may be measured by thermodilation methods and then used to derive systemic vascular resistance (SVR), pulmonary vascular resistance (PVR) and a host of other cardiac performance measures. Additionally, intra-operative blood samples are often taken and used to observe gas concentrations, blood-sugar levels, electrolyte concentrations and coagulation parameters.

The determination of when a patient is properly anesthetized thus is made by the anesthesiologist based on knowledge and experience of individual drug dose-response effects and synergistic effects of various drug combinations, combined with the observation of a number of indicators of patient status, such as those noted above. Vital signs such as mean arterial pressure (MAP), HR, and exhaled gases are commonly used to monitor patient status, but measurements of these quantities do not provide adequate information on the patient's anesthetic depth, and additional indicators must be considered. The electroencephalogram (EEG) may be monitored as one measure of the depth of sedation, or hypnosis, however, difficulties have been encountered in quantifying hypnotic depth in terms of EEG measurements, including varying effects on EEG by different anesthetic agents [58]. At present, there does not exist amongst anesthesiologists a single widely accepted indicator for anesthetic adequacy, and in fact it is obvious that a single indicator will not suffice for describing adequate levels of the three main components of anesthesia. As a result, anesthesiologists perform the role of a multivariable feedback controller during surgery, observing multiple patient indicators while simultaneously adjusting and controlling dosing and delivery of a number of anesthetic agents as well as respiration system parameters, in order to guarantee the health and safety of the patient.

Closed-loop administration of anesthetics during surgery promises to provide a number of benefits, such as tailoring and minimizing the overall amount of anesthetics required for individuals, and allowing the anesthesiologist to focus on critical safety tasks as necessitated by surgical demands on the patient and events that are both expected and unexpected. The main advantages of implementing closed-loop drug delivery, most obviously would include reduced pharmaceutical costs, but would also include reduced recovery time, and improved long-term patient outcomes [1, 32, 31], However, in order to design and implement feedback control schemes, mathematical models of the patient/drug delivery system are required.

The standard modeling paradigm that has been commonly used to describe the relationships between anesthetic inputs and patient output indicators (or effects) is that of *compartment models*. Pharmacokinetic (PK) compartment models are widely used as a means of predicting the disposition of drug in the body, by modeling the simultaneous diffusion of drug through body tissues and the flow of drug in blood. Most drugs are characterized by models containing a *central compartment*, which typically has a drug concentration corresponding to that of the blood, and peripheral compartments that represent groupings of internal organs and tissues of the body. As well, a theoretical *effect compartment* may be included which typically consists of a nonlinear pharmacodynamic (PD) model plus a first order linear time invariant system that is used to reflect the time-lag in the patient response to anesthesia (see [28, 48, 43, 67, 29] for details). The resulting mathematical models are inherently single-input single-output (SISO) and consist of a system of ordinary differential equations plus a nonlinear function, representing the relations between the drug input function, the concentration of drug in the various compartments, and the effect of the drug on specific patient endpoints. Unfortunately, as these models are strictly SISO, they are incapable of capturing the effects of disturbances, drug synergies, and the interrelation among effects in the human body.

In our research we have addressed this shortcoming directly by focusing on (1) the development of control-relevant multivariable models to describe patient response to anesthetic agents, ventilation controls and external stimuli, and (2) the development and implementation of control strategies for which patient safety and postoperative outcomes are improved. Although closed-loop control of anesthesia delivery has been studied for over 50 years [12], prior efforts have all essentially been SISO, thus, many important issues for MIMO modeling and control remain open. In our work we have targeted some of these problems, on which we elaborate in the sequel. The remainder of this chapter is organized as follows. In Section 2, we present an overview of traditional modeling approaches, i.e., compartmental modeling, and outline the MIMO piecewise-linear approach we have introduced. Simulation results and comparisons from these modeling efforts are given. In Section 3, we provide an overview of our control design approach, which is based on linear parameter-varying (LPV) techniques. A discussion of future directions is given in Section 4.

2 Modeling

In this section we provide a more in-depth discussion of compartment modeling. We then introduce our approach to modeling the response to anesthesia, which uses system identification-based piecewise-linear models. We focus primarily on our contributions, but provide some review of other work in the literature that motivates the primary problems of interest.

2.1 Compartment Modeling

As noted earlier, the standard modeling paradigm that has been used to describe the relationships between anesthetic inputs and patient outputs such as blood pressure or heart rate is that of compartment models. Combining both pharmacokinetic and pharmacodynamic properties, these models offer a continuous profile of a drug's concentration versus time in the body, which can then be related to the time course of the drug's effect by incorporating a theoretical effect compartment.

Pharmacokinetic (PK) compartment models are widely used as a means of predicting the disposition of drug in the body, by modeling the simultaneous diffusion of drug through body tissues and the flow of drug in the blood. That is, PK models address *"what the body does to the drug"*. PK models are derived from the consideration of mass balances of the drug distribution, relying on the anatomical relation of tissues to circulating blood for the derivation of models. A series of conceptual compartments which represents the body's tissues and organs are used to describe the interchange of drug within the body. At each compartment, drug flows into the compartment from external anesthetic inputs, or via transfer from other compartments, or both. Drug flows out of the compartment by transfer to other compartments and/or by elimination via metabolic clearance.

As an example, a three compartment model, which is shown in Figure 1, would be represented mathematically by the state equations

$$\frac{dx_1}{dt} = I + x_2 k_{21} + x_3 k_{31} - x_1 k_{10} - x_1 k_{12} - x_1 k_{13}$$

$$\frac{dx_2}{dt} = x_1 k_{12} - x_2 k_{21} \tag{1}$$

$$\frac{dx_3}{dt} = x_1 k_{13} - x_3 k_{31}$$

where x_i is the amount of drug in the ith compartment, k_{ij} is the distribution transfer rate from the ith compartment to the jth compartment, k_{10} is the clearance transfer rate out of the central compartment, and I represents the anesthetic infusion rate into the central compartment.

The peripheral compartment with volume V_2 may represent a *vessel rich* grouping of tissues and organs (e.g., brain, liver, etc.), which is assumed to reach steady-state equilibrium quickly. The peripheral compartment with volume V_3 may then correspond to fatty tissues and other *vessel poor* tissues and organs which equilibrate slowly (e.g., bones). It is assumed that the infused drug will mix immediately in the volume space of the central compartment (e.g., the heart for the intravenous injections and the lung for inhaled anesthetics). The drug concentration in the central compartment then decreases due to metabolic clearance and distribution to other compartments. The associated transfer rate constants, k_{ij}, are determined empirically.

Pharmacodynamic (PD) compartment models are used to describe the relationship between drug concentration and the observed or measured clinical effect. That is, PD models address *"what the drug does to the body"*. These models are typically given by static nonlinear functions, which are used to describe the equilibrium relationship between the drug concentration and drug effect. Measurements of specific effects and either measured or predicted values of concentration levels are required to construct the models. To be precise, concentrations at the actual effect site are needed to accurately construct PD models. However, effect site concentrations are not always readily available.

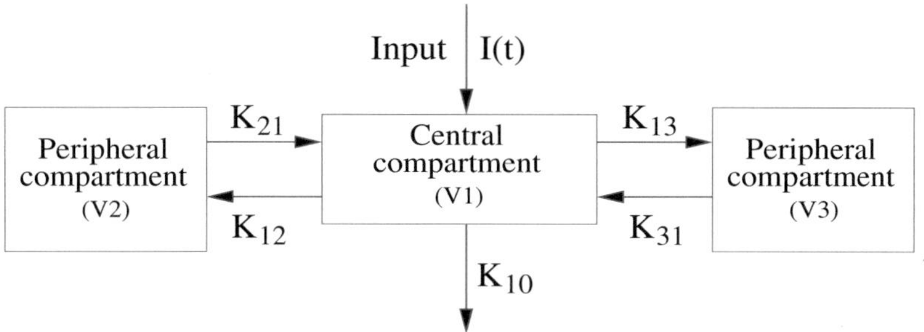

Fig. 1. Three-Compartment pharmacokinetic model

Alternatively, predicted concentrations in the blood captured from the PK models may be used in order to capture the response from the input source of anesthetics to the gross effect of the drug on specific endpoints. In this case, an *effect compartment* is attached to the central compartment of the PK model to capture the transport time to the effect site; see Figure 2. It is assumed that the amount of drug in the effect compartment itself is extremely small and has no impact on the pharmacokinetic steady-state condition. To model the time to equilibration of drug concentration at the effect site from the central compartment, an additional linear first order differential equation is derived, given by

$$\frac{dC_\epsilon}{dt} = K_{\epsilon 0}(C_c - C_\epsilon),\qquad(2)$$

where C_c and C_ϵ represent concentrations in the central and effect compartments, respectively, and $K_{\epsilon 0}$ is the associated equilibration rate constant.

Commonly used PD model structures are (1) fixed-effect models, (2) linear models, (3) log-linear models, (4) $E_{\max}$ models, and (5) sigmoidal $E_{\max}$ models [28, 48]. Sigmoidal $E_{\max}$ functions are those most frequently used to characterize the observed relationship between concentration and effect of anesthetic agents. The general form of this function is

$$E = \frac{E_{max} \cdot C^\gamma}{EC_{50}^\gamma + C^\gamma},\qquad(3)$$

where E is the measured effect (e.g, HR, or MAP), EC_{50} is the concentration of the drug at which half of the maximum achievable effect is observed in the patient, and the exponent γ is used to better fit observed data. Sigmoidal $E_{\max}$ functions have been used in our work and the previous work of others that is most relevant to our studies [52, 50, 42, 23, 24, 4, 37].

To evaluate the equilibration rate constant $K_{\epsilon 0}$ in (2), a plot of measured effect values versus blood plasma concentrations is considered. If the time course of the effect value is considered, a hysteresis-like response becomes evident. Since blood plasma only transports the drug and is not the intended site of action, the hysteresis loop represents the time delay from equilibrium drug concentration in the plasma to equilibrium drug concentration at the effect site to be attained.

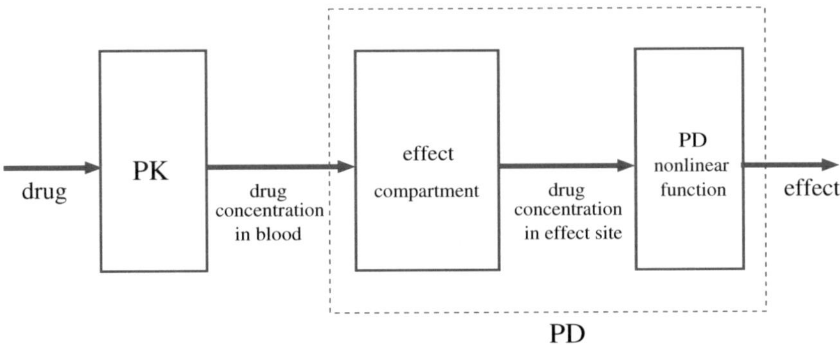

Fig. 2. Block diagram of a traditional pharmacokinetic-pharmacodynamic (PK-PD) model

This rate constant K_{e0} can thus be determined graphically by collapsing this hysteresis loop to minimize the circumscribed area [48]. The PD parameters EC_{50} and γ may also be determined in this manner. Alternatively, in [23, 24], a nested nonlinear regression algorithm to obtain optimal values of K_{e0}, EC_{50} and γ is proposed, which is essentially equivalent to the loop collapsing process. We have used the latter approach in our studies.

As noted before, the models one obtains from applying the traditional compartmental modeling framework are SISO. Such models have been used in a variety of research studies investigating the development of a controllers to automate anesthesia delivery. We outline a few of those studies here.

Closed-loop control has also been considered in a number of studies. For example, Schwilden and colleagues have used median frequencies from EEG power spectra as one measure of hypnotic effect to develop PK-PD model-based adaptive feedback control of propofol, methohexital, and alfentanil delivery during both clinical studies and for surgery [52, 50, 51]. A number of model-based closed-loop anesthesia control studies have been published by Gentilini and colleagues [21, 20, 23, 24, 25]. In [21], physiological models and rule-based controllers for the regulation of respiratory functions and MAP under administration of isoflurane are described. The application of model predictive control schemes to regulate MAP during delivery of isoflurane is investigated in [20]. In one of the most comprehensive control implementations completed to date, Gentilini *et al.* proposed a control scheme for the regulation of MAP and sedation level using PK-PD models for the anesthetic agent isoflurane [23, 24]. In this work, the authors discuss the design of a cascaded IMC controller to regulate the hypnotic affects of anesthesia via the bispectral index (BIS) level of the patient, and a three-observer-based state feedback controller to regulate MAP; the control designs are implemented in a loop-at-a-time manner. Mortier had also earlier considered control of sedation level via BIS monitoring in [42], where PK-PD model-based adaptive control of propofol is implemented in surgeries. More recently, Bailey *et al.* have completed adaptive and neural network based control designs for the regulation of unconciousness under administration of propofol. A relevant and thorough discussion of control system technology for clinical pharmacology, and references to additional studies are presented in a recent publication by Bailey and Haddad [4]. Although this by no means represents an exhaustive discussion of prior work on closed-loop control of anesthesia, it presents the work most closely related to our proposal. The main point to consider here is that all of the prior and ongoing work discussed above involves the use of SISO models, and SISO control design, for what is clearly (and widely agreed upon to be) a MIMO system.

In the next few subsections, we review our work on modeling and control of anesthetic pharmacodynamics.

2.2 Multivariable Piecewise-Linear Models

The modeling and control studies previously completed and published by the PIs were based on data collected from a clinical study of 10 volunteers, completed under the supervision of Dr. Bloom. For completeness, in the initial modeling stage we evaluated the use of standard MIMO linear state-space models to capture and predict human response to isoflurane and external stimuli [38]. Although these models captured the overall trend in patient response, the results, not surprisingly, did not provide an adequate fit due to the fact that the system being modelled is inherently non-linear. More specifically, one can view the response of the patient (or volunteer in our case) as transitioning from one set of dynamic behaviors to another as the course of anesthesia takes the subject from the alert state to the sedated state. We therefore proposed the use of linear switching systems, where the underlying subsystems are linear state-space models over which the volunteers' responses switch based on their sedative state. We briefly describe our data before discussing our methods and the simulation results we have obtained.

Input and Output Data

The original (IRB-approved) study was designed to define the relation between clinical evaluation of the state of conciousness, explicit recall, drug concentrations and BIS effects of the anesthetic agent isoflurane when administered alone to healthy volunteers under controlled conditions. Additionally, a series of external stimuli, or disturbances, were applied to the volunteers throughout the administration of anesthesia. These stimuli included: laryngeal mask insertion and removal, performed when the volunteer was considered completely sedated; evoked potential evaluations involving the application of short electrical stimulation signals to the wrist of the volunteer at a period of every 3 seconds and up to 100 μA and 100 V amplitude; and *alertness evaluations* which included yelling at, shaking, and squeezing the trapezius muscle of the volunteer. (See [27] for complete detail of the clinical protocols).

Time-synchronized measured volunteer outputs included BIS levels, MAP and HR. Note that BIS values range from $0 - 100$, where a BIS value near 100 corresponds to a completely alert state, a BIS value around 60 corresponds to a moderate hypnotic state, a BIS value around 40 corresponds to a deep hypnotic state, and very low BIS values are referred to as characterizing a profound anesthetic state [30, 53]. For healthy individuals, normal ranges for MAP are between 70 and 110 mmHg, and the average resting HR for normal adults is around 70 beats per minute [11].

An example of a set of data taken from one subject during that study is shown in Figures 3 and 4. Note that we developed quantitative models of the stimuli applied to the volunteers during the study, hence in Figure 3 the plot with vertical axis labeled EP represents evoked potentials stimuli, that labeled LMA represents the laryngeal mask insertion process, and that labeled EVAL respresent the alertness evaluation tests. Similar quantitative models were adopted by Spreckelsen and Bromm [65], who modeled the stimuli involved in evoked potential evaluations by a unit impulse, and by Derighetti, *et al.*, who used step-like excitations to represent intubation [15]. Frei, *et al.*, also used step-like functions of random heights to model surgical stimuli [20]. The maximum amplitude of the external stimuli has been normalized to one, with relative weightings determined implicitly by our identification algorithms.

Further, note the distinct transitions, or the effective *switching* that has been attained between the BIS levels in the plot presented in Figure 4.

Piecewise-Linear Modeling

We have found that piecewise-linear models effectively capture the response to anesthesia. The constituent subsystems in the piecewise-linear models we have proposed have been constructed using subspace identification methods applied to the data described above. We have identified two linear state-space subsystems, denoted

$$S_1 \begin{cases} x^1(k+1) = A_1 x^1(k) + B_1 u(k) + w(k) \\ y(k) \quad\;\; = C_1 x^1(k) + D_1 u(k) + v(k) \end{cases}$$

and

$$S_2 \begin{cases} x^2(k+1) = A_2 x^2(k) + B_2 u(k) + w(k) \\ y(k) \quad\;\; = C_2 x^2(k) + D_2 u(k) + v(k) \end{cases}$$

which model the patient response in the awake and sedated states, respectively. In this framework, $u(k)$ and $y(k)$ represent the sampled input and output data described in the preceding section, $x(k)$ is the state vector, $w(k)$ and $v(k)$ are assumed to be white-noise processes, and A_i, B_i, C_i, and D_i, $i = 1, 2$, are constant real-valued matrices estimated by the subspace identification process.

Observed BIS values have been used to choose between one of two models for a patient's response to anesthesia and stimuli (i.e., the alert models, and the sedated models which include both moderate and deeply sedated states). Switching between these two models occurs based on a BIS threshold value of 70; this choice of switching value is physiologically motivated, as it was noted in [53] that approximately 50% of the population will be unconscious at a BIS value of 70. From the data we used, it is clear that this value always lies in the transition region from alert to sedate states. Upon switching from one subsystem, S_j, to the other subsystem, S_i the initial state for the subsystem S_i is

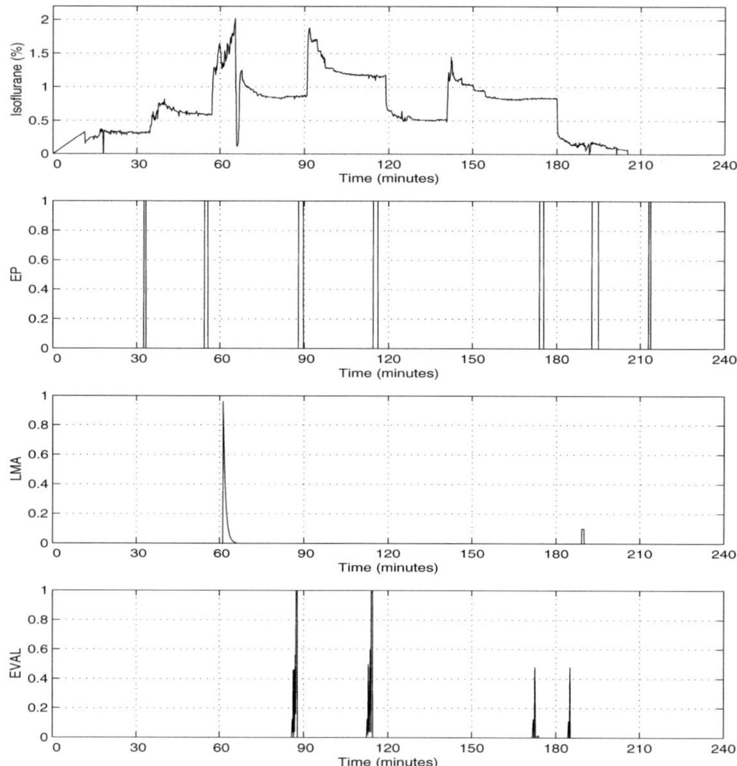

Fig. 3. Isoflurane and stimuli inputs versus time

calculated directly from the last output of S_j. We assume the outputs contain no feedforward information from the inputs, so that the D matrices are zero. Note that although the output will remain continuous, a jump in state values may result at the switching instant. Detailed descriptions of subspace algorithms may be found in [39, 64] and the references therein. A brief overview is given here.

Subspace Identification Methods

Subspace identification techniques have their origin in classical realization theory. However, a number of subspace identification algorithms have more been developed over the past 15 years that are numerically better conditioned than the classical approach. Basically, when the only information available is input-output data, impulse response matrices are difficult to estimate directly. The subspace identification algorithm employed in our work, the N4SID algorithm [59], avoids directly estimating the impulse response by utilizing projections of the data. These projections are computed directly from the given input-output data, and are then used both to estimate a low order state-space model

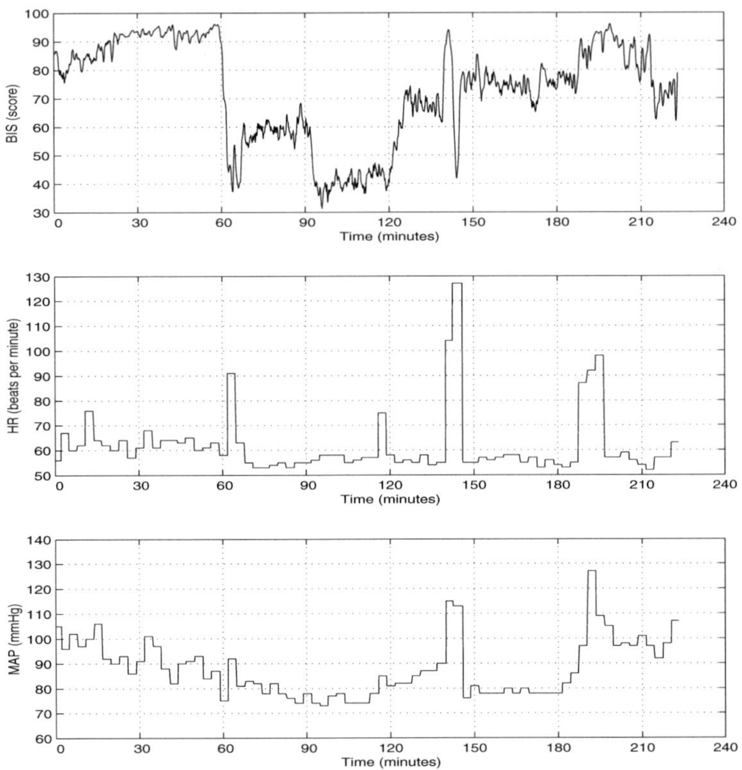

Fig. 4. BIS, HR and MAP outputs versus time

associated with the true system, as well as covariance matrices associated with measurement and process noise inherently present in the data.

Briefly, suppose we consider the state-space model structure

$$x(k+1) = Ax(k) + Bu(k) + w(k)$$
$$y(k) = Cx(k) + Du(k) + v(k),$$
(4)

where, as before, u is our input vector, y is our output vector, x is the state vector, w and v are assumed to be white noise processes that are not necessarily independent, and A, B, C and D are constant real-valued matrices. These state equations can alternatively be written in the following matrix notation:

$$Y(k) = \theta\phi(k) + e(k)$$
(5)

where

$$Y(k) = \begin{bmatrix} x(k+1) \\ y(k) \end{bmatrix}, \; \theta = \begin{bmatrix} A & B \\ C & D \end{bmatrix}, \; \phi(k) = \begin{bmatrix} x(k) \\ u(k) \end{bmatrix}, \; \text{and } e(k) = \begin{bmatrix} w(k) \\ v(k) \end{bmatrix}.$$

Clearly, if the sequence of state vectors $x(k)$ is known in addition to the measured input-output sequences $u(k)$ and $y(k)$, then (5) sets up a standard linear regression from which the elements of θ can be directly estimated.

The projection-based subspace algorithms thus begin by considering the following matrix formed directly from output data

$$
Y = \begin{bmatrix}
y(1) & y(2) & \cdots & y(N) \\
y(2) & y(3) & \cdots & y(N+1) \\
\vdots & \vdots & \ddots & \vdots \\
y(r) & y(r+1) & \cdots & y(N+r)
\end{bmatrix}
$$

where it is assumed that the number of data points available is greater than $N + r$. Note that based on the relationship of the output $y(k)$ to the impulse response coefficients, the rank of the matrix Y can be directly related to the order n of a minimal state-space representation for the system under certain assumptions. Thus, a singular value decomposition of Y may be used to estimate the system order. The main step in the N4SID subspace algortihm is then to construct an orthogonal projection from the input data. This projection is used to eliminate the direct dependence of the $y(k)$ terms on $u(k)$. The resulting "projected" output terms can then be seen to be dependent only on the states $x(k)$ and the noise terms (consider (5) above). The states may thus be estimated at this point. Standard least-squares estimations are then applied to identify the state matrices comprising θ. A number of projection-based subspace algorithms, similar to the N4SID algorithm, have been developed; a few of these are described in [33, 63, 62].

2.3   PK-PD Models from Data

In order to complete a comparative analysis, we also modelled the dosing and related effects of isoflurane using the standard pharmacological approaches, i.e., PK-PD models. For the PK model, we used the mammillary compartment model identified by Yasuda, *et al* [68]; this model has been determined based on data collected from seven healthy male volunteers.

The Yasuda model is given by the system of equations

$$
\dot{x}_1 = I + x_2 k_{21} + x_3 k_{31} + x_4 k_{41} + x_5 k_{51} - x_1 k_{10} - x_1 k_{12} - x_1 k_{13} - x_1 k_{14} - x_1 k_{15}
$$
$$
\dot{x}_2 = x_1 k_{12} - x_2 k_{21} - x_2 k_{20}
$$
$$
\dot{x}_3 = x_1 k_{13} - x_3 k_{31}
$$
$$
\dot{x}_4 = x_1 k_{14} - x_4 k_{41}
$$
$$
\dot{x}_5 = x_1 k_{15} - x_5 k_{51}
$$

where I is the inspired concentration, x_i is the amount of drug in the ith compartment, k_{ij} is the transfer rate constant from the ith compartment to the jth compartment, and k_{10} and k_{20} are the drug elimination or clearance rate for the first and the second compartments, respectively. Numerical values for the constants k_{ij} and k_{i0} have been determined empirically, as described in [68]. As isoflurane is a volatile anesthetic, the central, or first compartment represents the lungs. Note that one benefit of using volatile anesthetic agents is that *exhaled* end-tidal gas concentrations, which are easily measured, correlate well with anesthetic partial pressure at the intended site of action for the healthy volunteers. The second compartment represents the vessel-rich group (including the brain, heart and liver), and the third compartment represents muscle tissue. The fourth and fifth compartments represent fatty tissues, with the fourth compartment specifically used to represent the layer of fat receiving anesthesia from adjacent (vessel-rich) organs via intertissue diffusion. Note that in this model elimination of anesthetics occurs at two sites, namely the central compartment and the second compartment, due to the fact that volatile anesthetics are eliminated from the body mainly by exhalation and metabolism occuring via the liver. We evaluated predicted results from the Yasuda model versus measured responses for all volunteers in our data set and found the average MSE to be %3.0.

SISO effect compartment models resulting from utilizing the more standard PD compartment concepts have also been identified, in which the only input used is the anesthetic agent. Note that the PD models are individualized and not mean-based (as is the Yasuda PK model). That is, individual PD models are constructed for each patient for each effect. To model the BIS response to the isoflurane anesthetic the specific $E_{\max}$ function we have used is

$$BIS = BIS_0 \left(1 - \frac{C^\gamma}{EC_{50}^\gamma + C^\gamma} \right),$$

where BIS_0 represents the base point, or no-drug, patient BIS value (maximum of 100).

In order to find the parameters EC_{50} and γ, the previously described nested two stage nonlinear optimization algorithm based on the nonlinear regression algorithm proposed by Genitilini, *et. al*, in [24, 23], has been used. This algorithm is outlined below. Estimated values of $K_{\epsilon 0}$, γ and EC_{50} forming the BIS effect compartment models for all volunteers have been determined; these values are comparable to those found in [24] and [44], given the protocols of the respective studies. A more detailed discussion is given in [37].

PD Nonlinear Regression Algorithm

(1) The effect site concentration C_ϵ is computed for a given rate constant $K_{\epsilon 0}$.

(2) A least squares error minimization criterion is then used to obtain optimal EC_{50} and γ values, as well as the resulting optimal least square error term ε_{opt}^2, for the given $K_{\epsilon 0}$.

(3) Steps (1) and (2) are repeated to determine an optimal $K_{\epsilon 0}$ value, which is based on the minimal ε_{opt}^2 value evaluated with respect to a range of $K_{\epsilon 0}$ values. $K_{\epsilon 0}$ is constrained to be between 0 and 5.

Note that construction of PD models of the MAP and HR responses to isoflurane were also attempted. However we did not consider these results directly in

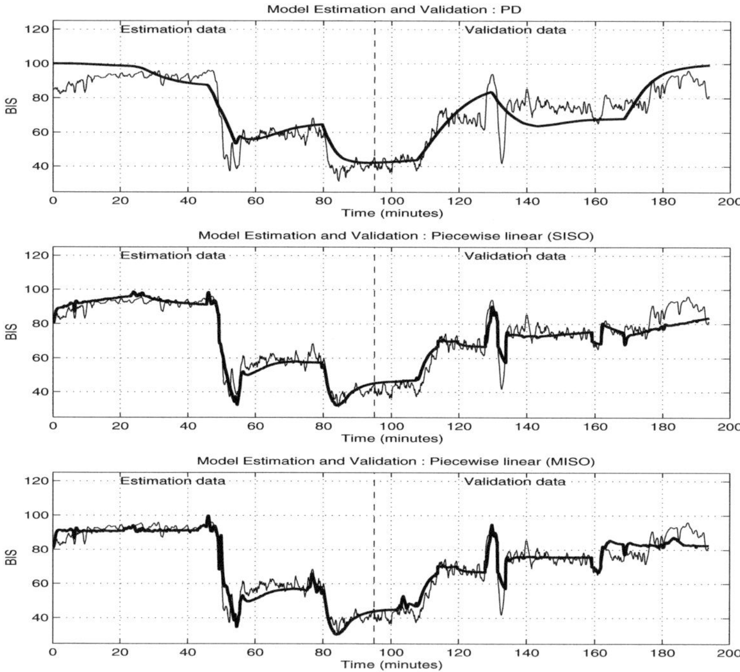

Fig. 5. Clinical data (thin line) and simulated responses (thick line) for Patient 1

our comparative analyses for the following reasons. First, note that in computing the optimal K_{e0} value in the first stage, the drug concentration in the blood versus the drug effect should graphically present itself as a hysteresis loop, as previously discussed. The optimal K_{e0} value is determined when the hysteresis loop collapses. However, using the clinical data available to us, a hysteresis loop could not be generated for HR and MAP effects. Additionally prior studies and discussions have indicated that the overall trend of the heart rate response to isoflurane is neither consistent, nor clearly understood (see [20, 18, 14, 19]).

2.4   Model Simulation Results

We first constructed SISO piecewise-linear models to compare their effectiveness to the standard SISO PK-PD models. We then constructed multi-input piecewise-linear BIS response models (with isoflurane and the external stimuli inputs), and compared these to the PK-PD model responses also. Finally, we constructed MIMO piecewise-linear models capturing the MAP and HR responses. Examples of measured and predicted outputs for the piecewise-linear multi-input BIS response models are shown for one volunteer in Figure 5, along with SISO piecewise-linear and PK-PD responses. The thin solid line represents the measured data, and the thick solid line represents the simulated responses. Note the dashed verticle line in each of the plots; this line represents the point

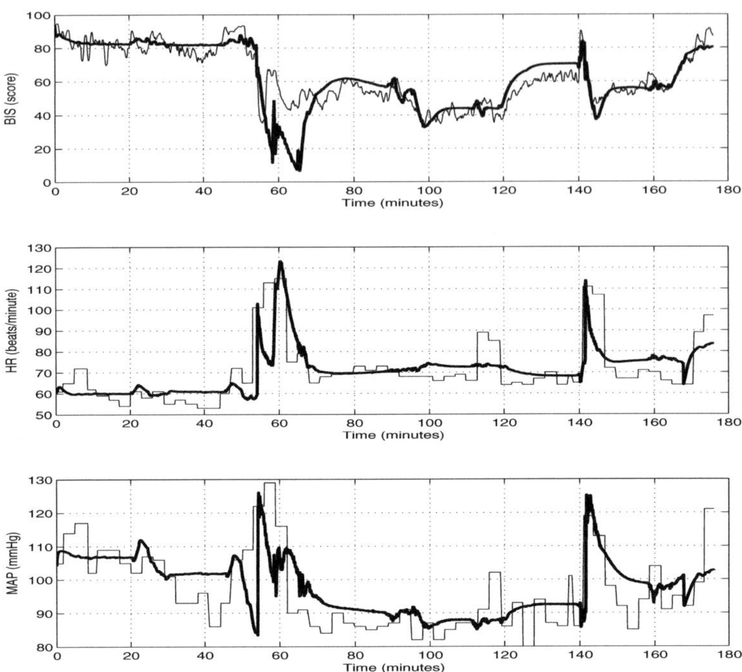

Fig. 6. Clinical data (thin line) and simulated responses (thick line) for Patient 3 using Patient 1 piecewise-linear model

at which we separated the data record for estimation and validation purposes, for the piecewise-linear models. The entire data record is used to construct the PK-PD models. To compare the performance of the PK-PD and the piecewise-linear models we considered (1) predictive capabilities, (2) error signal means and variances over pooled data, (3) normalized errors for individual patient data sets, and (4) computational effort required for model estimation.

Based on direct observation of prediction/estimation capabilities, the piecewise-linear models give improved responses over the PK-PD models (see [36] for a complete set of simulation results). We also found that the piecewise-linear models for one individual provide reasonable *central* models, i.e., models applicable to a group of subjects, in the sense that simulated output responses obtained utilizing the input data set for one volunteer (e.g., patient 3) applied to the piecewise-linear model for another volunteer (e.g., patient 1) produces an acceptable fit to the output data for the first patient; for one example see Figure 6. An evaluation of all combinations of BIS piecewise-linear models and data sets have been completed. The results have indicated that the models for volunteers 1, 6, 7 and 8 provide quite reasonable fits to all other data sets, and the models for volunteers 2 and 5 provide reasonable fits for a subset of the remaining data sets.

Quantitative comparisons of the results from the PK-PD and both the SISO and MISO piecewise-linear modeling efforts has also been completed, with error signal means and variances evaluated. The mean over individual data sets of the error signal $e(t) = y(t) - \hat{y}(t)$ was first computed, and population means over all data sets compared via a t-test ($p < 0.05$). Using the full data set to estimate the PK-PD model parameters, values for the population mean and variance of the error signal are obtained. Comparing the PK-PD error mean and variance values to those from the MISO piecewise-linear models, the F-test indicated there is no significant difference between the two population variances (F test: $p < 0.05$). However, the t-test indicated the error mean from the PK-PD models is significantly larger than that from the piecewise-linear models (one-tailed paired t test: $p = 0.011$). Thus in both the SISO and MISO cases, the piecewise-linear models we propose outperform the PK-PD models.

Normalized error tabulations may be found in [36]. On average, the normalized error for the piecewise-linear models was 0.0944, and the normalized error for the PK-PD models was 0.1341. The piecewise-linear models resulted in a better fit to the data, as measured by normalized errors, even though the entire data set was required to estimate the PK-PD model parameters.

Multivariable MAP and HR Models

Multi-input multi-output piecewise-linear models for MAP and HR responses also were constructed using the N4SID algorithm, from which a good fit was obtained. Representative examples of these model responses are shown in Figure 2.4. Results for all data sets are provided in [36].

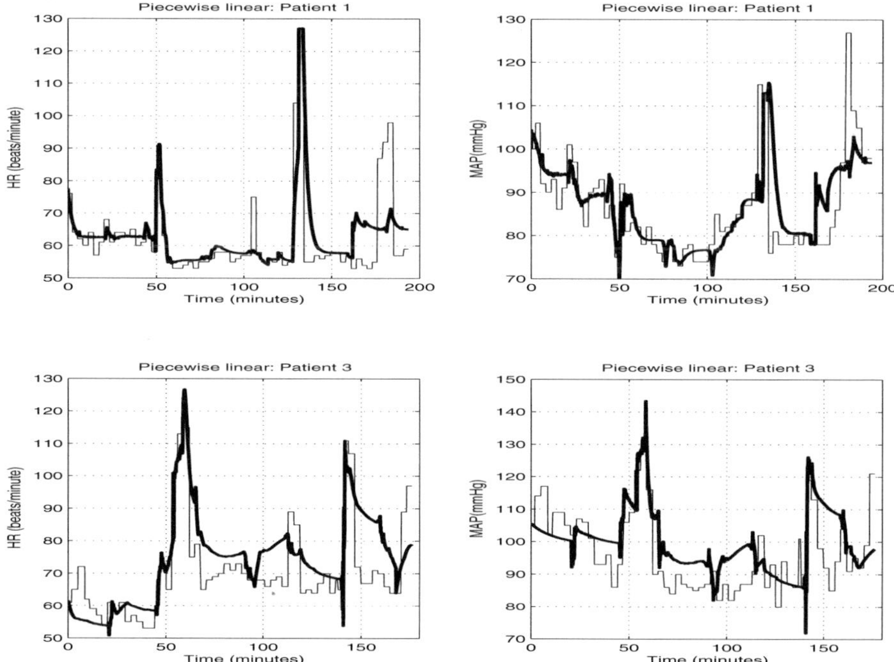

Computational efficiency was also evaluated for the PK-PD and piecewise-linear approaches described. For the input-to-BIS models, the PD models required approximately a factor of 5 times more computation time than the piecewise-linear models; for the input-to-MAP/HR models, the PD models required a factor of 5 to 20 times more computation time than did the piecewise-linear models.

3   Preliminary Control Designs

The use of piecewise-linear models in our work has been based, intuitively, on the course that the patient response to anesthesia takes from the alert state to the sedated state. Switching is based directly on knowledge of BIS values. However, current piecewise-linear synthesis strategies are based directly on *states*, i.e., on transitions from one partition in state space to another, and require either direct knowledge or estimation of both the current state and reasonable partitions of the state space. In the identification-based models we use, the states in our state-space models have no direct physical relevance, hence state-space partitions and transitions between such partitions are not practical to affect. Alternatively, we have considered linear parameter-varying (LPV) methods to complete the control designs [45], in which BIS is viewed as a measurable time-varying system parameter. The benefit of utilizing LPV models is that these models are able to capture the transition from alert to sedate and back in a continuous manner. In order to transform the piecewise-linear models into LPV models, a curve fitting process is used. The functions used essentially produce a "smooth piecewise-linear" type of response. A brief overview of LPV control methods is given here, followed by our control simulation results.

3.1   Linear Parameter-Varying Control Methods

The study of LPV systems has been largely motivated by the gain-scheduling philosophy [7, 46]. The state-space entries of LPV systems are linear fractional functions of one or more exogenous parameters, which are assumed to vary with time. These time-varying parameters are assumed to be bounded, and in most cases, have bounded measurable time derivative. The trajectories of these time-varying parameters are *a priori* unknown other than the range of variations [57, 2]. However, it is assumed these parameters may be measured or estimated upon operation of the system. Based on the *small-gain theorem* [69], a systematic LPV control design method has been developed [45, 2]. In this type of control design, state-space entries of both plant and controller are linear fractional functions of the time-varying parameters. The stability of the closed-loop system is then guaranteed using constant quadratic Lyapunov functions [2, 9]. A brief overview of the LPV control framework is now presented, which is excerpted from [2, 16].

Figure 7 shows the structure of the LPV control design paradigm. The upper two blocks in this figure, denoted M and Θ, represent a system whose dynamics are assumed to evolve relative to the set of time-varying parameters denoted by $\theta(t)$, as well as in time, that is, with state equations

$$x(t+1) = \tilde{A}(\theta(t))x(t) + \tilde{B}(\theta(t))\begin{bmatrix} w(t) \\ u(t) \end{bmatrix} \tag{6}$$

$$\begin{bmatrix} z(t) \\ y(t) \end{bmatrix} = \tilde{C}(\theta(t))x(t) + \tilde{D}(\theta(t))\begin{bmatrix} w(t) \\ u(t) \end{bmatrix}, \tag{7}$$

where $\tilde{A}(\cdot)$, $\tilde{B}(\cdot)$, $\tilde{C}(\cdot)$, $\tilde{D}(\cdot)$ are matrix-valued functions of appropriate dimensions, dependent on the vector-valued parameter function $\theta(t) = (\theta_1(t), \ldots, \theta_k(t))$. In the standard setup of [45, 2] we have that the parameter functions are known only to satisfy $-1 \leq \theta_i(t) \leq 1$, and that functions $\tilde{A}$, $\tilde{B}$, $\tilde{C}$, $\tilde{D}$ are linear fractional functions of the matrix $\Theta(t) = \mathrm{diag}\left[\theta_1(t)I_{m_1}, \ldots, \theta_{p-1}(t)I_{m_{p-1}}\right]$, where the dimensions m_i are appropriately defined. The lower two blocks in this figure, denoted K and Θ_K, represent an $\mathcal{H}_\infty$ control synthesis constructed to satisfy stability and performance specifications over the range of variations assumed for the θ_i; such controllers also have dynamics that evolve with respect to the set of time-varying parameters $\theta(t)$, with state equations

$$x_K(t+1) = \tilde{A}_K(\theta(t))x_K(t) + \tilde{B}_K(\theta(t))u_K(t) \tag{8}$$
$$y_K(t) = \tilde{C}_K(\theta(t))x_K(t) + \tilde{D}_K(\theta(t))u_K(t), \tag{9}$$

where in feedback $y_K(t) = u(t)$ and $u_K(t) = y(t)$ from (6).

The resulting closed-loop transfer function from disturbance input $w(t)$ to controlled output $z(t)$ is denoted by

$$T(M, K, \Theta) = (\Theta \star M) \star (K \star \Theta_K), \tag{10}$$

where for a general system realization $M = \begin{bmatrix} M_{11} & M_{12} \\ M_{21} & M_{22} \end{bmatrix}$ and $\Theta = diag[\delta_1(t)I_{n_1},$
$\ldots, \delta_p(t)I_{n_p}]$ we define

$$\Theta \star M = (M_{22} + M_{21}\Theta(I - M_{11}\Theta)^{-1}M_{12})$$

$$\text{and } M \star \Theta = M_{11} + M_{12}\Theta(I - M_{22}\Theta)^{-1}M_{21}.$$

The state dimensions and the dimensions of $\Theta(t)$ for the plant and controller indicate the dimensions of the constant realization matrices associated with the respective mappings $\Theta \star M$ and $K \star \Theta$. These matrices are partitioned as

$$\begin{bmatrix} A & B \\ C & D \end{bmatrix} \text{ and } \begin{bmatrix} A_K & B_K \\ C_K & D_K \end{bmatrix}$$

with the dimensions of A being $n \times n$ and of A_K being $n_K \times n_K$, where $\sum_{i=1}^{p} n_i = n$ and $\sum_{i=1}^{p} n_{Ki} = n_K$. The LPV H_∞ problem is formulated as finding a realization $M_K = \begin{bmatrix} A_K & B_K \\ C_K & D_K \end{bmatrix}$ such that the resulting LPV controller satisfies

- the closed-loop system given by (10) is internally stable for all assumed parameter variations

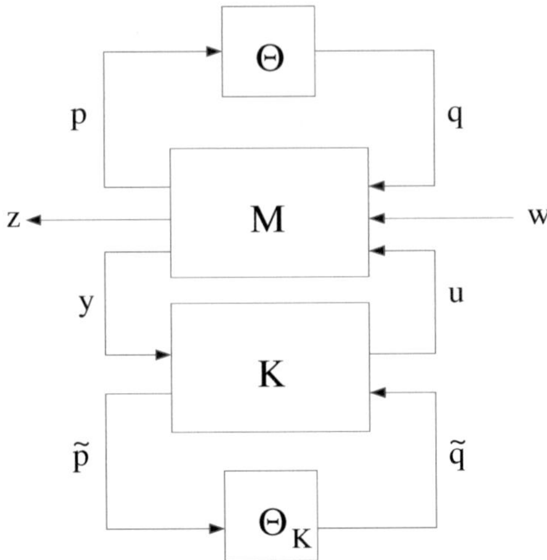

Fig. 7. Structure of LPV control

- the induced L_2 norm of the operator $T(M, K, \Theta)$ satisfies

$$\max_{\|\Theta\| \leq \frac{1}{\gamma}} \|T(M, K, \Theta)\|_\infty < \gamma. \tag{11}$$

Controllers satisfying our design objectives are found by applying algorithms resulting directly from the following theorem.

Theorem 1. *Consider an LPV plant defined by (6) and shown by the interconnection of upper blocks of Figure 7, where M is a proper discrete-time plant with minimal realization*

$$M(\lambda) = \begin{bmatrix} D_{11} & D_{12} \\ D_{21} & D_{22} \end{bmatrix} + \begin{bmatrix} C_1 \\ C_2 \end{bmatrix} (\lambda I - A)^{-1} \begin{bmatrix} B_1 & B_2 \end{bmatrix},$$

Θ is the parameter operator given by $\Theta = diag[\delta_1(t)I_{n_1}, \ldots, \delta_p(t)I_{n_p}]$, and λ is the usual shift or delay operator. Let L_Θ denote the set of scaling matrices defined by

$$L_\Theta = \{L \text{ positive definite} : L\Theta = \Theta L, \|\Theta\| < \frac{1}{\gamma}\} \subset R^{n \times n}.$$

Assume

- *(A, B_2, C_2) is stabilizable and detectable, and*
- *$D_{22} = 0$.*

Let $ImN_R = Ker\begin{bmatrix} B_2^T & D_{12}^T & 0 \end{bmatrix}$ and $ImN_S = Ker\begin{bmatrix} C_2 & D_{21} & 0 \end{bmatrix}$. Then, the parameter-dependent H_∞ control synthesis is solvable if there exist pairs of symmetric matrices (R, S) in $R^{n \times n}$ and (L, J) in $R^{n \times n}$ such that

$$N_R^T \begin{bmatrix} ARA^T - R & ARC_1^T & B_1 \\ C_1RA^T & C_1RC_1^T - \gamma J & D_{11} \\ B_1^T & D_{11}^T & -\gamma L \end{bmatrix} N_R < 0 \tag{12}$$

$$N_S^T \begin{bmatrix} ASA^T - S & A^TSB_1 & C_1^T \\ B_1^TSA & B_1^TSB_1 - \gamma L & D_{11}^T \\ C_1 & D_{11} & -\gamma J \end{bmatrix} N_R < 0 \tag{13}$$

$$\begin{bmatrix} R & I \\ I & S \end{bmatrix} \geq 0 \tag{14}$$

$$\begin{bmatrix} L & I \\ I & J \end{bmatrix} \geq 0 \tag{15}$$

Moreover, there exist γ-suboptimal controllers of order k if (12)-(15) hold for some (R, S, L, J) where R and S further satisfy

$$rank(I - RS) \leq k. \tag{16}$$

See [2] for a proof.

The equations (12)-(15) are linear matrix inequalities (LMIs) and can be solved for (R, S, L, J) [13]. When solutions R and S are found, an explicit controller can be computed by the algorithm described in [2, 16] which requires solving one additional LMI.

3.2   Control Design and Simulation Results

To transform the piecewise-linear models describing patient response to anesthetic and stimuli inputs into the LPV framework, the BIS signal is viewed as a measurable time-varying parameter. We applied a curve-fitting process, similar in approach to that described in [10], using the system realizations obtained for the patients in the awake (SA) and sedated (SS) states, $(A_{SA}, B_{SA}, C_{SA}, 0)$ and $(A_{SS}, B_{SS}, C_{SS}, 0)$ respectively. We defined the following relationships for the multi-input to BIS output models:

$$P_{BIS} \begin{cases} A(\delta) = \frac{\delta(\delta-1)}{2} A_{SA} + \frac{\delta+1}{2} A_{SS} \\ B(\delta) = \frac{\delta(\delta-1)}{2} B_{SA} + \frac{\delta+1}{2} B_{SS} \\ C(\delta) = \frac{\delta(\delta-1)}{2} C_{SA} + \frac{\delta+1}{2} C_{SS} \end{cases} \tag{17}$$

where the value of δ at any time t is based on patient measured BIS values. In particular we used

$$\delta(t) = 1 - \frac{2}{1 + \exp^{\eta*(70-BIS(t))}}. \tag{18}$$

This choice of the function for $\delta(t)$ is based on the fact that a BIS value around 100 corresponds to a completely alert state and a BIS value around 40 corresponds to a deep hypnotic state. As a result, $\delta = -1$ represents patients being

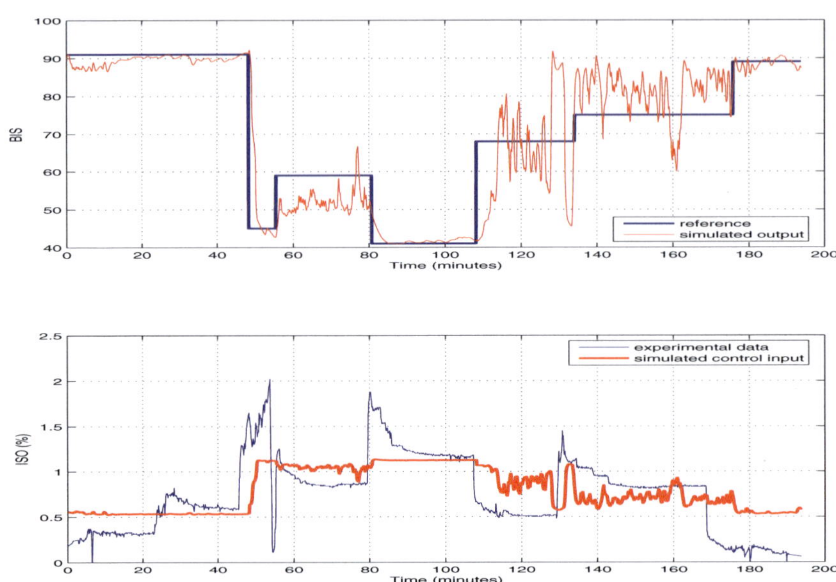

Fig. 8. Control simulation result of BIS reference tracking for patient 1

in a totally awake state. In contrast, $\delta = 1$ represents patients being in a deeply sedated state. The slope η was selected as 0.2, providing a "smooth piecewise-linear" type of response.

For the multi-input to MAP/HR output models, the LPV curve fitting equations are defined as

$$P_{MAP/HR} \begin{cases} A(\delta) = \frac{1-\delta}{2}A_{SA} + \frac{1+\delta}{2}A_{SS} \\ B(\delta) = \frac{1-\delta}{2}B_{SA} + \frac{1+\delta}{2}B_{SS} \\ C(\delta) = \frac{1-\delta}{2}C_{SA} + \frac{1+\delta}{2}C_{SS} \end{cases} \tag{19}$$

The resulting LPV systems have fairly high dimensions, for example for one of the patient models resulting dimensions are $n_1 = 36$ in the δ parameter and

Table 1. Normalized error ($\frac{\|\hat{y}-ref\|_2}{\|ref\|_2}$) for BIS tracking

Patient No.	Normalized error
1	0.1119
2	0.2161
3	0.1095
5	0.1194
6	0.1086
7	0.1309
8	0.2515

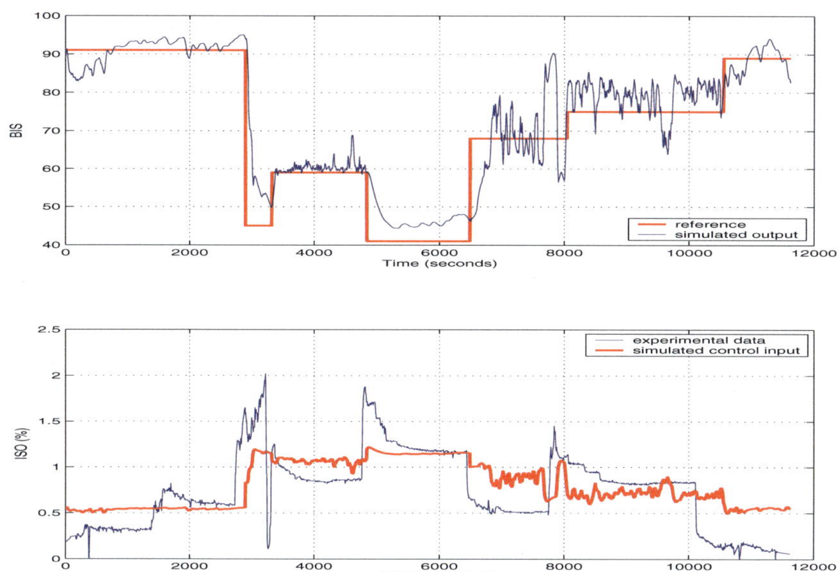

Fig. 9. Patient 3 controller with patient 1 model/data for BIS reference tracking

$n_2 = 4$ in the state. To facilitate control design, existing multidimensional model reduction methods [6] were first applied to the LPV models, resulting in reduced dimensions for example, of $n_{r_1} = 15$ in the δ parameter and $n_{r_2} = 4$ in the state. Model reduction errors are on the order of 5% or less.

Control Objectives

Desired reference trajectories for patient BIS values are determined by the attending anesthesiologist. The control goals are to track the BIS reference signal, attenuate the effect of disturbances, and for obvious safety reasons, maintain the HR and MAP within prespecified bounds. At the same time, the overall consumption of anesthetics should be minimized; these goals lead to a multi-objective control problem. In this discussion, we focus on tracking of the BIS reference trajectory when subjected to external disturbances (i.e., external stimuli). LPV control synthesis algorithms such as those proposed in [56, 2, 45, 8, 55, 9] have been implemented in MATLAB.

Control Results

Controllers were completed for each individual data set. Figure 8 shows one example of the control simulation results for the BIS reference tracking. In this 2-by-1 plot, the upper plot represents the desired BIS reference trajectory (thick line) and the control simulation output (thin line); the lower plot represents the isoflurane control input (thick line) (in %volume concentration) and the isoflurane measurements from the associated clinical trial (thin line).

Normalized error tabulations are given in Table 1, where *ref* represents the desired BIS reference trajectory and $\hat{y}$ represents the simulated control output. In this table, low values represent lower tracking error and thus better controller performance.

Controller cross-validations amongst patients for the BIS reference tracking also have been evaluated. Controllers generated from the estimated model for one subject have been implemented on other subjects' data and associated models. An example is shown in Figure 9. A number of such combinations were evaluated. The results indicate that the controller generated for patient 1 can be used as a controller for patients 2 and 6 with little degradation in performance, and similarly the controller generated for patient 3 can be used as a controller for patients 5, 7, and 8 without experiencing degradation in performance. This indicates that the use of *central* models for population subgroups (which could be based on covariate analyses, for example) combined with robust control techniques is an approach that should be investigated.

4   Conclusions

We have introduced and evaluated the use of multivariable piecewise-linear models for predicting and controlling patient response to the anesthetic agent isoflurane, and to external stimuli. These are the first explicit multivariable models of anesthetic pharmacodynamics, to our knowledge. Our results indicate that the proposed piecewise-linear models yield improved performance over traditional single-input, single-output PK-PD models, in comparisons made with respect to single output effects, and further the proposed models perform surprisingly well in cross-validation studies. Additionally, reduced computational effort is required to construct the models we propose, and the approach we have taken may be readily adapted to include additional or alternative inputs and outputs.

As a means of completing control synthesis for anesthetic pharmacodynamics, we have used $\mathcal{H}_\infty$ gain-scheduling techniques, also commonly known as linear parameter-varying methods. The primary advantage of using the LPV control methodology for this research, versus direct piecewise-linear design methods, is the straightforward manner in which the time-varying parameters (BIS) can be incorporated into the problem formulation.

A number of directions for additional studies are suggested by our results, including:

- **Pharmacodynamic Modeling of Multi-agent Anesthetic Strategies:** The construction of multi-drug MIMO piecewise-linear models, that is, an extension of the approach described herein to a more realistic surgical scenario that incorporates multiple anesthetic agents could be readily completed using the identification-based techniques described in this paper. In particular, the construction and evaluation of models capturing the synergistic effects of combining opioid and hypnotic agents would be relevant and timely.

- **Multi-objective Control Formulation:**
  The use of multi-output models directly allow for the design of multi-objective control strategies [66, 47, 3], for example, feedback control strategies aimed at optimizing dosing and administration of anesthetic agents affecting hypnosis and neuromuscular blockade levels while simultaneously regulating hemodynamic functions. Additionally, as the multivariable models we propose provide a means for disturbance modeling, robustness analysis and design objectives could be readily incorporated.
- **Population-specific Model Development:**
  The evaluation of whether *central* models exist for specific segments of the population, for example, examining to what extent individual covariates such as age extremes, gender, body weight, ethnicity or coloration [35] are correlated with how well models based on one patient's data set predict the responses exhibited by another patient's data sets, is warranted based on preliminary cross-validation results, as well as existing data in the anesthetic pharmacodynamics literature [54, 49, 17, 41].
- **Implementation of Identification-based Approaches:**
  An investigation into the feasibility of applying system identification methods that allow for updating models on-line (i.e., implementing recursive or semi-recursive identification procedures during surgery), combined with on-line model validation and real-time updating of robust control algorithms would be instructive for actual application of the control methods proposed herein. Model predictive control methods may also be considered in this context, extending those discussed in [22, 40].
- **Direct Identification of LPV Models:**
  System identification of state-space models has been used to form piecewise-linear models, which are then transformed into LPV systems. Algorithms for direct identification of LPV systems should be evaluated [34, 5, 26, 61, 60].
- **Extension of Piecewise-Linear Control Synthesis Methods:**
  The development of piecewise-linear control synthesis methods that rely on switching based on measurable system parameters, rather than on knowledge of state-space partitions and transitions, would provide a viable alternative controller design approach.

References

1. J.P. Abenstein and M.A. Warner. Anesthesia providers, patient outcomes, and costs. *Anesth. Analg.*, 82:1273–1283, 1996.
2. P. Apkarian and P. Gahinet. A convex characterization of gain-scheduled H-infinity controllers. *IEEE Trans. on Automatic Control*, 1995. Vol. 40, No. 5.
3. Pierre Apkarian, Paulo C. Pellanda, and Hoang Duong Tuan. Mixed H_2/H_∞ multi-channel linear parameter-varying control in discrete time. *System and Control Letters*, 41:333–346, 2000.
4. J. M. Bailey and W. M. Haddad. Drug dosing control in clinical pharmacology. *IEEE Control Systems Magazine*, pages 35–51, April 2005.
5. B. Bamieh and L. Giarré. Identification of linear paramerter-varying models. *International Journal of Robust and Nonlinear Control*, 12:841–853, 2002.

6. C. L Beck, J.C. Doyle, and K. Glover. Model reduction of multi-dimensional and uncertain systems. *IEEE Trans. on Automatic Control*, 41(10), 1996.

7. G. Becker. *Quadratic Stabiliby and Performance of Linear Parameter Dependent Systems*. PhD thesis, University of California at Berkeley, Dec. 1993.

8. G. Becker and A. Packard. Robust performance of linear parametrically varying systems using parametrically-depend linear feedback. *Systems and Control Letters*, 23(3):205–215, Sept. 1994.

9. G. S. Becker. *Quadratic Stability and Performance of Linear Parameter Dependent Systems*. PhD thesis, University of California at Berkeley, Nov. 1993.

10. P. Bendotti and C. L. Beck. On the role of LFT model reduction methods in robust controller synthesis for a pressurized water reactor. *IEEE Trans. on Control Systems Technology*, pages 248 – 257, 1999.

11. R. Berne, M. N. Levy, B. M. Koeppen, and B. A. Stanton. *Physiology*. Mosby, Inc., St. Louis, MI, 4th edition, 1998.

12. R.G. Bickford. Automatic electroencephalographic control of anesthesia (servo-anesthesia). *Electroenceph. Clin. Neurophys.*, 3:83–86, 1951.

13. S. P. Boyd, L. El Ghaoui, and E. Feron. *Linear Matrix Inequalities in System and Control Theory*. Philadelphia, Pa: SIAM, 1994.

14. O. Dale and B. R. Brown Jr. Clinical pharmacokinetics of the inhalational anaesthetics. *Clinical Pharmacokinetics*, 12:145–167, 1987.

15. M. Derighetti, C. W. Frei, M. Buob, A. M. Zbinden, and T. W. Schnider. Modeling the effect of surgical stimulation on mean arterial blood pressure. In *Proceedings of the 1997 19th Annual International Conference of the IEEE Engineering in Medicine and Biology Society*, volume 5, pages 2172–2175, 1997.

16. G. E. Dullerud and Fernando Paganini. *A Course in Robust Control Theory: A Convex Approach*. Springer-Verlag, New York, 2000.

17. T.D. Egan, B. Huizinga, S.K. Gupta, R.L. Jaarsma, R.J. Sperry, J.B. Yee, and K.T. Muir. Remifentanil pharmacokinetics in obese versus lean patients. *Anesthesiology*, 89(3):562–573, 1998.

18. E. I. Eger II. Isoflurance: A review. *Anesthesiology*, 55(5):559–576, Nov. 1981.

19. E. I. Eger II. Current and future perspectives on inhaled anesthetics. *Pharmacotherapy*, 18(5):895–910, 1998.

20. C. W. Frei, M. Derighetti, M. Morari, A. H. Glattfelder, and A. M. Zbinden. Improving regulation of mean arterial blood pressure during anesthesia through estimates of surgery effects. *IEEE Trans. on Biomed. Eng.*, 47(11):1456–1464, Nov. 2000.

21. C. W. Frei, A. Gentilini, M. Derighetti, A. H. Glattfelder, M. Morari, T. W. Schnider, and A. M. Zbinden. Automation in anesthesia. In *Proc. American Control Conference*, pages 1258–1263, 1999.

22. C. E. Garcia, D. M. Prett, and M. Morari. Model predictive control: Theory and practice - a survey. *Automatica*, 25(3):335–348, 1989.

23. A. Gentilini, C. W. Frei, A. H. Glattfelder, M. Morari, T. J. Sieber, R. Wymann, T. W. Schnider, and A. M. Zbinden. Multi-tasked closed loop control in anesthesia. *IEEE Eng. in Med. and Bio. Magazine*, 20(1):39–53, Jan. 2001.

24. A. Gentilini, M. Rossoni-Gerosa, C. Frei, R. Wymann, M. Morari, A. Zbinden, and T. Schnider. Modeling and closed loop control of hypnosis by means of bispectral index (BIS) with isoflurane. *IEEE Trans. on Biomed. Eng.*, 48(8):874–889, Aug. 2001.

25. A. Gentilini, C. Schaniel, M. Morari, C. Bieniok, R. Wymann, and T. Schnider. A new paradigm for the closed-loop intraoperative administration of analgesics in humans. *IEEE Trans. on Biomed. Eng.*, 49(4):289–299, April 2002.

26. L. Giarré, D. Bauso, P. Falugi, and B. Bamieh. LPV model identification for gain scheduling control: An application to the rotating stall and surge control problem. *Control Engineering Practice*, 14(4):351–361, 2006.

27. P. S. Glass, M. Bloom, L. Kearse, C. Rosow, P. Sebel, and P. Manberg. Bispectral analysis measures sedation and memory effects of propofol, midazolam, isoflurane, and alfentanil in healthy volunteers. *Anesthesiology*, 86(4):836–847, April 1997.

28. N. H. G. Holford and L. B. Sheiner. Understanding the dose-effect relationship: Clinical application of pharmacokinetic-pharmacodynamic models. *Clinical Pharmacokinetics*, 6:429–453, 1981.

29. J. R. Jacobs and E. A. Williams. Algorithm to control effect compartment drug concentrations in pharmocokinetic model-driven drug delivery. *IEEE Trans. on Biomed. Eng.*, 40(10):993–999, Oct. 1993.

30. J. O. Johnson and S. E. Kern. Bispectral electroencephalographic analysis for patient monitoring during anesthesia. *Advances in Anesthesia*, pages 61–81, 1999.

31. B. Kabon and A. Kurz. Optimal perioperative oxygen administration. *Current Opinion in Anesth.*, 19:11–18, 2006.

32. A. Kurz and D. Sessler. Opioid-induced bowel dysfunction: pathophysiology and potential new therapies. *Drugs*, 63(7):649–671, 2003.

33. W.E. Larimore. Canonical variate analysis in identification, filtering, and adaptive control. In *Proc. IEEE Conference on Decision and Control*, pages 596–604, 1990.

34. L. H. Lee and K. Poolla. Identification of linear parameter-varying systems. In *35th IEEE CDC*, pages 1545–1550, 1996.

35. E. B. Liem, C-M. Lin, M-I. Suleman, A.G. Doufas, and D. I. Sessler. Increased anesthetic requirement in subjects with naturally red hair. *Anesthesiology*, 2002. 97:A-77.

36. H. H. Lin. *Multivariable Modeling and Control of the Response to Anesthesia*. PhD thesis, University of Illinois at Urbana-Champaign, May 2006.

37. H. H. Lin, C. L. Beck, and M. J. Bloom. On the use of multivariable piecewise-linear models for predicting human response to anesthesia. *IEEE Trans. on Biomedical Engineering*, 51(11):1876–1887, 2004.

38. H. H. Lin, C. L. Beck, J. Nedumgottil, and M. Bloom. An investigation of multivariable models for human response to anesthesia. In *Proc. IEEE Conference on Decision and Control*, pages 867–868, 2001.

39. L. Ljung. *System Identification: Theory for the User*. Prentice Hall PTR, Upper Saddle River, NJ, 2nd edition, 1999.

40. D.Q. Mayne, J.B. Rawlings, C.V. Rao, and P.O.M. Scokaert. Contrained model predictive control: Stability and optimality. *Automatica*, 36(6):789–814, 2000.

41. C.F. Minto, T.W. Schnider, T.D. Egan, E.J. Youngs, H.J. Lemmens, P.L. Gambus, V. Billard, J.F. Hoke, K.H. Moore, D.J. Hermann, K.T. Muir, J.W. Mandema, and S.L. Shafer. Influence of age and gender of the pharmacokinetics and pharmacodynamics of remifentanil. I. model development. *Anesthesiology*, 86(1):10–23, 1997.

42. E. Mortier, M. Struys, T. De Smet, L. Versichelen, and G. Rolly. Closed-loop controlled administration of propofol using bispectral analysis. *Anesthesia*, 53(8):749–754, 1998.

43. KT. Olkkola and H. Schwilden. Adaptive closed-loop feedback control of vecuronium-induced neuromuscalar relaxation. *European Journal of Anaesthesiology*, 8(1):7–12, Jan. 1991.

44. E. Olofsen and A. Dahan. The dynamic relationship between end-tidal sevoflurane and isoflurane concentrations and bispectral index and spectral edge frequency of the electroencephalogram. *Anesthesiology*, 90(5):1345–1353, May 1999.

45. A. Packard. Gain scheduling via linear fraction transformations. *System and Control Letters*, 22(2):79–92, Feb. 1994.

46. W. J. Rugh and J. S. Shamma. Research on gain scheduling. *Automatica*, 36:1401–1425, 2000.

47. C. Scherer and P. Gahinet. Multiobjective output-feedback control via LMI optimization. *IEEE Transaction on Automatic Control*, 42(7):896–910, 1997.

48. T. W. Schnider, C. F. Minto, and D. R. Stanski. The effect compartment concept in pharmacodynamic modelling. *Anesthetic Pharmacology Review*, 2(3):204–213, 1994.

49. T.W. Schnider, C.F. Minto, S. L. Shafer, P.L. Gambus, C. Andresen, D.B. Goodale, and E.J. Youngs. The influence of age on propofol pharmacodynamics. *Anesthesiology*, 90(6):1502–1516, 1999.

50. H. Schwilden and H. Stoeckel. Effective therapeutic infusions produced by closed-loop feedback control of methohexital administration during total intravenous anesthesia with fentanyl. *Anesthesiology*, 73(2):225–229, Aug. 1990.

51. H. Schwilden and H. Stoeckel. Closed-loop feedback controlled administration of alfentanil during alfentanil-nitrous oxide anaesthesia. *British Journal of Anaesthesia*, pages 389–393, 1993.

52. H. Schwilden, H. Stoeckel, and J. Schüttler. Closed-loop feedback control of propofol anaesthesia by quantitative EEG analysis in humans. *British Journal of Anaesthesia*, pages 290–296, 1989.

53. P. S. Sebel. Bispectral monitoring technology: Clinical applications. In T.H. Stanley and T.D. Egan, editors, *Anesthesia for the New Millennium - Modern Anesthetic Clinical Pharmacology*, pages 99–104. Kluwer Academic Publishers, 1999.

54. S. L. Shafer. The pharmacology of anesthetic drugs in elderly patients. *Anesthesiol. Clin. North America*, 18(1):1–29, 2000.

55. S. Shahruz and S. Behtash. Design of controllers for linear parameter varying systems by the gain scheduling technique. *Journal of Mathematical Analysis and Applications*, 168(1):195–217, 1992.

56. J. Shamma and M. Athans. Guaranteed properties of gain scheduled control of linear parameter-varying plants. *Automatica*, 27(3):559–564, 1991.

57. J.S. Shamma and M. Athans. Gain scheduling: Potential hazards and possible remedies. *IEEE Control System Magazine*, 12:101–107, 1992.

58. D. R. Stanksi. *Monitoring Depth of Anesthesia*. Churchill Livingston, New York, 3rd edition, 1990. in *Anesthesia*, ED R. D. Miller.

59. P. Van Overschee and B. De Moor. N4SID: Subspace algorithms for identification of combined deterministic-stochastic systems. *Automatica*, 30(1):75–93, 1994.

60. V. Verdult, M. Lovera, and M. Verhaegen. Identification of linear parameter-varying state-space models with application to helicopter rotor dynamics. *International Journal of Control*, 77(13):1149–1159, 2004.

61. V. Verdult and M. Verhaegen. Subspace identification of multivariable linear paramerter-varying systems. *Automatica*, 38:805–814, 2002.

62. M. Verhaegen. Identification of the deterministic part of MIMO state space models given in innovations form from input-output data. *Automatica*, 30(1):61–74, 1994.

63. M. Viberg. Subspace-based methods for the identification of the linear time-invariant systems. *Automatica*, 31(12):1835–1852, 1995.

64. M. Viberg, B. Wahlberg, and B. Ottersten. Analysis of state space system identification methods based on instrumental variables and subspace fitting. *Automatica*, 33(9):1603–1616, 1997.

65. M. von Spreckelsen and B. Bromm. Estimation of single-evoked cerebral potentials by means of parametric modeling and kalman filtering. *IEEE Trans. on Biomed. Eng.*, 35(9):691–700, Sept. 1988.

66. B. Vroemen and B. de Jager. Multiobjective control: An overview. In *36th IEEE CDC*, pages 440–445, 1997.

67. D. R. Wada and D. S. Ward. The hybrid model: A new pharmacokinetic model for computer-controlled infusion pumps. *IEEE Trans. on Biomed. Eng.*, 41(2):134–142, Feb. 1994.

68. N. Yasuda, S. H. Lockhart, E. I. Eger II, R. B. Weiskopf, J. Liu, M. Laster, S. Taheri, and N. A. Peterson. Comparison of kinetics of sevoflurane and isoflurane in humans. *Anesthesia Analgesia*, pages 316–324, 1991.

69. G. Zames. On the input-output stability of nonlinear time-varying feedback systems, parts i and ii. *IEEE Trans. on Automatic Control*, pages 228–238 and 465–476, 1966.

Direct Adaptive Control of Nonnegative and Compartmental Dynamical Systems with Time Delay

VijaySekhar Chellaboina[1], Wassim M. Haddad[2], Jayanthy Ramakrishnan[3], and Qing Hui[4]

[1] Mechanical, Aerospace, and Biomedical Engineering, University of Tennessee, Knoxville, TN 37996-2210
chellaboina@utk.edu
[2] School of Aerospace Engineering, Georgia Institute of Technology, Atlanta, GA 30332-0150
wm.haddad@aerospace.gatech.edu
[3] Mechanical, Aerospace, and Biomedical Engineering, University of Tennessee, Knoxville, TN 37996-2210
jayanthy@utk.edu
[4] School of Aerospace Engineering, Georgia Institute of Technology, Atlanta, GA 30332-0150
qing_hui@ae.gatech.edu

Summary. In this paper, we develop a direct adaptive control framework for uncertain linear nonnegative and compartmental dynamical systems with unknown time delay. The specific focus of the paper is on compartmental pharmacokinetic models and their applications to drug delivery systems. In particular, we develop a Lyapunov-Krasovskii-based direct adaptive control framework for guaranteeing set-point regulation of the closed-loop system in the nonnegative orthant in the presence of unknown system time delay. The framework additionally guarantees nonnegativity of the control signal. Finally, we demonstrate the framework on a drug delivery model for general anesthesia involving system time delays.

1 Introduction

Nonnegative and compartmental models play a key role in the understanding of many processes in biological and medical sciences [1, 2, 3, 4, 5, 6, 7, 8, 9, 10, 11, 12, 13, 14]. Compartmental systems are modeled by interconnected subsystems (or compartments) which exchange variable nonnegative quantities of material with conservation laws describing transfer, accumulation, and elimination between compartments and the environment. In many compartmental pharmacokinetic system models, transfers between compartments are assumed to be instantaneous, that is, the model does not account for material in transit. Even though this is a valid assumption for certain biological and physiological systems, it is not true in general, especially in pharmacokinetic and pharmacodynamic models. For example, if a bolus (impulsive) dose of drug is injected,

I. Queinnec et al. (Eds.): Bio. & Ctrl. Theory: Current Challenges, LNCIS 357, pp. 291–316, 2007.
springerlink.com

there is a time lag before the drug is detected in the extracellular and intercellular space of an organ [15, 16, 11]. In this case, assuming instantaneous mass transfer between compartments will yield erroneous models. Hence, to accurately describe the distribution of pharmacological agents in the human body, it is necessary to include in any mathematical compartmental pharmacokinetic model some information of the past system states. In this case, the state of the system at any given time involves a *piece of trajectories* in the space of continuous functions defined on an interval in the nonnegative orthant. This of course leads to (infinite-dimensional) delay dynamical systems [17, 18, 19, 20].

In a recent paper [21], the authors present a direct adaptive control framework for set-point regulation of linear nonnegative and compartmental systems with applications to clinical pharmacology. In this paper, we extend the results of [21] to the case of nonnegative and compartmental dynamical systems with unknown system time delay. Specifically, we develop a Lyapunov-Krasovskii-based direct adaptive control framework for guaranteeing set-point regulation for linear uncertain nonnegative and compartmental dynamical systems with unknown time delay. The specific focus of the paper is on pharmacokinetic models and their applications to drug delivery systems. In particular, we develop direct adaptive controllers with nonnegative control inputs as well as adaptive controllers with the absence of such a restriction. Finally, we demonstrate the framework on a drug delivery model for general anesthesia that involves system time delays.

2 Mathematical Preliminaries

In this section we introduce notation, several definitions, and some key results concerning linear nonnegative dynamical systems with time delay [22, 23] that are necessary for developing the main results of this paper. Specifically, for $x \in \mathbb{R}^n$ we write $x \geq\geq 0$ (resp., $x >> 0$) to indicate that every component of x is nonnegative (resp., positive). In this case, we say that x is *nonnegative* or *positive*, respectively. Likewise, $A \in \mathbb{R}^{n \times m}$ is *nonnegative*[1] or *positive* if every entry of A is nonnegative or positive, respectively, which is written as $A \geq\geq 0$ or $A >> 0$, respectively. Furthermore, for $A \in \mathbb{R}^{n \times n}$ we write $A \geq 0$ (resp., $A > 0$) to indicate that A is a nonnegative-definite (resp., positive-definite) matrix. In addition, spec (A) denotes the spectrum of A, $(\cdot)^{\mathrm{T}}$ denotes transpose, $(\cdot)^{-1}$ denotes inverse, and $(\cdot)^{\#}$ denotes group generalized inverse. Let $\overline{\mathbb{R}}_+^n$ and $\mathbb{R}_+^n$ denote the nonnegative and positive orthants of $\mathbb{R}^n$, that is, if $x \in \mathbb{R}^n$, then $x \in \overline{\mathbb{R}}_+^n$ and $x \in \mathbb{R}_+^n$ are equivalent, respectively, to $x \geq\geq 0$ and $x >> 0$. Finally, let $\mathbf{e} \in \mathbb{R}^n$ denote the ones vector of order n, that is, $\mathbf{e} = [1, \ldots, 1]^{\mathrm{T}}$.

The following definition introduces the notion of a nonnegative (resp., positive) function.

Definition 1. *Let $T > 0$. A real function $u : [0, T] \to \mathbb{R}^n$ is a* nonnegative *(resp.,* positive*) function if $u(t) \geq\geq 0$ (resp., $u(t) >> 0$) on the interval $[0, T]$.*

---

[1] In this paper it is important to distinguish between a square nonnegative (resp., positive) matrix and a nonnegative-definite (resp., positive-definite) matrix.

The next definition introduces the notion of essentially nonnegative matrices and compartmental matrices.

Definition 2 ([24]). *Let $A \in \mathbb{R}^{n \times n}$. A is* essentially nonnegative *if $A_{(i,j)} \geq 0$, $i, j = 1, \ldots, n$, $i \neq j$. A is* compartmental *if A is essentially nonnegative and $A^{\mathrm{T}} \mathbf{e} \leq\leq 0$.*

The following results are necessary for the development of the main results of this paper.

Lemma 1 ([14]). *Let $A \in \mathbb{R}^{n \times n}$ be essentially nonnegative and assume there exists a vector $q \in \mathbb{R}_+^n$ such that $A^{\mathrm{T}} q \leq\leq 0$. Then A is semistable, that is, $\mathrm{Re}\,\lambda < 0$, or $\lambda = 0$ and λ is semisimple, where $\lambda \in \mathrm{spec}\,(A)$. Alternatively, A is Hurwitz if and only if $A^{\mathrm{T}} q << 0$.*

Corollary 1. *Let $A \in \mathbb{R}^{n \times n}$ be an essentially nonnegative matrix such that $A = A^{\mathrm{T}}$. If there exists $q \in \mathbb{R}_+^n$ such that $A^{\mathrm{T}} q \leq\leq 0$, then $A \leq 0$.*

Proof. The proof is a direct consequence of Lemma 1 by noting that if A is symmetric, then semistability implies that $A \leq 0$. $\qquad\square$

Lemma 2. *Let $X \in \mathbb{R}^{n \times n}$ and $Z \in \mathbb{R}^{m \times m}$ be such that $X = X^{\mathrm{T}}$ and $Z = Z^{\mathrm{T}}$, and let $Y \in \mathbb{R}^{n \times m}$ be such that $Y - Y Z^{\#} Z$. Then*

$$M \triangleq \begin{bmatrix} X & Y \\ Y^{\mathrm{T}} & Z \end{bmatrix} \leq 0 \tag{1}$$

if and only if $Z \leq 0$ and $X - Y Z^{\#} Y^{\mathrm{T}} \leq 0$.

Proof. Let $T = \begin{bmatrix} I_n & -Y Z^{\#} \\ 0 & I_m \end{bmatrix}$ and note that $\det T \neq 0$. Now, noting that $T M T^{\mathrm{T}} \leq 0$ if and only if $M \leq 0$, and

$$\begin{aligned} T M T^{\mathrm{T}} &= \begin{bmatrix} I_n & -Y Z^{\#} \\ 0 & I_m \end{bmatrix} \begin{bmatrix} X & Y \\ Y^{\mathrm{T}} & Z \end{bmatrix} \begin{bmatrix} I_n & 0 \\ -Z^{\#} Y^{\mathrm{T}} & I_m \end{bmatrix} \\ &= \begin{bmatrix} X - Y Z^{\#} Y^{\mathrm{T}} & 0 \\ 0 & Z \end{bmatrix} \\ &\leq 0, \end{aligned}$$

the result follows immediately. $\qquad\square$

Lemma 3. *Let $A \in \mathbb{R}^{n \times n}$ be an essentially nonnegative matrix and let $A_{\mathrm{d}i} \in \mathbb{R}^{n \times n}$, $i = 1, \ldots, p$, be nonnegative matrices such that $\hat{A} \mathbf{e} \leq\leq 0$ and $\hat{A}^{\mathrm{T}} \mathbf{e} \leq\leq 0$, where $\hat{A} \triangleq A + \sum_{i=1}^{p} A_{\mathrm{d}i}$. Then there exist diagonal matrices $Q_i \geq 0$, $i = 1, \ldots, p$, such that*

$$A^{\mathrm{T}} + A + \sum_{i=1}^{p} (Q_i + A_{\mathrm{d}i}^{\mathrm{T}} Q_i^{\#} A_{\mathrm{d}i}) \leq 0. \tag{2}$$

Proof. For each $i \in \{1, \ldots, p\}$, let Q_i be the diagonal matrix defined by

$$Q_{i(l,l)} \triangleq \sum_{m=1}^{n} A_{\mathrm{d}i(l,m)},$$

and note that $(A + \sum_{i=1}^{p} Q_i)\mathbf{e} = (A + \sum_{i=1}^{p} A_{\mathrm{d}i})\mathbf{e} \leq\leq 0$, $(A_{\mathrm{d}i} - Q_i)\mathbf{e} = 0$, and $Q_i Q_i^{\#} A_{\mathrm{d}i} = A_{\mathrm{d}i}$, $i = 1, \ldots, p$. Hence, $M\mathbf{e} \leq\leq 0$, where

$$M \triangleq \begin{bmatrix} A^{\mathrm{T}} + A + \sum_{i=1}^{p} Q_i & A_{\mathrm{d}1}^{\mathrm{T}} & A_{\mathrm{d}2}^{\mathrm{T}} & \cdots & A_{\mathrm{d}p}^{\mathrm{T}} \\ A_{\mathrm{d}1} & -Q_1 & 0 & \cdots & 0 \\ \vdots & \vdots & \vdots & \vdots & \vdots \\ A_{\mathrm{d}p} & 0 & 0 & \cdots & -Q_p \end{bmatrix}. \tag{3}$$

Now, it follows from Corollary 1 that $M \leq 0$ and since $Q_i Q_i^{\#} A_{\mathrm{d}i} = A_{\mathrm{d}i}$, $i = 1, \ldots, p$, it follows from Lemma 2 that $M \leq 0$ if and only if (2) holds. $\qquad\square$

Theorem 1. *Let $A \in \mathbb{R}^{n \times n}$ be an essentially nonnegative matrix and let $A_{\mathrm{d}i} \in \mathbb{R}^{n \times n}$, $i = 1, \ldots, p$, be nonnegative matrices such that $\hat{A} \triangleq A + \sum_{i=1}^{p} A_{\mathrm{d}i}$ is Hurwitz. Then there exist diagonal matrices P and $Q_i \in \mathbb{R}^{n \times n}$, with $P > 0$ and $Q_i \geq 0$, $i = 1, \ldots, p$, such that*

$$A^{\mathrm{T}}P + PA + \sum_{i=1}^{p}(Q_i + A_{\mathrm{d}i}^{\mathrm{T}}PQ_i^{\#}PA_{\mathrm{d}i}) < 0. \tag{4}$$

Proof. Since $\hat{A} \triangleq A + \sum_{i=1}^{p} A_{\mathrm{d}i}$ is essentially nonnegative and Hurwitz, it follows from Lemma 1 that there exists $l \in \mathbb{R}_+^n$, $l = [l_1, \ldots, l_n]^{\mathrm{T}}$, such that $\hat{A}^{\mathrm{T}}l << 0$, which implies that $\hat{A}^{\mathrm{T}}L\mathbf{e} << 0$, where $L = \mathrm{diag}[l_1, \ldots, l_n]$. This further implies that $L\hat{A}$ is compartmental and Hurwitz.

Since $L\hat{A}$ is compartmental and Hurwitz, it follows that $\hat{A}^{\mathrm{T}}L$ is essentially nonnegative and Hurwitz, and by Lemma 1, there exists $m \in \mathbb{R}_+^n$, $m = [m_1, \ldots, m_n]^{\mathrm{T}}$ such that $L\hat{A}m << 0$, which further implies that $L\hat{A}M\mathbf{e} << 0$, where $M = \mathrm{diag}[m_1, \ldots, m_n]$. Furthermore, since $\hat{A}^{\mathrm{T}}L\mathbf{e} << 0$, it follows that $M\hat{A}^{\mathrm{T}}L\mathbf{e} << 0$, and hence, $L\hat{A}M$ is compartmental and Hurwitz.

Next, define $\tilde{A} \triangleq LAM$ and $\tilde{A}_{\mathrm{d}i} \triangleq LA_{\mathrm{d}i}M$, $i = 1, \ldots, p$. By construction $\tilde{A}$ is an essentially nonnegative matrix and $\tilde{A}_{\mathrm{d}i}$, $i = 1, \ldots, p$, are nonnegative matrices. Hence, $\tilde{A} + \sum_{i=1}^{p} \tilde{A}_{\mathrm{d}i}$ is essentially nonnegative and Hurwitz since $\tilde{A} + \sum_{i=1}^{p} \tilde{A}_{\mathrm{d}i} = L\hat{A}M$. Now, since $(\tilde{A} + \sum_{i=1}^{p} \tilde{A}_{\mathrm{d}i})\mathbf{e} \leq\leq 0$ and $(\tilde{A} + \sum_{i=1}^{p} \tilde{A}_{\mathrm{d}i})^{\mathrm{T}}\mathbf{e} \leq\leq 0$, it follows from Lemma 3 that there exist diagonal matrices $\tilde{Q}_i$, $i = 1, \ldots, p$, such that

$$\tilde{A}^{\mathrm{T}} + \tilde{A} + \sum_{i=1}^{p}(\tilde{Q}_i + \tilde{A}_{\mathrm{d}i}^{\mathrm{T}}\tilde{Q}_i^{\#}\tilde{A}_{\mathrm{d}i}) \leq 0,$$

which further implies that

$$MA^{\mathrm{T}}L + LAM + \sum_{i=1}^{p}(\tilde{Q}_i + MA_{\mathrm{d}i}^{\mathrm{T}}L\tilde{Q}_i^{\#}LA_{\mathrm{d}i}M) \leq 0,$$

and hence,

$$A^{\mathrm{T}}LM^{-1} + M^{-1}LA + \sum_{i=1}^{p}(M^{-1}\tilde{Q}_i M^{-1} + A_{\mathrm{d}i}^{\mathrm{T}} L\tilde{Q}_i^{\#} LA_{\mathrm{d}i}) \leq 0. \tag{5}$$

Next, it can be shown that (5) holds with a strict inequality by noting that the proof remains valid by replacing A with $A + \varepsilon I_n$ for sufficiently small $\varepsilon > 0$. Now, (4) follows with $Q_i = M^{-1}\tilde{Q}_i M^{-1}$, $i = 1,\ldots,p$, and $P = LM^{-1}$. $\square$

In this paper, we consider a controlled linear time-delay dynamical system $\mathcal{G}$ of the form

$$\dot{x}(t) = Ax(t) + \sum_{i=1}^{p} A_{\mathrm{d}i}x(t - \tau_i) + Bu(t), \quad x(\theta) = \eta(\theta), \quad -\bar{\tau} \leq \theta \leq 0, \ t \geq 0, \tag{6}$$

where $x(t) \in \mathbb{R}^n$, $u(t) \in \mathbb{R}^m$, $t \geq 0$, $A \in \mathbb{R}^{n \times n}$ is an unknown essentially nonnegative matrix, $A_{\mathrm{d}i} \in \mathbb{R}^{n \times n}$, $i = 1,\ldots,p$, and $B \in \mathbb{R}^{n \times m}$ are unknown nonnegative matrices, $\bar{\tau} = \max_{i \in \{1,\ldots,p\}} \tau_i$, $\eta(\cdot) \in \mathcal{C} = \mathcal{C}([-\bar{\tau},0],\mathbb{R}^n)$ is a continuous vector valued function specifying the initial state of the system, and $\mathcal{C}([-\bar{\tau},0],\mathbb{R}^n)$ denotes a Banach space of continuous functions mapping the interval $[-\bar{\tau},0]$ into $\mathbb{R}^n$ with the topology of uniform convergence. Note that the state of (6) at time t is the *piece of trajectories* x between $t - \bar{\tau}$ and t or, equivalently, the *element* x_t in the space of continuous functions defined on the interval $[-\bar{\tau},0]$ and taking values in $\mathbb{R}^n$, that is, $x_t \in \mathcal{C}([-\bar{\tau},0],\mathbb{R}^n)$, where $x_t(\theta) \triangleq x(t + \theta)$, $\theta \in [-\bar{\tau},0]$. Furthermore, since for a given time t the piece of the trajectories x_t is defined on $[-\bar{\tau},0]$, the uniform norm $\|x(t)\| = \sup_{\theta \in [-\bar{\tau},0]} \|x(t+\theta)\|$, where $\|\cdot\|$ denotes the Euclidean vector norm, is used for the definitions of Lyapunov and asymptotic stability of (6) with $u(t) \equiv 0$. For further details see [18, 17]. Finally, note that since $\eta(\cdot)$ is continuous it follows from Theorem 2.1 of [18, p. 14] that there exists a unique solution $x(\eta)$ defined on $[-\bar{\tau},\infty)$ that coincides with η on $[-\bar{\tau},0]$ and satisfies (6) for all $t \geq 0$.

Definition 3. *The linear time-delay dynamical system given by (6) is nonnegative if for every $\eta(\cdot) \in \mathcal{C}_+$, and $u(t) \geq\geq 0$, $t \geq 0$, where $\mathcal{C}_+ \triangleq \{\psi(\cdot) \in \mathcal{C} : \psi(\theta) \geq\geq 0, \ \theta \in [-\bar{\tau},0]\}$, the solution $x(t)$, $t \geq 0$, to (6) is nonnegative.*

Theorem 2 ([22, 23]). *Consider the linear nonnegative dynamical system $\mathcal{G}$ given by (6) where $A \in \mathbb{R}^{n \times n}$ is essentially nonnegative, $A_{\mathrm{d}i} \in \mathbb{R}^{n \times n}$, $i = 1,\ldots,p$, is nonnegative, and $u(t) \equiv 0$. Then, $\mathcal{G}$ is asymptotically stable for all $\bar{\tau} \in [0,\infty)$ if and only if there exist $p, r \in \mathbb{R}^n$ such that $p \gg 0$ and $r \gg 0$ satisfy*

$$0 = \left(A + \sum_{i=1}^{p} A_{\mathrm{d}i}\right)^{\mathrm{T}} p + r. \tag{7}$$

Remark 1. It follows from Lemma 1 and Theorems 1 and 2 that $\mathcal{G}$ is asymptotically stable for all $\bar{\tau} \in [0,\infty)$ if and only if there exist diagonal matrices P and $Q_i \in \mathbb{R}^{n \times n}$, with $P > 0$ and $Q_i \geq 0$, $i = 1,\ldots,p$, such that

$$A^{\mathrm{T}}P + PA + \sum_{i=1}^{p}(Q_i + A_{\mathrm{d}i}^{\mathrm{T}}PQ_i^{\#}PA_{\mathrm{d}i}) < 0.$$

The proof of this fact follows from standard arguments using the Lyapunov-Krasovskii functional

$$V(\psi(\cdot)) = \psi^{\mathrm{T}}(0)P\psi(0) + \sum_{i=1}^{p}\int_{-\tau_i}^{0}\psi^{\mathrm{T}}(\theta)A_{\mathrm{d}i}^{\mathrm{T}}PQ_i^{\#}PA_{\mathrm{d}i}^{\mathrm{T}}\psi(\theta)\mathrm{d}\theta.$$

The following proposition is needed for the main results of the paper.

Proposition 1 ([22, 23]). *The linear dynamical system $\mathcal{G}$ given by (6) is non-negative if and only if $A \in \mathbb{R}^{n \times n}$ is essentially nonnegative, $A_{\mathrm{d}i} \in \mathbb{R}^{n \times n}$ is nonnegative, and $B \in \mathbb{R}^{n \times m}$ is nonnegative.*

It follows from Proposition 1 that the control input signal $Bu(t)$, $t \geq 0$, needs to be nonnegative to guarantee the nonnegativity of the state of (6). This is due to the fact that when the initial state of (6) belongs to the boundary of the nonneg-ative orthant, a negative input can destroy the nonnegativity of the state of (6). Alternatively, however, if the initial state is in the interior of the nonnegative orthant, then it follows from continuity of solutions with respect to the system initial conditions that, over a small interval of time, nonnegativity of the state of (6) is guaranteed *irrespective* of the sign of each element of the control input $Bu(t)$ over this time interval. However, unlike open-loop control wherein lack of coordination between the input and the state necessitates nonnegativity of the control input, a *feedback* control signal predicated on the system state variables allows for the anticipation of loss of nonnegativity of the state. Hence, state feedback control signals can take negative values while assuring nonnegativity of the system states. For further discussion of the above fact see [25, 21].

Next, we present a time-varying extension to Proposition 1 needed for the main theorems of this paper. Specifically, we consider the linear time-varying delay dynamical system

$$\dot{x}(t) = A(t)x(t) + \sum_{i=1}^{p}A_{\mathrm{d}i}(t)x(t-\tau_i) + Bu(t), \quad x(\theta) = \eta(\theta), \ \theta \in [-\bar{\tau}, 0], \ t \geq 0, \quad (8)$$

where $A : [0, \infty) \to \mathbb{R}^{n \times n}$ and $A_{\mathrm{d}i} : [0, \infty) \to \mathbb{R}^{n \times n}$, $i = 1, \ldots, p$ are continuous. For the following result the definition of nonnegativity holds with (6) replaced by (8).

Proposition 2. *Consider the time-varying delay dynamical system (8) where $A : [0, \infty) \to \mathbb{R}^{n \times n}$ and $A_{\mathrm{d}i} : [0, \infty) \to \mathbb{R}^{n \times n}$, $i = 1, \ldots, p$, are continuous. If for every $t \in [0, \infty)$, $A : [0, \infty) \to \mathbb{R}^{n \times n}$ is essentially nonnegative, $A_{\mathrm{d}i} : [0, \infty) \to \mathbb{R}^{n \times n}$, $i = 1, \ldots, p$, is nonnegative, $B \in \mathbb{R}^{n \times m}$ is nonnegative, and $u(t)$ is nonnegative, then the solution $x(t)$, $t \geq 0$, to (8) is nonnegative.*

Proof. The result is a direct consequence of the nonlinear analogue to Propo-sition 1 by equivalently representing the time-varying delay dynamical system

(8) as an autonomous nonlinear time-delay system by appending another state to represent time. See [21] for a similar proof. $\square$

Since stabilization of nonnegative systems naturally deals with equilibrium points in the interior of the nonnegative orthant $\overline{\mathbb{R}}_+^n$, the following proposition provides necessary conditions for the existence of an interior equilibrium state $x(\theta) = x_e \in \mathbb{R}_+^n$, $\theta \in [-\bar{\tau}, 0]$, of (6) in terms of the stability properties of the system matrices A and A_{di}. For this result, recall that a matrix $M \in \mathbb{R}^{n \times n}$ is *semistable* if and only if $\lim_{t \to \infty} e^{Mt}$ exists.

Proposition 3. *Consider the nonnegative time-delay dynamical system (6) and assume there exist $x_e \in \mathbb{R}_+^n$ and $u_e \in \overline{\mathbb{R}}_+^m$ such that*

$$0 = \left(A + \sum_{i=1}^{p} A_{di} \right) x_e + B u_e. \tag{9}$$

Then, $A + \sum_{i=1}^{p} A_{di}$ is semistable.

Proof. The proof is a direct consequence of *ii*) of Theorem 3.2 in [14] with A replaced by $A + \sum_{i=1}^{p} A_{di}$, $p = x_e$, and $r = B u_e$. $\square$

It follows from Proposition 3 that the existence of an equilibrium state $x(\theta) = x_e \in \mathbb{R}_+^n$, $\theta \in [-\bar{\tau}, 0]$, for (6) implies that the matrix $A + \sum_{i=1}^{p} A_{di}$ is semistable. Hence, if (9) holds for $x_e \in \mathbb{R}_+^n$ and $u_e \in \overline{\mathbb{R}}_+^m$, then $A + \sum_{i=1}^{p} A_{di}$ is Hurwitz or $0 \in \operatorname{spec}(A + \sum_{i=1}^{p} A_{di})$ is a simple eigenvalue of $A + \sum_{i=1}^{p} A_{di}$, and all other eigenvalues of $A + \sum_{i=1}^{p} A_{di}$ have negative real parts since $-(A + \sum_{i=1}^{p} A_{di})$ is an M-matrix [24].

Next, we consider a subclass of nonnegative systems, namely, compartmental systems. As noted in the Introduction, linear compartmental dynamical systems are of major importance in biological and physiological systems. For example, almost the entire field of distribution of tracer labelled materials in steady state systems can be captured by linear compartmental dynamical systems [11].

Definition 4 ([22, 23]). *The linear time-delay dynamical system (6) is called a* compartmental dynamical system *if A and $A_d \triangleq \sum_{i=1}^{p} A_{di}$ are such that A is essentially nonnegative, A_{di}, $i = 1, \ldots, p$, is nonnegative, and $A + \sum_{i=1}^{p} A_{di}$ is a* compartmental matrix.

Note that if A and A_d are such that

$$A_{(i,j)} = \begin{cases} -\sum_{k=1}^{n} a_{ki}, & i = j, \\ 0, & i \neq j, \end{cases} \qquad A_{d(i,j)} = \begin{cases} 0, & i = j, \\ a_{ij}, & i \neq j, \end{cases} \tag{10}$$

where $a_{ii} \geq 0$, $i \in \{1, \ldots, n\}$, denote the loss coefficients of the ith compartment and $a_{ij} \geq 0$, $i \neq j$, $i, j \in \{1, \ldots, n\}$, denote the transfer coefficients from the jth compartment to the ith compartment, then the linear dynamical system (6) is a compartmental system.

3 Adaptive Control for Linear Nonnegative Uncertain Dynamical Systems with Time Delay

In this section, we consider the problem of characterizing adaptive feedback control laws for nonnegative and compartmental uncertain dynamical systems with time delay to achieve *set-point* regulation in the nonnegative orthant. Specifically, consider the following controlled linear uncertain time-delay dynamical system $\mathcal{G}$ given by

$$\dot{x}(t) = Ax(t) + \sum_{i=1}^{p} A_{\mathrm{d}i}x(t - \tau_i) + Bu(t), \quad x(\theta) = \eta(\theta), \quad -\bar{\tau} \le \theta \le 0, \quad t \ge 0,$$

(11)

where $x(t) \in \mathbb{R}^n$, $t \ge 0$, is the state of the system, $u(t) \in \mathbb{R}^m$, $t \ge 0$, is the control input, $A \in \mathbb{R}^{n \times n}$ is an *unknown* essentially nonnegative matrix, $A_{\mathrm{d}i} \in \mathbb{R}^{n \times n}$ and $B \in \mathbb{R}^{n \times m}$ are *unknown* nonnegative matrices, $\eta(\cdot) \in \{\psi(\cdot) \in \mathcal{C}_+([-\bar{\tau}, 0], \mathbb{R}^n) : \psi(\theta) \ge\ge 0, \theta \in [-\bar{\tau}, 0]\}$, and $\bar{\tau} = \max_{i \in \{1,\ldots,p\}} \tau_i$, where $\tau_i \ge 0$, $i = 1, \ldots, p$, are *unknown* system delays. The control input $u(\cdot)$ in (11) is restricted to the class of admissible controls consisting of measurable functions such that $u(t) \in \mathbb{R}^m$, $t \ge 0$.

It follows from Proposition 1 that the state trajectories of nonnegative and compartmental dynamical systems remain in the nonnegative orthant of the state space for nonnegative initial conditions and nonnegative inputs. However, even though active control of drug delivery systems for physiological applications requires control (source) inputs to be nonnegative, in many applications of nonnegative systems such as biological systems, population dynamics, and ecological systems, the positivity constraint on the control input is not natural. Hence, in this section we do not place any restriction on the sign of the control signal and design an adaptive controller that guarantees that the system states remain in the nonnegative orthant and converge to a desired equilibrium state.

Specifically, for a given desired set point $x_{\mathrm{e}} \in \overline{\mathbb{R}}_+^n$, our aim here is to design a control input $u(t)$, $t \ge 0$, such that $\lim_{t \to \infty} \|x(t) - x_{\mathrm{e}}\| = 0$, where $\|\cdot\|$ denotes any vector norm on $\mathbb{R}^n$. However, since in many applications of nonnegative systems and in particular, compartmental systems, it is often necessary to regulate a subset of the nonnegative state variables which usually include a central compartment, here we require that $\lim_{t \to \infty} x_i(t) = x_{\mathrm{d}i} \ge 0$ for $i = 1, \ldots, m \le n$, where $x_{\mathrm{d}i}$ is a desired set point for the ith state $x_i(t)$. Furthermore, we assume that control inputs are injected directly into m separate compartments such that the input matrix is given by

$$B = \begin{bmatrix} B_u \\ 0_{(n-m) \times m} \end{bmatrix},$$

(12)

where $B_u \triangleq \mathrm{diag}[b_1, \ldots, b_m]$ and $b_i \in \mathbb{R}_+$, $i = 1, \ldots, m$. For compartmental systems this assumption is not restrictive since control inputs correspond to control inflows to each individual compartment. Here, we assume that for $i \in \{1, \ldots, m\}$, b_i is *unknown*. For the statement of our main result define $x_{\mathrm{e}} \triangleq [x_{\mathrm{d}}^{\mathrm{T}}, x_{\mathrm{u}}^{\mathrm{T}}]^{\mathrm{T}}$, where $x_{\mathrm{d}} \triangleq [x_{\mathrm{d}1}, \ldots, x_{\mathrm{d}m}]^{\mathrm{T}}$ and $x_{\mathrm{u}} \triangleq [x_{\mathrm{u}1}, \ldots, x_{\mathrm{u}(n-m)}]^{\mathrm{T}}$.

Theorem 3. *Consider the linear uncertain time-delay dynamical system $\mathcal{G}$ given by (11) where A is essentially nonnegative, $A_{\mathrm{d}i}$, $i = 1,\ldots,p$, is nonnegative, and B is nonnegative and given by (12). Assume there exist nonnegative vectors $x_{\mathrm{u}} \in \overline{\mathbb{R}}_+^{n-m}$ and $u_{\mathrm{e}} \in \overline{\mathbb{R}}_+^m$ such that*

$$0 = \left(A + \sum_{i=1}^{p} A_{\mathrm{d}i} \right) x_{\mathrm{e}} + B u_{\mathrm{e}}. \tag{13}$$

Furthermore, assume there exists a diagonal matrix $K_{\mathrm{g}} = \mathrm{diag}[k_{\mathrm{g}1},\ldots,k_{\mathrm{g}m}]$, such that $A_{\mathrm{s}} + \sum_{i=1}^{p} A_{\mathrm{d}i}$ is Hurwitz, where $A_{\mathrm{s}} \triangleq A + B\tilde{K}_{\mathrm{g}}$ and $\tilde{K}_{\mathrm{g}} \triangleq [K_{\mathrm{g}} \ 0_{m\times(n-m)}]$. Finally, let q_i and $\hat{q}_i$, $i = 1,\ldots,m$, be positive constants. Then the adaptive feedback control law

$$u(t) = K(t)(\hat{x}(t) - x_{\mathrm{d}}) + \phi(t), \tag{14}$$

where $K(t) = \mathrm{diag}[k_1(t),\ldots,k_m(t)]$, $\hat{x}(t) = [x_1(t),\ldots,x_m(t)]^{\mathrm{T}}$, and $\phi(t) \in \mathbb{R}^m$, $t \geq 0$, or, equivalently,

$$u_i(t) = k_i(t)(x_i(t) - x_{\mathrm{d}i}) + \phi_i(t), \quad i = 1,\ldots,m, \tag{15}$$

where $k_i(t) \in \mathbb{R}$, $t \geq 0$, and $\phi_i(t) \in \mathbb{R}$, $t \geq 0$, $i = 1,\ldots,m$, with update laws

$$\dot{k}_i(t) = -q_i(x_i(t) - x_{\mathrm{d}i})^2, \quad k_i(0) \leq 0, \quad i = 1,\ldots,m, \tag{16}$$

$$\dot{\phi}_i(t) = \begin{cases} 0, \ \text{if } \phi_i(t) = 0 \ \text{and } x_i(t) \geq x_{\mathrm{d}i}, \ \phi_i(0) \geq 0, \quad i = 1,\ldots,m, \\ -\hat{q}_i(x_i(t) - x_{\mathrm{d}i}), \ \text{otherwise,} \end{cases}$$

$$\tag{17}$$

guarantees that the solution $(x(t), K(t), \phi(t)) \equiv (x_{\mathrm{e}}, K_{\mathrm{g}}, u_{\mathrm{e}})$ of the closed-loop system given by (11), (14), (16), and (17) is Lyapunov stable and $x_i(t) \to x_{\mathrm{d}i}$, $i = 1,\ldots,m$, as $t \to \infty$ for all $\eta(\cdot) \in \mathcal{C}_+$. Furthermore, $x(t) \geq\geq 0$, $t \geq 0$, for all $\eta(\cdot) \in \mathcal{C}_+$.

Proof. Note that with $u(t)$, $t \geq 0$, given by (14), it follows from (11) that

$$\dot{x}(t) = Ax(t) + \sum_{i=1}^{p} A_{\mathrm{d}i}x(t - \tau_i) + BK(t)(\hat{x}(t) - x_{\mathrm{d}}) + B\phi(t),$$

$$x(\theta) = \eta(\theta), \quad -\bar{\tau} \leq \theta \leq 0, \quad t \geq 0, \tag{18}$$

or, equivalently, using (13),

$$\dot{x}(t) = A_{\mathrm{s}}(x(t) - x_{\mathrm{e}}) + \sum_{i=1}^{p} A_{\mathrm{d}i}(x(t - \tau_i) - x_{\mathrm{e}}) + B(K(t) - K_{\mathrm{g}})(\hat{x}(t) - x_{\mathrm{d}})$$

$$+B(\phi(t) - u_{\mathrm{e}}), \quad x(\theta) = \eta(\theta), \quad -\bar{\tau} \leq \theta \leq 0, \quad t \geq 0. \tag{19}$$

Since $A_{\mathrm{s}} + \sum_{i=1}^{p} A_{\mathrm{d}i}$ is essentially nonnegative and Hurwitz it follows from Theorem 1 that there exist diagonal matrices $P > 0$ and $\tilde{Q}_i \geq 0$, $i = 1,\ldots,p$, and a positive-definite matrix $R \in \mathbb{R}^{n\times n}$ such that

$$0 = A_{\mathrm{s}}^{\mathrm{T}} P + P A_{\mathrm{s}} + \sum_{i=1}^{p} (\tilde{Q}_i + A_{\mathrm{d}i}^{\mathrm{T}} P \tilde{Q}_i^{\#} P A_{\mathrm{d}i}) + R. \tag{20}$$

To show Lyapunov stability of the closed-loop system (16), (17), and (19) consider the Lyapunov-Krasovskii functional candidate $V : \mathcal{C}_+ \times \mathbb{R}^{m \times m} \times \mathbb{R}^m \to \mathbb{R}$ given by

$$\begin{aligned}
V(\psi, K, \phi) &= (\psi(0) - x_{\mathrm{e}})^{\mathrm{T}} P (\psi(0) - x_{\mathrm{e}}) \\
&\quad + \sum_{i=1}^{p} \int_{-\tau_i}^{0} (\psi(\theta) - x_{\mathrm{e}})^{\mathrm{T}} A_{\mathrm{d}i}^{\mathrm{T}} P \tilde{Q}_i^{\#} P A_{\mathrm{d}i} (\psi(\theta) - x_{\mathrm{e}}) \mathrm{d}\theta \\
&\quad + \mathrm{tr}(K - K_{\mathrm{g}})^{\mathrm{T}} Q^{-1} (K - K_{\mathrm{g}}) + (\phi - u_{\mathrm{e}})^{\mathrm{T}} \hat{Q}^{-1} (\phi - u_{\mathrm{e}}),
\end{aligned} \tag{21}$$

or, equivalently,

$$\begin{aligned}
V(\psi, K, \phi) &= \sum_{i=1}^{n} p_i (\psi_i(0) - x_{\mathrm{e}i})^2 + \sum_{i=1}^{p} \int_{-\tau_i}^{0} (\psi(\theta) - x_{\mathrm{e}})^{\mathrm{T}} A_{\mathrm{d}i}^{\mathrm{T}} P \tilde{Q}_i^{\#} P \\
&\quad \cdot A_{\mathrm{d}i}(\psi(\theta) - x_{\mathrm{e}}) \mathrm{d}\theta + \sum_{i=1}^{m} \frac{p_i b_i}{q_i} (k_i - k_{\mathrm{g}i})^2 + \sum_{i=1}^{m} \frac{p_i b_i}{\hat{q}_i} (\phi_i - u_{\mathrm{e}i})^2,
\end{aligned}$$

where $Q = \mathrm{diag}\left[\frac{q_1}{p_1 b_1}, \ldots, \frac{q_m}{p_m b_m}\right]$ and $\hat{Q} = \mathrm{diag}\left[\frac{\hat{q}_1}{p_1 b_1}, \ldots, \frac{\hat{q}_m}{p_m b_m}\right]$. Note that $V(\psi_{\mathrm{e}}, K_{\mathrm{g}}, u_{\mathrm{e}}) = 0$, where $\psi_{\mathrm{e}}(\theta) = x_{\mathrm{e}}, \theta \in [-\bar{\tau}, 0]$. Furthermore, note that there exist $\mathcal{K}$-class functions $\alpha_1(\cdot)$, $\alpha_2(\cdot)$, and $\alpha_3(\cdot)$ such that

$$V(\psi, K, \phi) \geq \alpha_1(\|\psi(0) - x_{\mathrm{e}}\|) + \alpha_2(\|K - K_{\mathrm{g}}\|_{\mathrm{F}}) + \alpha_3(\|\phi - u_{\mathrm{e}}\|),$$

where $\|\cdot\|$ denotes the Euclidean vector norm and $\|\cdot\|_{\mathrm{F}}$ denotes the Frobenius matrix norm.

Next, letting $x(t)$, $t \geq 0$, denote the solution to (19) and using (16) and (17), it follows that the Lyapunov-Krasovskii directional derivative of $V(x_t, K(t), \phi(t))$ along the closed-loop system trajectories is given by

$$\begin{aligned}
\dot{V}(x_t, K(t), \phi(t)) &= 2(x(t) - x_{\mathrm{e}})^{\mathrm{T}} P \dot{x}(t) \\
&\quad + \sum_{i=1}^{p} (x(t) - x_{\mathrm{e}})^{\mathrm{T}} A_{\mathrm{d}i}^{\mathrm{T}} P \tilde{Q}_i^{\#} P A_{\mathrm{d}i}(x(t) - x_{\mathrm{e}}) \\
&\quad - \sum_{i=1}^{p} (x(t - \tau_i) - x_{\mathrm{e}})^{\mathrm{T}} A_{\mathrm{d}i}^{\mathrm{T}} P \tilde{Q}_i^{\#} P A_{\mathrm{d}i}(x(t - \tau_i) - x_{\mathrm{e}}) \\
&\quad + 2 \sum_{i=1}^{m} \frac{p_i b_i}{q_i} (k_i(t) - k_{\mathrm{g}i}) \dot{k}_i(t) \\
&\quad + 2 \sum_{i=1}^{m} \frac{p_i b_i}{\hat{q}_i} (\phi_i(t) - u_{\mathrm{e}i}) \dot{\phi}_i(t),
\end{aligned} \tag{22}$$

which, using (19), implies

$$\dot{V}(x_t, K(t), \phi(t)) = (x(t) - x_{\mathrm{e}})^{\mathrm{T}}(A_{\mathrm{s}}^{\mathrm{T}}P + PA_{\mathrm{s}})(x(t) - x_{\mathrm{e}})$$

$$+2(x(t) - x_{\mathrm{e}})^{\mathrm{T}}P\sum_{i=1}^{p}A_{\mathrm{d}i}(x(t - \tau_i) - x_{\mathrm{e}})$$

$$+2(x(t) - x_{\mathrm{e}})^{\mathrm{T}}PB(K(t) - K_{\mathrm{g}})(\hat{x}(t) - x_{\mathrm{d}})$$

$$+2(x(t) - x_{\mathrm{e}})^{\mathrm{T}}PB(\phi(t) - u_{\mathrm{e}})$$

$$+\sum_{i=1}^{p}(x(t) - x_{\mathrm{e}})^{\mathrm{T}}A_{\mathrm{d}i}^{\mathrm{T}}P\tilde{Q}_i^{\#}PA_{\mathrm{d}i}(x(t) - x_{\mathrm{e}})$$

$$-\sum_{i=1}^{p}(x(t - \tau_i) - x_{\mathrm{e}})^{\mathrm{T}}A_{\mathrm{d}i}^{\mathrm{T}}P\tilde{Q}_i^{\#}PA_{\mathrm{d}i}(x(t - \tau_i) - x_{\mathrm{e}})$$

$$+2\sum_{i=1}^{m}\frac{p_i b_i}{q_i}(k_i(t) - k_{\mathrm{g}i})\dot{k}_i(t)$$

$$+2\sum_{i=1}^{m}\frac{p_i b_i}{\hat{q}_i}(\phi_i(t) - u_{\mathrm{e}i})\dot{\phi}_i(t). \tag{23}$$

Now, using (20), (23) yields

$$\dot{V}(x_t, K(t), \phi(t)) = -(x(t) - x_{\mathrm{e}})^{\mathrm{T}}R(x(t) - x_{\mathrm{e}})$$

$$-\sum_{i=1}^{p}[\tilde{Q}_i(x(t) - x_{\mathrm{e}}) - PA_{\mathrm{d}i}(x(t - \tau_i) - x_{\mathrm{e}})]^{\mathrm{T}}\tilde{Q}_i^{\#}$$

$$\cdot[\tilde{Q}_i(x(t) - x_{\mathrm{e}}) - PA_{\mathrm{d}i}(x(t - \tau_i) - x_{\mathrm{e}})]$$

$$+2\sum_{i=1}^{m}p_i b_i(k_i(t) - k_{\mathrm{g}i})(x_i(t) - x_{\mathrm{d}})^2$$

$$+2\sum_{i=1}^{m}p_i b_i(x_i(t) - x_{\mathrm{e}})(\phi_i(t) - u_{\mathrm{e}i})$$

$$+2\sum_{i=1}^{m}\frac{p_i b_i}{q_i}(k_i(t) - k_{\mathrm{g}i})\dot{k}_i(t) + 2\sum_{i=1}^{m}\frac{p_i b_i}{\hat{q}_i}(\phi_i(t) - u_{\mathrm{e}i})\dot{\phi}_i(t)$$

$$\leq -(x(t) - x_{\mathrm{e}})^{\mathrm{T}}R(x(t) - x_{\mathrm{e}})$$

$$+2\sum_{i=1}^{m}p_i b_i(\phi_i(t) - u_{\mathrm{e}i})\left[(x_i(t) - x_{\mathrm{d}i}) + \frac{1}{\hat{q}_i}\dot{\phi}_i(t)\right], \quad t \geq 0. \tag{24}$$

For the two cases given in (17), the last term on the right-hand side of (24) gives:

1. If $\phi_i(t) = 0$ and $x_i(t) \geq x_{\mathrm{d}i}$, $t \geq 0$, then $\dot{\phi}_i(t) = 0$, $t \geq 0$, and hence, for $t \geq 0$,

$$p_i b_i (\phi_i(t) - u_{\mathrm{e}i}) \left[(x_i(t) - x_{\mathrm{d}i}) + \frac{1}{\hat{q}_i} \dot{\phi}_i(t) \right] = -p_i b_i u_{\mathrm{e}i}(x_i(t) - x_{\mathrm{d}i}) \leq 0.$$

2. Otherwise, $\dot{\phi}_i(t) = -\hat{q}_i(x_i(t) - x_{\mathrm{d}i})$, and hence, for $t \geq 0$,

$$p_i b_i (\phi_i(t) - u_{\mathrm{e}i}) \left[(x_i(t) - x_{\mathrm{d}i}) + \frac{1}{\hat{q}_i} \dot{\phi}_i(t) \right] = 0.$$

Hence, it follows that in either case

$$\dot{V}(x_t, K(t), \phi(t)) \leq -(x(t) - x_{\mathrm{e}})^{\mathrm{T}} R(x(t) - x_{\mathrm{e}}) \leq 0, \quad t \geq 0, \tag{25}$$

which proves that the solution $(x(t), K(t), \phi(t)) \equiv (x_{\mathrm{e}}, K_{\mathrm{g}}, u_{\mathrm{e}})$ to (16), (17), and (19) is Lyapunov stable. Furthermore, since the positive orbit $\gamma^+(\eta(\theta), K_0, \phi_0)$ of the closed-loop system (16), (17), and (19) is bounded, $\gamma^+(\eta(\theta), K_0, \phi_0)$ belongs to a compact subset of $\mathcal{C}_+ \times \mathbb{R}^{m \times m} \times \mathbb{R}^m$ [26], and $R > 0$, it follows from the Krasovskii-LaSalle invariant set theorem for infinite-dimensional systems [18, p. 143] that $x(t) \to x_{\mathrm{e}}$ as $t \to \infty$ for all $\eta(\cdot) \in \mathcal{C}_+$.

Finally, to show that $x(t) \geq\geq 0$, $t \geq 0$, for all $\eta(\cdot) \in \mathcal{C}_+$, note that the closed-loop system (11), (14), (16), and (17) is given by

$$\dot{x}(t) = Ax(t) + \sum_{i=1}^{p} A_{\mathrm{d}i}x(t - \tau_i) + BK(t)(\hat{x}(t) - x_{\mathrm{d}}) + B\phi(t)$$

$$= (A + B[K(t), 0_{m \times (n-m)}])x(t) + \sum_{i=1}^{p} A_{\mathrm{d}i}x(t - \tau_i) - BK(t)x_{\mathrm{d}} + B\phi(t)$$

$$= \tilde{A}(t)x(t) + \sum_{i=1}^{p} A_{\mathrm{d}i}x(t - \tau_i) + v(t) + w(t), \tag{26}$$

where

$$\tilde{A}(t) \triangleq \begin{bmatrix} a_{11} + b_1 k_1(t) & a_{12} & \cdots & a_{1m} & a_{1\,m+1} & \cdots & a_{1n} \\ a_{21} & a_{22} + b_2 k_2(t) & & \vdots & \vdots & \ddots & a_{2n} \\ \vdots & & \ddots & & & & \vdots \\ a_{m1} & \cdots & & a_{mm} + b_m k_m(t) & a_{m\,m+1} & \cdots & a_{mn} \\ a_{m+1\,1} & \cdots & & a_{m+1\,m} & a_{m+1\,m+1} & \cdots & a_{m+1\,n} \\ \vdots & \ddots & & \vdots & \vdots & \ddots & \vdots \\ a_{n1} & \cdots & & a_{nm} & a_{n\,m+1} & \cdots & a_{nn} \end{bmatrix}, \tag{27}$$

$$v(t) \triangleq - \begin{bmatrix} b_1 k_1(t)x_{\mathrm{d}1} \\ \vdots \\ b_m k_m(t)x_{\mathrm{d}m} \\ 0 \\ \vdots \\ 0 \end{bmatrix}, \qquad w(t) \triangleq \begin{bmatrix} b_1 \phi_1(t) \\ \vdots \\ b_m \phi_m(t) \\ 0 \\ \vdots \\ 0 \end{bmatrix}. \tag{28}$$

Now, since by (16) and (17), $k_i(t) \leq 0$, $t \geq 0$, $i = 1, \ldots, m$, and $\phi_i(t) \geq 0$, $t \geq 0$, $i = 1, \ldots, m$, it follows that $v(t) \geq\geq 0$, $t \geq 0$, and $w(t) \geq\geq 0$, $t \geq 0$. Hence, since for every $t \in [0, \infty)$, $\tilde{A}(t)$ is essentially nonnegative and $A_{\mathrm{d}i}$, $i = 1, \ldots, p$, is nonnegative, it follows from Proposition 2 that $x(t) \geq\geq 0$, $t \geq 0$, for all $\eta(\cdot) \in \mathcal{C}_+$. $\square$

Remark 2. Note that the conditions in Theorem 3 imply that $x(t) \to x_{\mathrm{e}}$ as $t \to \infty$, and hence, it follows from (16) and (17) that $(x(t), K(t), \phi(t)) \to \mathcal{M} \triangleq \{(x_t, K, \phi) \in \mathcal{C}_+ \times \mathbb{R}^{m \times m} \times \mathbb{R}^m : x_t = x_{\mathrm{e}}, \dot{K} = 0, \dot{\phi} = 0\}$ as $t \to \infty$.

It is important to note that the adaptive control law (14), (16), and (17) does not require the explicit knowledge of the system matrices A, $A_{\mathrm{d}i}$, $i = 1, \ldots, p$, and B, the gain matrix K_{g}, and the nonnegative constant vector u_{e}, even though Theorem 3 requires the existence of K_{g} and nonnegative vectors x_{u} and u_{e} such that $A_{\mathrm{s}} + \sum_{i=1}^{p} A_{\mathrm{d}i}$ is asymptotically stable and that the conditions (13) holds. Furthermore, in the case where $A + \sum_{i=1}^{p} A_{\mathrm{d}i}$ is semistable and minimum phase with respect to the output $y = \hat{x}$, or $A + \sum_{i=1}^{p} A_{\mathrm{d}i}$ is asymptotically stable, then there always exists a diagonal matrix $K_{\mathrm{g}} \in \mathbb{R}^{m \times m}$ such that $A_{\mathrm{s}} + \sum_{i=1}^{p} A_{\mathrm{d}i}$ is asymptotically stable. In addition, note that for $i = 1, \ldots, m$, the control input signal $u_i(t)$, $t \geq 0$, can be negative depending on the values of $x_i(t)$, $k_i(t)$, and $\phi_i(t)$, $t \geq 0$. However, as is required in nonnegative and compartmental dynamical systems the closed-loop plant states remain nonnegative.

In the case where our objective is zero set-point regulation, that is, $\psi_{\mathrm{e}}(\theta) = x_{\mathrm{e}} = 0$, $\theta \in [-\bar{\tau}, 0]$, the adaptive controller given in Theorem 3 can be considerably simplified. Specifically, since in this case $x(t) \geq\geq x_{\mathrm{e}} = 0$, $t \geq 0$, and condition (13) is trivially satisfied with $u_{\mathrm{e}} = 0$, we can set $\phi(t) \equiv 0$ so that update law (17) is superfluous. Furthermore, since (13) is trivially satisfied, A can possess eigenvalues in the open right-half plane. Alternatively, exploiting a *linear* Lyapunov-Krasovskii functional construction for the plant dynamics, an even simpler adaptive controller can be derived. This result is given in the following theorem.

Theorem 4. *Consider the linear uncertain time-delay system $\mathcal{G}$ given by (11) where A is essentially nonnegative, $A_{\mathrm{d}i}$, $i = 1, \ldots, p$, is nonnegative, and B is nonnegative and given by (12). Assume there exists a diagonal matrix $K_{\mathrm{g}} = \mathrm{diag}[k_{\mathrm{g}1}, \ldots, k_{\mathrm{g}m}]$ such that $A_{\mathrm{s}} + \sum_{i=1}^{p} A_{\mathrm{d}i}$ is asymptotically stable, where $A_{\mathrm{s}} = A + B\tilde{K}_{\mathrm{g}}$ and $\tilde{K}_{\mathrm{g}} = [K_{\mathrm{g}}, 0_{m \times (n-m)}]$. Furthermore, let q_i, $i = 1, \ldots, m$, be positive constants. Then the adaptive feedback control law*

$$u(t) = K(t)\hat{x}(t), \tag{29}$$

where $K(t) = \mathrm{diag}[k_1(t), \ldots, k_m(t)]$ and $\hat{x}(t) = [x_1(t), \ldots, x_m(t)]^{\mathrm{T}}$ or, equivalently,

$$u_i(t) = k_i(t)x_i(t), \quad i = 1, \ldots, m, \tag{30}$$

where $k_i(t) \in \mathbb{R}$, $i = 1, \ldots, m$, with update law

$$\dot{K}(t) = -\mathrm{diag}[q_1 x_1(t), \ldots, q_m x_m(t)], \quad K(0) \leq\leq 0, \tag{31}$$

guarantees that the solution $(x(t), K(t)) \equiv (0, K_{\mathrm{g}})$ of the closed-loop system given by (11), (29), and (31) is Lyapunov stable and $x(t) \to 0$ as $t \to \infty$ for all $\eta(\cdot) \in \mathcal{C}_+$. Furthermore, $x(t) \geq\geq 0$, $t \geq 0$, for all $\eta(\cdot) \in \mathcal{C}_+$.

Proof. Note that with $u(t)$, $t \geq 0$, given by (29) it follows from (11) that

$$\dot{x}(t) = Ax(t) + \sum_{i=1}^{p} A_{\mathrm{d}i} x(t - \tau_i) + BK(t)\hat{x}(t),$$

$$x(\theta) = \eta(\theta), \quad -\bar{\tau} \leq \theta \leq 0, \quad t \geq 0. \tag{32}$$

Now, since for every $t \in [0, \infty)$, $\tilde{A}(t) \triangleq A + B[K(t), 0_{m \times (n-m)}]$ is essentially nonnegative and $A_{\mathrm{d}i}$, $i = 1, \ldots, p$, is nonnegative it follows from Proposition 2 that $x(t) \geq\geq 0$, $t \geq 0$, for all $\eta(\cdot) \in \mathcal{C}_+$. Next, using $A_{\mathrm{s}} = A + B\tilde{K}_{\mathrm{g}}$, note that (32) can be equivalently written as

$$\dot{x}(t) = A_{\mathrm{s}} x(t) + \sum_{i=1}^{p} A_{\mathrm{d}i} x(t - \tau_i) + B(K(t) - K_{\mathrm{g}})\hat{x}(t),$$

$$x(\theta) = \eta(\theta) \quad -\bar{\tau} \leq \theta \leq 0, \quad t \geq 0. \tag{33}$$

Furthermore, since A_{s} is essentially nonnegative, $A_{\mathrm{d}i}$ is nonnegative, and $A_{\mathrm{s}} + \sum_{i=1}^{p} A_{\mathrm{d}i}$ is asymptotically stable, it follows from Theorem 2 that there exist vectors $p >> 0$ and $r >> 0$ satisfying

$$0 = \left(A_{\mathrm{s}} + \sum_{i=1}^{p} A_{\mathrm{d}i} \right)^{\mathrm{T}} p + r. \tag{34}$$

To show Lyapunov stability of the closed-loop system (31) and (33) consider the Lyapunov-Krasovskii functional candidate $V : \mathcal{C}_+ \times \mathbb{R}^{m \times m} \to \mathbb{R}$ given by

$$V(\psi, K) = p^{\mathrm{T}} \psi(0) + \sum_{i=1}^{p} \int_{-\tau_i}^{0} p^{\mathrm{T}} A_{\mathrm{d}i} \psi(\theta) \mathrm{d}\theta + \frac{1}{2} \mathrm{tr}(K - K_{\mathrm{g}})^{\mathrm{T}} Q^{-1} (K - K_{\mathrm{g}}), \tag{35}$$

where $Q = \mathrm{diag}[\frac{q_1}{p_1 b_1}, \ldots, \frac{q_m}{p_m b_m}]$. Furthermore, note that $V(\psi_{\mathrm{e}}, K_{\mathrm{g}}) = 0$, where $\psi_{\mathrm{e}}(\theta) = 0$, $\theta \in [-\bar{\tau}, 0]$, and there exist $\mathcal{K}$-class functions $\alpha_1(\cdot)$ and $\alpha_2(\cdot)$ such that

$$V(\psi, K) \geq \alpha_1(\|\psi(0)\|) + \alpha_2(\|K - K_{\mathrm{g}}\|_{\mathrm{F}}).$$

Now, letting $x(t)$, $t \geq 0$, denote the solution to (33) and using (31), it follows that the Lyapunov-Krasovskii directional derivative of $V(x_t, K(t))$ along the closed-loop system trajectories is given by

$$\dot{V}(x_t, K(t)) = p^{\mathrm{T}} A_{\mathrm{s}} x(t) + p^{\mathrm{T}} \sum_{i=1}^{p} A_{\mathrm{d}i} x(t - \tau_i) + p^{\mathrm{T}} B(K(t) - K_{\mathrm{g}})\hat{x}(t)$$

$$+ p^{\mathrm{T}} \sum_{i=1}^{p} A_{\mathrm{d}i} x(t) - p^{\mathrm{T}} \sum_{i=1}^{p} A_{\mathrm{d}i} x(t - \tau_i)$$

$$+ \mathrm{tr}(K(t) - K_{\mathrm{g}})^{\mathrm{T}} Q^{-1} \dot{K}(t)$$

$$= -r^{\mathrm{T}} x(t)$$

$$\leq 0, \quad t \geq 0,$$

which proves that the solution $(x(t), K(t)) \equiv (0, K_{\mathrm{g}})$ to (31) and (33) is Lyapunov stable. Furthermore, since the positive orbit $\gamma^+(\eta(\theta), K_0)$ of the closed-loop system (31) and (33) is bounded, $\gamma^+(\eta(\theta), K_0)$ belongs to a compact subset of $\mathcal{C}_+ \times \mathbb{R}^{m \times m}$ [26], and $r >> 0$, it follows from Corollary 3.1 of [18, p. 143] that $x(t) \to 0$ as $t \to \infty$ for all $\eta(\cdot) \in \mathcal{C}_+$. $\square$

4   Adaptive Control for Linear Nonnegative Dynamical Systems with Nonnegative Control and Time Delay

In drug delivery systems for physiological processes, control (source) inputs are usually constrained to be nonnegative as are the system states. Hence, in this section we develop adaptive control laws for nonnegative retarded systems with nonnegative control inputs. However, since condition (9) is required to be satisfied for $x_{\mathrm{e}} \in \overline{\mathbb{R}}_+^n$ and $u_{\mathrm{e}} \in \overline{\mathbb{R}}_+^m$, it follows from Brockett's necessary condition for asymptotic stabilizability [25] that there does not exist a continuous stabilizing *nonnegative* feedback if $0 \in \mathrm{spec}(A + \sum_{i=1}^p A_{\mathrm{d}i})$ and $x_{\mathrm{e}} \in \mathbb{R}_+^n$ (see [21] for further details). Hence, in this section we assume that $A + \sum_{i=1}^p A_{\mathrm{d}i}$ is an asymptotically stable compartmental matrix. Thus, we proceed with the aforementioned assumptions to design adaptive controllers for uncertain time-delay compartmental systems that guarantee that $\lim_{t \to \infty} x_i(t) = x_{\mathrm{d}i} \geq 0$ for $i - 1, \ldots, m \leq n$, where $x_{\mathrm{d}i}$ is a desired set point for the ith compartmental state while guaranteeing a nonnegative control input.

Theorem 5. *Consider the linear uncertain time-delay system $\mathcal{G}$ given by (11), where A is essentially nonnegative, $A_{\mathrm{d}i}$, $i = 1, \ldots, p$, is nonnegative, $A + \sum_{i=1}^p A_{\mathrm{d}i}$ is asymptotically stable, and B is nonnegative and given by (12). For a given $x_{\mathrm{d}} \in \mathbb{R}^m$, assume there exist vectors $x_{\mathrm{u}} \in \mathbb{R}_+^{n-m}$ and $u_{\mathrm{e}} \in \overline{\mathbb{R}}_+^m$ such that (13) holds. Furthermore, let q_i and $\hat{q}_i$, $i = 1, \ldots, m$, be positive constants. Then, the adaptive feedback control law*

$$u_i(t) = \max\{0, \hat{u}_i(t)\}, \quad i = 1, \ldots, m, \tag{36}$$

where

$$\hat{u}_i(t) = k_i(t)(x_i(t) - x_{\mathrm{d}i}) + \phi_i(t), \quad i = 1, \ldots, m, \tag{37}$$

$k_i(t) \in \mathbb{R}$, $t \geq 0$, and $\phi_i(t) \in \mathbb{R}$, $t \geq 0$, $i = 1, \ldots, m$, *with update laws*

$$\dot{k}_i(t) = \begin{cases} 0, & \text{if } \hat{u}_i(t) < 0, \\ -q_i(x_i(t) - x_{\mathrm{d}i})^2, & \text{otherwise,} \end{cases} \quad k_i(0) \leq 0, \quad i = 1, \ldots, m, \tag{38}$$

$$\dot{\phi}_i(t) = \begin{cases} 0, & \text{if } \phi_i(t) = 0 \text{ and } x_i(t) > x_{\mathrm{d}i}, \text{ or if } \hat{u}_i(t) \leq 0, \\ -\hat{q}_i(x_i(t) - x_{\mathrm{d}i}), & \text{otherwise,} \end{cases}$$

$$\phi_i(0) \geq 0, \quad i = 1, \ldots, m, \tag{39}$$

guarantees that the solution $(x(t), K(t), \phi(t)) \equiv (x_{\mathrm{e}}, 0, u_{\mathrm{e}})$ of the closed-loop system given by (11), (36), (38), and (39) is Lyapunov stable and $x_i(t) \to x_{\mathrm{d}i}$, $i = 1, \ldots, m$, as $t \to \infty$ for all $\eta(\cdot) \in \mathcal{C}_+$. Furthermore, $u(t) >> 0$, $t \geq 0$, and $x(t) >> 0$, $t \geq 0$, for all $\eta(\cdot) \in \mathcal{C}_+$.

Proof. First, define $K_{\mathrm{u}}(t) \triangleq \operatorname{diag}[k_{\mathrm{u}1}(t), \ldots, k_{\mathrm{u}m}(t)]$ and $\phi_{\mathrm{u}}(t) \triangleq [\phi_{\mathrm{u}1}(t), \ldots, \phi_{\mathrm{u}m}(t)]^{\mathrm{T}}$, where

$$k_{\mathrm{u}i}(t) = \begin{cases} 0, & \text{if } \hat{u}_i(t) < 0, \\ k_i(t), & \text{otherwise,} \end{cases} \qquad i = 1, \ldots, m,$$

$$\phi_{\mathrm{u}i}(t) = \begin{cases} 0, & \text{if } \hat{u}_i(t) < 0, \\ \phi_i(t), & \text{otherwise,} \end{cases} \qquad i = 1, \ldots, m.$$

Now, note that with $u(t)$, $t \geq 0$, given by (36), it follows from (11) that

$$\dot{x}(t) = Ax(t) + \sum_{i=1}^{p} A_{\mathrm{d}i}x(t - \tau_i) + BK_{\mathrm{u}}(t)(\hat{x}(t) - x_{\mathrm{d}}) + B\phi_{\mathrm{u}}(t),$$

$$x(\theta) = \eta(\theta), \quad -\bar{\tau} \leq \theta \leq 0, \quad t \geq 0, \quad (40)$$

or, equivalently, using (13),

$$\dot{x}(t) = A(x(t) - x_{\mathrm{e}}) + \sum_{i=1}^{p} A_{\mathrm{d}i}(x(t - \tau_i) - x_{\mathrm{e}}) + BK_{\mathrm{u}}(t)(\hat{x}(t) - x_{\mathrm{d}})$$

$$+ B(\phi_{\mathrm{u}}(t) - u_{\mathrm{e}}), \quad x(\theta) = \eta(\theta), \quad -\bar{\tau} \leq \theta \leq 0, \quad t \geq 0. \quad (41)$$

Since $A + \sum_{i=1}^{p} A_{\mathrm{d}i}$ is essentially nonnegative and Hurwitz, it follows from Theorem 1 that there exists diagonal matrices $P > 0$ and $\tilde{Q}_i \geq 0$, $i = 1, \ldots, p$, and a positive-definite matrix $R \in \mathbb{R}^{n \times n}$ such that

$$0 = A^{\mathrm{T}}P + PA + \sum_{i=1}^{p} (\tilde{Q}_i + A_{\mathrm{d}i}^{\mathrm{T}}P\tilde{Q}_i^{\#}PA_{\mathrm{d}i}) + R. \quad (42)$$

To show Lyapunov stability of the closed-loop system (38), (39), and (41) consider the Lyapunov-Krasovskii functional candidate $V : \mathcal{C}_+ \times \mathbb{R}^{m \times m} \times \mathbb{R}^m \to \mathbb{R}$ given by

$$V(\psi, K, \phi) = (\psi(0) - x_{\mathrm{e}})^{\mathrm{T}}P(\psi(0) - x_{\mathrm{e}})$$

$$+ \sum_{i=1}^{p} \int_{-\tau_i}^{0} (\psi(\theta) - x_{\mathrm{e}})^{\mathrm{T}}A_{\mathrm{d}i}^{\mathrm{T}}P\tilde{Q}_i^{\#}PA_{\mathrm{d}i}(\psi(\theta) - x_{\mathrm{e}})\mathrm{d}\theta$$

$$+ \operatorname{tr} K^{\mathrm{T}}Q^{-1}K + (\phi - u_{\mathrm{e}})^{\mathrm{T}}\hat{Q}^{-1}(\phi - u_{\mathrm{e}}), \quad (43)$$

or, equivalently,

$$V(\psi, K, \phi) = \sum_{i=1}^{n} p_i(\psi_i(0) - x_{\mathrm{e}i})^2$$

$$+ \sum_{i=1}^{p} \int_{-\tau_i}^{0} (\psi(\theta) - x_{\mathrm{e}})^{\mathrm{T}}A_{\mathrm{d}i}^{\mathrm{T}}P\tilde{Q}_i^{\#}PA_{\mathrm{d}i}(\psi(\theta) - x_{\mathrm{e}})\mathrm{d}\theta$$

$$+ \sum_{i=1}^{m} \frac{p_i b_i}{q_i} k_i^2 + \sum_{i=1}^{m} \frac{p_i b_i}{q_i}(\phi_i - u_{\mathrm{e}i})^2,$$

where $Q = \mathrm{diag}[\frac{q_1}{p_1 b_1}, \ldots, \frac{q_m}{p_m b_m}]$ and $\hat{Q} = \mathrm{diag}[\frac{\hat{q}_1}{p_1 b_1}, \ldots, \frac{\hat{q}_m}{p_m b_m}]$. Note that $V(\psi_\mathrm{e}, 0, u_\mathrm{e}) = 0$, where $\psi_\mathrm{e}(\theta) = x_\mathrm{e}$, $\theta \in [-\bar{\tau}, 0]$. Furthermore, there exist $\mathcal{K}$-class functions $\alpha_1(\cdot)$, $\alpha_2(\cdot)$, and $\alpha_3(\cdot)$ such that

$$V(\psi, K, \phi) \geq \alpha_1(\|\psi(0) - x_\mathrm{e}\|) + \alpha_2(\|K\|_\mathrm{F}) + \alpha_3(\|\phi - u_\mathrm{e}\|).$$

Next, letting $x(t)$, $t \geq 0$, denote the solution to (41) and using (38) and (39), it follows that the Lyapunov-Krasovskii directional derivative of $V(x_t, K(t), \phi(t))$ along the closed-loop system trajectories is given by

$$
\begin{aligned}
\dot{V}(x_t, K(t), \phi(t)) = {} & -(x(t) - x_\mathrm{e})^\mathrm{T} R(x(t) - x_\mathrm{e}) \\
& - \sum_{i=1}^{p} [\tilde{Q}_i(x(t) - x_\mathrm{e}) - PA_{\mathrm{d}i}(x(t - \tau_i) - x_\mathrm{e})]^\mathrm{T} \tilde{Q}_i^{\#} \\
& \cdot [\tilde{Q}_i(x(t) - x_\mathrm{e}) - PA_{\mathrm{d}i}(x(t - \tau_i) - x_\mathrm{e})] \\
& + 2 \sum_{i=1}^{m} p_i b_i k_{\mathrm{u}i}(t)(x_i(t) - x_\mathrm{d})^2 \\
& + 2 \sum_{i=1}^{m} p_i b_i (x_i(t) - x_\mathrm{e})(\phi_{\mathrm{u}i}(t) - u_{\mathrm{e}i}) \\
& + 2 \sum_{i=1}^{m} \frac{p_i b_i}{q_i} (k_i(t) - k_{\mathrm{g}i})\dot{k}_i(t) + 2 \sum_{i=1}^{m} \frac{p_i b_i}{\hat{q}_i} (\phi_i(t) - u_{\mathrm{e}i})\dot{\phi}_i(t) \\
\leq {} & -(x(t) - x_\mathrm{e})^\mathrm{T} R(x(t) - x_\mathrm{e}) \\
& + 2 \sum_{i=1}^{m} p_i b_i \Big[k_{\mathrm{u}i}(t)(x_i(t) - x_{\mathrm{d}i})^2 + \frac{1}{q_i} k_i(t)\dot{k}_i(t) \Big] \\
& + 2 \sum_{i=1}^{m} p_i b_i \Big[(x_i(t) - x_{\mathrm{d}i})(\phi_{\mathrm{u}i}(t) - u_{\mathrm{e}i}) \\
& + \frac{1}{\hat{q}_i}(\phi_i(t) - u_{\mathrm{e}i})\dot{\phi}_i(t) \Big], \quad t \geq 0. \tag{44}
\end{aligned}
$$

For the two cases given in (38) and (39), the last two terms on the right-hand side of (44) give:

i) If $\hat{u}_i(t) < 0$, $t \geq 0$, then $k_{\mathrm{u}i}(t) = 0$, $\phi_{\mathrm{u}i}(t) = 0$, $\dot{k}_i(t) = 0$, and $\dot{\phi}_i(t) = 0$, $t \geq 0$. Furthermore, since $\phi_i(t) \geq 0$ and $k_i(t) \leq 0$ for all $t \geq 0$ and $i = 1, \ldots, m$, it follows from (37) that $\hat{u}_i(t) < 0$ only if $x_i(t) > x_{\mathrm{d}i}$, $t \geq 0$, and hence, for $t \geq 0$,

$$k_{\mathrm{u}i}(t)(x_i(t) - x_{\mathrm{d}i})^2 + \tfrac{1}{q_i} k_i(t)\dot{k}_i(t) = 0,$$

$$(x_i(t) - x_{\mathrm{d}i})(\phi_{\mathrm{u}i}(t) - u_{\mathrm{e}i}) + \tfrac{1}{q_i}(\phi_i(t) - u_{\mathrm{e}i})\dot{\phi}_i(t) = -(x_i(t) - x_{\mathrm{d}i})u_{\mathrm{e}i} \leq 0.$$

ii) Otherwise, $k_{\mathrm{u}i}(t) = k_i(t)$ and $\phi_{\mathrm{u}i}(t) = \phi_i(t)$, and hence, for $t \geq 0$,

$$k_{\mathrm{u}i}(x_i(t) - x_{\mathrm{d}i})^2 + \frac{1}{q_i}k_i(t)\dot{k}_i(t) = 0,$$

$$(x_i(t) - x_{\mathrm{d}i})(\phi_{\mathrm{u}i}(t) - u_{\mathrm{e}i}) + \frac{1}{\hat{q}_i}(\phi_i(t) - u_{\mathrm{e}i})\dot{\phi}_i(t)$$

$$= \begin{cases} -(x_i(t) - x_{\mathrm{d}i})u_{\mathrm{e}i} \leq 0, & \text{if } \phi_i(t) = 0 \text{ and } x_i(t) \geq x_{\mathrm{d}i}, \\ 0, & \text{otherwise.} \end{cases}$$

Hence, it follows that in either case

$$\dot{V}(x_t, K(t), \phi(t)) \leq -(x(t) - x_{\mathrm{e}})^{\mathrm{T}} R(x(t) - x_{\mathrm{e}}) \leq 0, \quad t \geq 0, \tag{45}$$

which proves that the solution $(x(t), K(t), \phi(t)) \equiv (x_{\mathrm{e}}, 0, u_{\mathrm{e}})$ to (38), (39), and (41) is Lyapunov stable. Furthermore, since the positive orbit $\gamma^+(\eta(\theta), K_0, \phi_0)$ of the closed-loop system (38), (39), and (41) is bounded, $\gamma^+(\eta(\theta), K_0, \phi_0)$ belongs to a compact subset of $\mathcal{C}_+ \times \mathbb{R}^{m \times m} \times \mathbb{R}^m$ [26], and $R > 0$, it follows from the Krasovskii-LaSalle invariant set theorem for infinite-dimensional systems [18, p. 143] that $x(t) \to x_{\mathrm{e}}$ as $t \to \infty$ for all $\eta(\cdot) \in \mathcal{C}_+$.

Finally, $u(t) \geq\geq 0$, $t \geq 0$, is a restatement of (36). Now, since $B \geq\geq 0$ and $u(t) \geq\geq 0$, $t \geq 0$, it follows from Proposition 1 that $x(t) \geq\geq 0$, $t \geq 0$, for all $\eta(\cdot) \in \mathcal{C}_+$. $\qquad\square$

As in the case of Theorem 3, it is important to note that the adaptive control law (36), (38), and (39) does not require the explicit knowledge of the nonnegative constant vector u_{e}, even though Theorem 5 requires the existence of nonnegative vectors x_{u} and u_{e} such that the condition (13) holds.

5 Adaptive Control for General Anesthesia

Almost all anesthetics are *myocardial* depressants, that is, they decrease the strength of the contraction of the heart and lower *cardiac output* (i.e., the volume of blood pumped by the heart per unit time). As a consequence, decreased cardiac output slows down the transfer of blood from the central compartments comprising the heart, brain, kidney, and liver to the peripheral compartments of muscle and fat. In addition, decreased cardiac output can increase drug concentrations in the central compartments, compounding side effects. This instability can lead to overdosing which, at the very least, can delay recovery from anesthesia and, in the worst case, can result in respiratory and cardiovascular collapse. Alternatively, underdosing can cause psychological trauma from awareness and pain during surgery.

Control of drug effect is clinically important since overdosing or underdosing incur risk for the patient. To illustrate the adaptive control framework developed in this paper for general anesthesia we consider a hypothetical model for the intravenous anesthetic propofol. The pharmacokinetics of propofol are described by the three-compartment model [21, 27] shown in Figure 1, where x_1

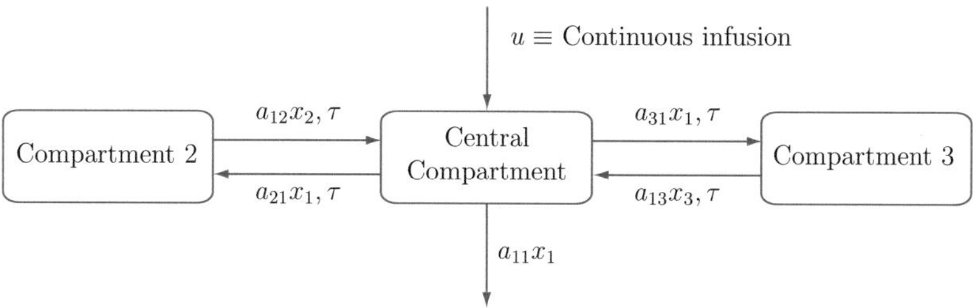

Fig. 1. Three-compartment mammillary model for disposition of propofol

denotes the mass of drug in the central compartment, which is the site for drug administration and includes the *intravascular blood* (blood within arteries or veins) volume as well as *highly perfused* organs, that is, organs with high ratios of blood flow to weight such as the heart, brain, kidney, and liver, which receive a large fraction of the cardiac output. The remainder of the drug in the body is assumed to reside in two peripheral compartments, one identified with muscle and one with fat; the masses in these compartments are denoted by x_2 and x_3, respectively. These compartments receive less than 20% of the cardiac output. Finally, the transfer time between the central compartment and peripheral compartment 2 (muscle) is given by $\tau > 0$, and transfer time between the central compartment and peripheral compartment 3 (fat) is given by $\tau > 0$.

A mass balance for the whole compartmental system yields

$$\dot{x}_1(t) = -(a_{11} + a_{21} + a_{31})x_1(t) + a_{12}x_2(t - \tau) + a_{13}x_3(t - \tau) + u(t),$$
$$x_1(\theta) = \eta_1(\theta), \quad -\tau \le \theta \le 0, \quad t \ge 0, \tag{46}$$
$$\dot{x}_2(t) = -a_{12}x_2(t) + a_{21}x_1(t - \tau), \quad x_2(\theta) = \eta_2(\theta), \quad -\tau \le \theta \le 0, \tag{47}$$
$$\dot{x}_3(t) = -a_{13}x_3(t) + a_{31}x_1(t - \tau), \quad x_3(\theta) = \eta_3(\theta), \quad -\tau \le \theta \le 0, \tag{48}$$

where $x_1(t)$, $x_2(t)$, $x_3(t)$, $t \ge 0$, are the masses in grams of propofol in the central compartment and compartments 2 and 3, respectively, $u(t)$, $t \ge 0$, is the infusion rate in grams/min of the anesthetic drug propofol into the central compartment, $a_{ij} > 0$, $i \ne j$, $i, j = 1, 2, 3$, are the rate constants in min^{-1} for drug transfer between compartments, and $a_{11} > 0$ is the rate constant in min^{-1} of drug metabolism and elimination (metabolism typically occurs in the liver) from the central compartment. Even though the transfer and loss coefficients are positive, they can be uncertain due to patient gender, weight, pre-existing disease, age, and concomitant medication. Hence, adaptive control for propofol set-point regulation can significantly improve the outcome for drug administration over manual (open-loop) control.

It has been reported in [28] that a 2.5–6 μg/ml blood concentration level of propofol is required during the maintenance stage in general anesthesia depending on patient fitness and extent of surgical stimulation. Hence, continuous infusion control is required for maintaining this desired level of anesthesia. Here we assume that the transfer and loss coefficients a_{11}, a_{12}, a_{21}, a_{13}, and a_{31} are unknown and our objective is to regulate the propofol concentration level of the central compartment to the desired level of 3.4 μg/ml in the face of system uncertainty. Furthermore, since propofol mass in the blood plasma cannot be measured directly, we measure the concentration of propofol in the central compartment, that is, x_1/V_c, where V_c is the volume in liters of the central compartment. As noted in [29], V_c can be approximately calculated by $V_c = (0.159\,\mathrm{l/kg})(M\,\mathrm{kg})$, where M is the mass in kilograms of the patient.

Next, note that (46)–(48) can be written in the state space form (11) with state vector $x = [x_1,\ x_2,\ x_3]^{\mathrm{T}}$,

$$A = \begin{bmatrix} -(a_{11}+a_{21}+a_{31}) & 0 & 0 \\ 0 & -a_{12} & 0 \\ 0 & 0 & -a_{13} \end{bmatrix}, \quad A_{\mathrm{d}} = \begin{bmatrix} 0 & a_{12} & a_{13} \\ a_{21} & 0 & 0 \\ a_{31} & 0 & 0 \end{bmatrix}, \quad B = \begin{bmatrix} 1 \\ 0 \\ 0 \end{bmatrix}. \tag{49}$$

Now, it can be shown that for $x_{\mathrm{d}1}/V_c = 3.4$ μg/ml, all the conditions of Theorem 5 are satisfied. Even though propofol concentration levels in the blood plasma will lead to the desired depth of anesthesia, they cannot be measured in *real-time* during surgery. Furthermore, we are more interested in drug *effect* (depth of hypnosis) rather than drug *concentration*. Hence, we consider a more realistic model involving pharmacokinetics (drug concentration as a function of time) and pharmacodynamics (drug effect as a function of concentration) for control of anesthesia. Specifically, we use an electroencephalogram (EEG) signal as a measure of drug effect of anesthetic compounds on the brain [30]. Since electroencephalography provides real-time monitoring of the central nervous system activity, it can be used to quantify levels of consciousness and hence is amenable for feedback (closed-loop) control in general anesthesia. Furthermore, we use the Bispectral Index (BIS), an EEG indicator, as a measure of anesthetic effect [31]. This index quantifies the nonlinear relationships between the component frequencies in the electroencephalogram, as well as analyzing their phase and amplitude.

The BIS signal is related to drug concentration by the empirical relationship

$$\mathrm{BIS}(c_{\mathrm{eff}}) = \mathrm{BIS}_0 \left(1 - \frac{c_{\mathrm{eff}}^{\gamma}}{c_{\mathrm{eff}}^{\gamma} + \mathrm{EC}_{50}^{\gamma}} \right), \tag{50}$$

where BIS_0 denotes the base line (awake state) value and, by convention, is typically assigned a value of 100, c_{eff} is the propofol concentration in grams/liter in the effect site compartment (brain), EC_{50} is the concentration at half maximal effect and represents the patient's sensitivity to the drug, and γ determines the degree of nonlinearity in (50). Here, the effect-site compartment is introduced to account for finite equilibration time between the central compartment concentration and the central nervous system concentration [32].

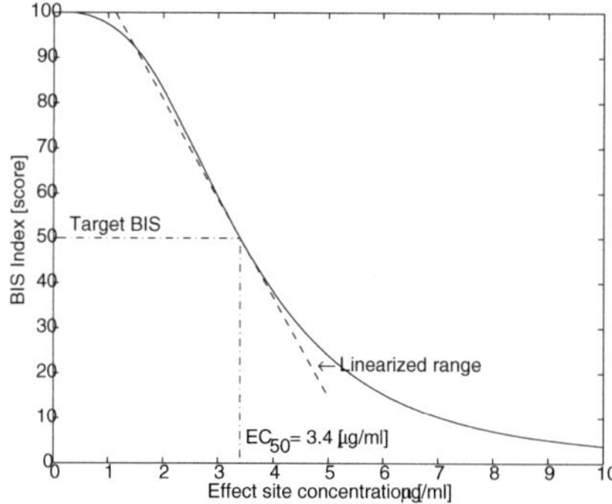

Fig. 2. BIS index versus effect site concentration

The effect-site compartment concentration is related to the concentration in the central compartment by the first-order model

$$\dot{c}_{\text{eff}}(t) = a_{\text{eff}}(x_1(t)/V_{\text{c}} - c_{\text{eff}}(t)), \quad c_{\text{eff}}(0) = x_1(0), \quad t \geq 0, \tag{51}$$

where a_{eff} in min^{-1} is an unknown positive time constant. In reality, the effect-site compartment equilibrates with the central compartment in a matter of a few minutes. The parameters a_{eff}, EC$_{50}$, and γ are determined by data fitting and vary from patient to patient. BIS index values of 0 and 100 correspond, respectively, to an *isoelectric* EEG signal (no cerebral electrical activity) and an EEG signal of a fully conscious patient; the range between 40 and 60 indicates a moderate hypnotic state [33].

In the following numerical simulation we set EC$_{50}$ = 3.4 μg/ml, $\gamma = 3$, and BIS$_0$ = 100, so that the BIS signal is shown in Figure 2. The values for the pharmacodynamic parameters (EC$_{50}$, γ) are within the typical range of those observed for ligand-receptor binding [34, 35]. The target (desired) BIS value, BIS$_{\text{target}}$, is set at 50. In this case, the linearized BIS function about the target BIS value is given by

$$\text{BIS}(c_{\text{eff}}) \simeq \text{BIS}(\text{EC}_{50}) - \text{BIS}_0 \cdot \text{EC}_{50}^{\gamma} \cdot \left. \frac{\gamma c_{\text{eff}}^{\gamma-1}}{(c_{\text{eff}}^{\gamma} + \text{EC}_{50}^{\gamma})^2} \right|_{c_{\text{eff}}=\text{EC}_{50}} \cdot c_{\text{eff}}$$

$$= 125 - 22.06 c_{\text{eff}}. \tag{52}$$

Furthermore, for simplicity of exposition, we assume that the effect-site compartment equilibrates instantaneously with the central compartment, that is, we assume that $c_{\text{eff}}(t) = x_1(t)/V_{\text{c}}$ for all $t \geq 0$.

Now, using the adaptive feedback controller

$$u_1(t) = \max\{0, \hat{u}_1(t)\}, \tag{53}$$

where

$$\hat{u}_1(t) = -k_1(t)(\text{BIS}(t) - \text{BIS}_{\text{target}}) + \phi_1(t), \tag{54}$$

Table 1. Pharmacokinetic parameters [36]

Set	a_{11} (min^{-1})	a_{21} (min^{-1})	a_{12} (min^{-1})	a_{31} (min^{-1})	a_{13} (min^{-1})
A	0.152	0.207	0.092	0.040	0.0048
B	0.119	0.114	0.055	0.041	0.0033

and $k_1(t) \in \mathbb{R}$ and $\phi_1(t) \in \mathbb{R}$ are scalars for $t \geq 0$, with update laws

$$\dot{k}_1(t) = \begin{cases} 0, & \text{if } \hat{u}_1(t) < 0, \\ -q_{\mathrm{BIS}_1}(\mathrm{BIS}(t) - \mathrm{BIS}_{\mathrm{target}})^2, & \text{otherwise,} \end{cases} \quad k_1(0) \leq 0, \qquad (55)$$

$$\dot{\phi}_1(t) = \begin{cases} 0, & \begin{aligned} &\text{if } \phi_1(t) = 0 \text{ and } \mathrm{BIS}(t) > \mathrm{BIS}_{\mathrm{target}}, \\ &\text{or if } \hat{u}_1(t) \leq 0, \end{aligned} \\ \hat{q}_{\mathrm{BIS}_1}(\mathrm{BIS}(t) - \mathrm{BIS}_{\mathrm{target}}), & \text{otherwise,} \end{cases}$$

$$\phi_1(0) \geq 0, \qquad (56)$$

where q_{BIS_1} and $\hat{q}_{\mathrm{BIS}_1}$ are positive constants, it follows from Theorem 5 that the control input (anesthetic infusion rate) satisfies $u(t) \geq 0$ for all $t \geq 0$ and $\mathrm{BIS}(t) \to \mathrm{BIS}_{\mathrm{target}}$ as $t \to \infty$ for all nonnegative values of the transfer and loss coefficients in the range of c_{eff} where the linearized BIS equation (52) is valid. It is important to note that during actual surgery or intensive care unit sedation the BIS signal is obtained directly from the EEG and not (50). Furthermore, since our adaptive controller only requires the error signal $\mathrm{BIS}(t) - \mathrm{BIS}_{\mathrm{target}}$ over the linearized range of (50), we do not require knowledge of the slope of the linearized equation (52), nor do we require knowledge of the parameters γ and EC_{50}.

To numerically illustrate the efficacy of the proposed adaptive control law, we use the average set of pharmacokinetic parameters given in [36] for 29 patients requiring general anesthesia for noncardiac surgery. For our design we assume $M = 70$ kg and we switch from Set A to Set B given in Table 1 at $t = 25$ min. Furthermore, we assume that at $t = 25$ min the pharmacodynamic parameters EC_{50} and γ are switched from 3.4 μg/ml and 3 to 4.0 μg/ml and 4, respectively. Here, we consider noncardiac surgery since cardiac surgery often utilizes hypothermia which itself changes the BIS signal. With $\tau = 1$ min, $q_{\mathrm{BIS}_1} = 1 \times 10^{-6}$ g/min^2, $\hat{q}_{\mathrm{BIS}_1} = 1 \times 10^{-3}$ g/min^2, and initial conditions $x(0) = [0, 0, 0]^{\mathrm{T}}$ g, $k_1(0) = 0$ min^{-1}, and $\phi_1(0) = 0.01$ g/min^{-1}, Figure 3 shows the masses of propofol in all three compartments versus time. Figure 4 shows the BIS index versus time. Figure 5 shows the propofol concentration in the central compartment and the control signal (propofol infusion rate) versus time. Finally, Figure 6 shows the adaptive gain history versus time.

The adaptive controller (53)–(56) does not require knowledge of the pharmacokinetic and pharmacodynamic parameters, in contrast to previous algorithms for closed-loop control of anesthesia [37, 38]. However, the adaptive controller (53)–(56) does not account for time delays due to equilibration between the central circulation and the effect-site compartment as well as due to the proprietary

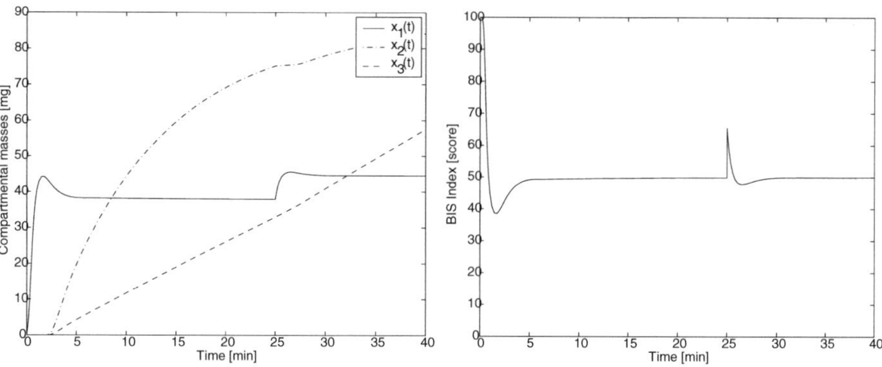

Fig. 3. Compartmental masses versus time

Fig. 4. BIS index versus time

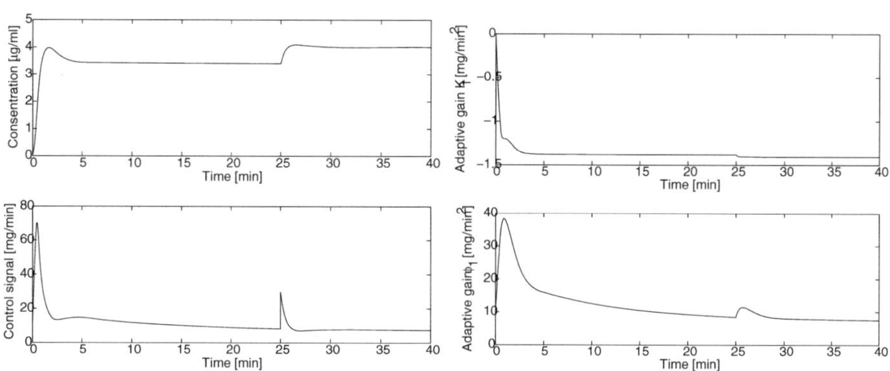

Fig. 5. Drug concentration in the central compartment and control signal (infusion rate) versus time

Fig. 6. Adaptive gain history versus time

signal-averaging algorithm within the BIS monitor. The adaptive controller also ignores measurement noise. Extensive clinical testing is needed to access the significance of these assumptions and approximations. Since there is often a substantial delay between observed changes in patient status and a change in the BIS signal, other measures of depth of anesthesia may be needed [39, 40, 41].

6 Conclusion

In this paper, we developed a direct adaptive control framework for linear uncertain nonnegative and compartmental dynamical systems with unknown time delay. In particular, a Lyapunov-Krasovskii-based direct adaptive control framework for guaranteeing set-point regulation for nonnegative and compartmental time-delay systems with specific applications to mammillary pharmacokinetic models was developed. Finally, we numerically demonstrated the framework on a drug delivery pharmacokinetic/pharmacodynamic model with time delay.

An important issue for future research is sensor measurement noise. In particular, EEG signals may have as much as 10% variation due to noise. For example, the BIS signal may be corrupted by *electromyographic noise*, that is, signals emanating from muscle rather than the central nervous system. Even though electromyographic noise can be minimized by muscle paralysis, there are other sources of measurement noise, such as electrocautery, that are stochastic in nature and need to be accounted for within the control design processes. Extensions of the proposed adaptive control framework to systems with exogenous disturbances will be addressed in a future paper.

Acknowledgement

This research was supported in part by the National Science Foundation under Grants ECS-0551947 and ECS-0601311, and the Air Force Office of Scientific Research under Grant FA9550-06-1-0240.

References

1. R. R. Mohler, "Biological modeling with variable compartmental structure," *IEEE Trans. Autom. Control*, vol. 19, pp. 922–926, 1974.
2. H. Maeda, S. Kodama, and F. Kajiya, "Compartmental system analysis: Realization of a class of linear systems with physical constraints," *IEEE Trans. Circuits Syst.*, vol. 24, pp. 8–14, 1977.
3. W. Sandberg, "On the mathematical foundations of compartmental analysis in biology, medicine and ecology," *IEEE Trans. Circuits Syst.*, vol. 25, pp. 273–279, 1978.
4. H. Maeda, S. Kodama, and Y. Ohta, "Asymptotic behavior of nonlinear compartmental systems: Nonoscillation and stability," *IEEE Trans. Circuits Syst.*, vol. 25, pp. 372–378, 1978.
5. A. Berman, M. Neumann, and R. J. Stern, *Nonnegative Matrices in Dynamical Systems*. New York: Wiley, 1979.
6. R. E. Funderlic and J. B. Mankin, "Solution of homogeneous systems of linear equations arising from compartmental models," *SIAM J. Sci. Statist. Comput.*, vol. 2, pp. 375–383, 1981.
7. D. H. Anderson, *Compartmental Modeling and Tracer Kinetics*. New York: Springer-Verlag, 1983.
8. J. W. Nieuwenhuis, "About nonnegative realizations," *Syst. Control Lett.*, vol. 1, pp. 283–287, 1982.
9. K. Godfrey, *Compartmental Models and their Applications*. New York: Academic, 1983.
10. Y. Ohta, H. Maeda, and S. Kodama, "Reachability, observability and realizability of continuous-time postive systems," *SIAM J. Control Optim.*, vol. 22, pp. 171–180, 1984.
11. J. A. Jacquez, *Compartmental Analysis in Biology and Medicine*. Ann Arbor, MI: Univ. Michigan Press, 1985.

12. D. S. Bernstein and D. C. Hyland, "Compartmental modeling and second-moment analysis of state space systems," *SIAM J. Matrix Anal. Appl.*, vol. 14, pp. 880–901, 1993.

13. J. A. Jacquez and C. P. Simon, "Qualitative theory of compartmental systems," *SIAM Rev.*, vol. 35, pp. 43–79, 1993.

14. W. M. Haddad and V. Chellaboina, "Stability and dissipativity theory for nonnegative and dynamical systems: a unified analysis framework for biological and physiological systems," *Nonlinear Analysis: Real World Applications*, vol. 6, pp. 35–65, 2005.

15. I. Gyori, "Delay differential and integro-differential equations in biological compartment models," *Syst. Sci.*, vol. 8, no. 2–3, pp. 167–187, 1982.

16. H. Maeda, S. Kodama, and T. Konishi, "Stability theory and existence of periodic solutions of time delayed compartmental systems," *Electron. Commun. Jpn.*, vol. 65, no. 1, pp. 1–8, 1982.

17. N. N. Krasovskii, *Stability of Motion*. Stanford, CA: Stanford Univ. Press, 1963.

18. J. K. Hale and S. M. V. Lunel, *Introduction to Functional Differential Equations*. New York: Springer-Verlag, 1993.

19. L. Dugard and E. Verriest, Eds., *Stability and Control of Time-Delay Systems*. New York: Springer-Verlag, 1998.

20. S. I. Niculescu, *Delay Effects on Stability: A Robust Control Approach*. New York: Springer-Verlag, 2001.

21. W. M. Haddad, T. Hayakawa, and J. M. Bailey, "Adaptive control for nonnegative and compartmental dynamical systems with applications to general anesthesia," *Int. J. Adapt. Control Signal Process.*, vol. 17, pp. 209–235, 2003.

22. W. M. Haddad, V. Chellaboina, and T. Rajpurohit, "Dissipativity theory for nonnegative and compartmental dynamical systems with time delay," *IEEE Trans. Autom. Control*, vol. 49, pp. 747–751, 2004.

23. W. M. Haddad and V. Chellaboina, "Stability theory for nonnegative and compartmental dynamical systems with time delay," *Syst. Control Lett.*, vol. 51, pp. 355–361, 2004.

24. A. Berman and R. J. Plemmons, *Nonnegative Matrices in the Mathematical Sciences*. New York: Academic, 1979.

25. P. D. Leenheer and D. Aeyels, "Stabilization of positive linear systems," *Syst. Control Lett.*, vol. 44, pp. 259–271, 2001.

26. J. K. Hale, "Dynamical systems and stability," *J. Math. Anal. Appl.*, vol. 26, pp. 39–59, 1969.

27. B. Marsh, M. White, N. Morton, and G. N. Kenny, "Pharmacokinetic model driven infusion of propofol in children," *Brit. J. Anaesth.*, vol. 67, no. 1, pp. 41–48, 1991.

28. M. White and G. N. C. Kenny, "Intravenous propofol anaesthesia using a computerised infusion system," *Anaesthesia*, vol. 45, pp. 204–209, 1990.

29. D. A. Linkens, M. F. Abbod, and J. E. Peacock, "Clinical implementation of advanced control in anaesthesia," *Trans. Inst. Meas. Control*, vol. 22, pp. 303–330, 2000.

30. J. C. Sigl and N. G. Chamoun, "An introduction to bispectral analysis for the electroencephalogram," *J. Clin. Monit.*, vol. 10, pp. 392–404, 1994.

31. E. Mortier, M. Struys, T. De Smet, L. Versichelen, and G. Rolly, "Closed-loop controlled administration of propofol using bispectral analysis," *Anaesthesia*, vol. 53, pp. 749–754, 1998.

32. T. W. Schnider, C. F. Minto, and D. R. Stanski, "The effect compartment concept in pharmacodynamic modelling," *Anaes. Pharmacol. Rev.*, vol. 2, pp. 204–219, 1994.

33. A. Gentilini, M. Rossoni-Gerosa, C. W. Frei, R. Wymann, M. Morari, A. M. Zbinden, and T. W. Schnider, "Modeling and closed-loop control of hypnosis by means of bispectral index (BIS) with isoflurane," *IEEE Trans. Biomed. Eng.*, vol. 48, pp. 874–889, 2001.

34. R. G. Eckenhoff and J. S. Johansson, "On the relevance of 'clinically relevant concentrations' of inhaled anesthetics in vitro experiments," *Anesthesiology*, vol. 91, pp. 856–860, 1999.

35. T. Kazama, K. Ikeda, K. Morita, M. Kikura, M. Doi, T. Ikeda, and T. Kurita, "Comparison of the effect site KeOs of propofol for blood pressure and EEG bispectral index in elderly and young patients," *Anesthesiology*, vol. 90, pp. 1517–1527, 1999.

36. P. S. Glass, D. K. Goodman, B. Ginsberg, J. G. Reves, and J. R. Jacobs, "Accuracy of pharmacokinetic model-driven infusion of propofol," *Anesthesiology*, vol. 71, p. A277, 1989.

37. H. Schwilden, J. Schuttler, and H. Stoeckel, "Closed-loop feedback control of methohexital anesthesia by quantitative EEG analysis in humans," *Anesthesiology*, vol. 67, no. 3, pp. 341–347, 1987.

38. M. Struys, T. De Smet, L. Versichelen, S. Van de Vilde, R. Van den Broecke, and E. Mortier, "Comparison of closed-loop controlled administration of propofol using BIS as the controlled variable versus 'standard practice' controlled administration," *Anesthesiology*, vol. 95, no. 1, pp. 6–17, 2001.

39. X.-S. Zhang, R. J. Roy, and E. W. Jensen, "EEG complexity as a measure of depth of anesthesia for patients," *IEEE Trans. Biomed. Eng.*, vol. 48, no. 12, pp. 1424–1433, 2001.

40. S. Bibian, T. Zikov, G. A. Dumont, C. R. Ries, E. Puil, H. Ahmadi, M. Huzmezan, and B. A. Macleod, "Estimation of the anesthetic depth using wavelet analysis of electroencephalogram," in *Proc. Conf. IEEE Eng. Medicine Biology Soc.*, (Istanbul, Turkey), pp. 951–955, October 2001.

41. T. Zikov, S. Bibian, G. A. Dumont, and M. Huzmezan, "A wavelet based denoising technique for ocular artifact correction of the electroencephalogram," in *Proc. Conf. IEEE Eng. Medicine Biology Soc.*, (Huston, TX), pp. 98–105, October 2002.

Analysis and Control of Dynamical Biological Systems in Presence of Limitations

Isabelle Queinnec, Sophie Tarbouriech, and Germain Garcia

LAAS-CNRS, University of Toulouse, 7 avenue du Colonel Roche,
31077 Toulouse cedex 4, France
`queinnec@laas.fr`, `tarbour@laas.fr`, `garcia@laas.fr`

Summary. This chapter concerns the use of control theory to analyze some characteristic properties of biological systems involving limitations such as non-negativeness of variables and physical limitations. A biological system may be studied at different levels from the genes expression to the cells, through proteins levels or pathways. In this chapter, some interesting features related to constrained systems are recast to analyze and control biological systems at the level of cells, which aggregate at a macroscopic level interactions among proteins, RNA, DNA... based on the key notion of positively invariant sets.

1 Introduction

One main aspect related to biological systems concerns the growth of bacteria in controlled environmental conditions. For large scale industrial processes such as cell mass and primary metabolites productions as well as at laboratory scale operation for modelling microbial growth, stress conditions like the depletion of a nutriment source prevent the growth of bacteria at fast rates. This justifies the strong incentive to develop control schemes that would enable rapid start-up and stabilization of steady states in continuous bioreactors. Since the seventies, numerous control schemes have been proposed for continuous bioreactors, but none of them involve constraint problems related to the limitations of inputs, neither for analysis of the closed-loop system behavior nor for synthesis aspects [3], [12], [14], [26].

The positive invariance property of some sets has been used as an interesting theoretical tool to treat the problem of constraints satisfaction (on the state and/or input vectors) for controlled uncertain linear systems [23], [6], [21]. For such systems, cones properties offer nice tools to construct some positive invariant and asymptotically stable domain of linear behavior for the closed-loop system [23]. The potential interests with the application of these properties to the constrained output feedback control of biological reactions involved in a chemostat have to be examined. Two different classes of physical constraints are present in such systems. Input limitations are related to the constraints on the interactions with the environment. State (or output) limitations are related to the fact that such systems are nonnegative (and compartmental) dynamical systems [10].

I. Queinnec et al. (Eds.): Bio. & Ctrl. Theory: Current Challenges, LNCIS 357, pp. 317–338, 2007.
springerlink.com   © Springer-Verlag Berlin Heidelberg 2007

318   I. Queinnec, S. Tarbouriech, and G. Garcia

The motivation of this chapter is then to investigate input constrained linear control and analysis strategies based on cone properties from a practical perspective. The control problem of the continuous bioprocess is expressed in terms of stabilization and disturbances attenuation of an input and state constrained uncertain linear system. A positively invariant set of linear behavior has to be determined for a linearized output feedback system subject to disturbances, input and state constraints. The notion of positively invariant sets is a key factor in the derivation of the main results of this chapter. The organization of the paper is as follows. The section 2 is related to the modelling aspects of the biological process under study while the control objectives are stated in section 3. Basic concepts are introduced in section 4, before establishing some results related to the control and analysis problem in section 5. Some numerical evaluations are provided in section 6 before concluding the chapter in order to draw some forthcoming works.

Nomenclature. The notation used in the chapter is standard. $\Re_+$ is the set of nonnegative real numbers. For any vector $x \in \Re^n$, $x_{(i)}$ denotes the ith component of x. For any vector $x \in \Re^n$, $x \succeq 0$ means that all the components of x, denoted $x_{(i)}$, are nonnegative. For two vectors x, y of $\Re^n$, the notation $x \succeq y$ means that $x_{(i)} - y_{(i)} \geq 0$, $\forall i = 1, \ldots, n$. A', $\sigma(A)$ and $rank(A)$ denote the transpose, the spectrum and the rank of A, respectively. A^{-1} denotes the inverse of the nonsingular square matrix A. I_m denotes the m-order identity matrix.

2 Modelling Aspects

Molecular biological models are mainly characterized by the time evolution of variables which represent concentrations of external ligand, of cellules, of enzymes affecting the reaction... As an example the rate of growth of a species c using some substrate s may be described by:

$$\begin{cases} \dot{c} = \frac{\mu_{max}s}{k_s+s}c - k_c c \\ \dot{s} = -\frac{\nu_{max}s}{k_s+s}c - k_s s \end{cases}$$

where the first term of the right part of both equations is a classical approximation, namely a Michaelis-Menten term, which mimics the formation or consumption reaction rate as its underlying enzymatic reaction rate [18]. Such a monotone law expresses the limitation of the reaction rate by the presence of the substance s. Many other kinetics expression may be used, especially to represent inhibition by this substance, inhibition by other substances (cross-inhibition) [19], competitive coexistence of species [16]... The term $k_c c$ may represent a degradation term, a death term or a dilution term relying the species c with its environment. In the same way, the term $k_s s$ may be viewed as a degradation or a dilution term of the substrate s. Steady-state responses of such models are clearly described in [20].

Other terms may be involved in such expressions which would represent some other links with the environment, such as addition of nutriment, or which could be interpreted as additive disturbance on the biological models. We are then concerned herein with a continuous-flow biological growth process of a bacteria population x on limiting substrate s described by the following state equations:

$$\begin{cases} \dot{c} = \mu c - dc \\ \dot{s} = -\frac{\mu}{Y}c + d(s_{in} - s) \end{cases} \tag{1}$$

where the specific growth rate is given by:

$$\mu = \frac{\mu_{max}s}{k_s + s} \tag{2}$$

Furthermore, in (1), d is the dilution rate, s_{in} is the influent substrate concentration and Y is the yield of cell mass. Although this is a rough model, it does represent the dynamical behavior of many important constant-volume stirred tank bioreactors characterized by the growth of a single cell population on a single limiting substrate.

For a given dilution rate $\tilde{d}$ and a given influent substrate concentration $\tilde{s}_{in}$, one can directly determine the steady state of the open-loop operation (chemostat), which usually corresponds to quality specifications or from a steady state optimization of some performance index:

$$\begin{cases} \tilde{\mu} = \tilde{d} \\ \tilde{c} = Y(\tilde{s}_{in} - \tilde{s}) \end{cases} \tag{3}$$

It is worth noting that transient perturbations on the influent substrate concentration s_{in} around a nominal value will not alter the steady state. On the contrary, a step on s_{in} will cause a new steady state $\begin{bmatrix} \tilde{c} & \tilde{s} \end{bmatrix}'$, after some transient evolution. At last, if s_{in} moves slowly inside a set of admissible values, no real steady state will be attained during an open-loop operation. In the following, we consider the presence of perturbations Δs_{in} on the influent substrate concentration around the nominal value $\tilde{s}_{in}$ where:

$$-\Delta s_{in2} \leq \Delta s_{in} \leq \Delta s_{in1} \tag{4}$$

What we intend to do in this chapter is to consider the influence of the limitations in the state-space and in the input-space to analyze some properties of such systems, mainly with a control perspective. There are two types of constraints, namely input constraints and output (or state) constraints. The input constraints are always present and are imposed by physical limitations of the actuators which cannot be exceeded under any circumstances. The control input, d, is subject to a linear constraint defined by:

$$0 \leq d(t) \leq d_{max}, \quad \text{for } t \geq 0 \tag{5}$$

In the same way, state variables c and s are subject to constraints provided by physical considerations. The substrate concentration is physically limited by the influent substrate concentration, and the biomass concentration is then physically constrained according to the substrate bounds. However, it is also desirable to keep specific variables within certain limits for reasons related to plant operation, e.g., a substrate concentration over a certain value corresponds to a waste of energy. We then obtain, for $t \geq 0$:

$$0 \leq Y(\tilde{s}_{in} - \Delta s_{in2} - s_{max}) \leq c(t) \leq Y(\tilde{s}_{in} + \Delta s_{in1}) \tag{6}$$

$$0 \leq s(t) \leq s_{max} \leq \tilde{s}_{in} - \Delta s_{in2} \tag{7}$$

Just remark that it is usually unavoidable to exceed "economic" , at least temporarily, for example when the system is subjected to unexpected disturbances. Some authors have then referred such constraints as soft constraints [27].

Considering a disturbance Δs_{in} on the influent substrate concentration satisfying (4), Taylor series expansion allows to derive a linear model which approximates the plant around a non-trivial nominal steady state $(\tilde{c}, \tilde{s})$. Let $x_e = x - \tilde{x} = \begin{bmatrix} c - \tilde{c} & s - \tilde{s} \end{bmatrix}'$ and $u_e = u - \tilde{u} = d - \tilde{d} = d - \tilde{\mu}$, we get:

$$\dot{x}_e = A x_e + B u_e - A \begin{bmatrix} Y \Delta s_{in} \\ 0 \end{bmatrix} \tag{8}$$

where

$$A = \begin{bmatrix} 0 & (\frac{\partial \tilde{\mu}}{\partial s})(\tilde{c} + Y \Delta s_{in}) \\ -\frac{\tilde{\mu}}{Y} & -\tilde{\mu} - (\frac{\partial \tilde{\mu}}{\partial s})(\frac{\tilde{c}}{Y} + \Delta s_{in}) \end{bmatrix} \tag{9}$$

$$B = \begin{bmatrix} -\tilde{c} - Y \Delta s_{in} \\ \frac{\tilde{c}}{Y} + \Delta s_{in} \end{bmatrix} \tag{10}$$

The input and state constraints are then defined by the following domains:

$$\mathcal{D}(u_e) = \{u_e \in \Re; -\tilde{\mu} \leq u_e \leq d_{max} - \tilde{\mu}\} \tag{11}$$

$$\mathcal{D}(x_e) = \left\{ x_e \in \Re^2; \begin{bmatrix} Y(\tilde{s} - s_{max} - \Delta s_{in2}) \\ -\tilde{s} \end{bmatrix} \preceq x_e \preceq \begin{bmatrix} Y(\tilde{s} + \Delta s_{in1}) \\ s_{max} - \tilde{s} \end{bmatrix} \right\} \tag{12}$$

From (8), the open-loop system $(u_e = 0)$ is described by

$$\dot{x}_e = A x_e - A \begin{bmatrix} Y \Delta s_{in} \\ 0 \end{bmatrix} \tag{13}$$

which is a non-autonomous system. Hence system (13) admits a set of equilibrium states depending on the value of Δs_{in}.

In the current case of mono-increasing cell growth rate (Monod model), $(\frac{\partial \tilde{\mu}}{\partial s})$ can never be less than zero. It can then be noticed that $\sigma(A) = \{-\tilde{\mu}; -(\frac{\partial \tilde{\mu}}{\partial s})(\frac{\tilde{c}}{Y} + \Delta s_{in})\}$ is an asymptotically stable spectrum for all admissible perturbations Δs_{in} satisfying (4). Nevertheless, it is well-known that although this condition is necessary for asymptotic stability, it is not a sufficient condition due to the fact that matrix A is time-varying [2]. The existence of a common polyhedral dissymmetrical Lyapunov function for all admissible perturbations Δs_{in} should permit to conclude about the stability of the open-loop system (13).

Furthermore, the following lemma gives a condition on the choice of s_{max} in order for the uncontrollable mode to be the slowest mode.

Lemma 1. *In the open-loop spectrum, the controllable eigenvalue is faster than the uncontrollable one, for all admissible perturbations Δs_{in} if and only if*

$$s_{max} \leq -k_s + k_s \sqrt{1 + \frac{\tilde{s}_{in} - \Delta s_{in2}}{k_s}} \tag{14}$$

Proof: The controllable eigenvalue of the open-loop perturbed matrix A, denoted $\lambda_c = -(\tilde{c} + Y \Delta s_{in}) \frac{1}{Y} (\frac{\partial \tilde{\mu}}{\partial s})$ is faster than the uncontrollable mode $\lambda_u = -\tilde{\mu}$ if and only if $\lambda_c < \lambda_u$. By replacing $\tilde{c}$ and $(\frac{\partial \tilde{\mu}}{\partial s})$ by their values, it follows: $\lambda_c < \lambda_u \Leftrightarrow \tilde{s}(2k_s + \tilde{s}) < k_s(\tilde{s}_{in} + \Delta s_{in})$, which must be satisfied for all admissible perturbations. Then from (4), it follows: $\tilde{s}(2k_s + \tilde{s}) - k_s \tilde{s}_{in} - k_s \Delta s_{in1} \leq \tilde{s}(2k_s + \tilde{s}) - k_s(\tilde{s}_{in} + \Delta s_{in}) \leq \tilde{s}(2k_s + \tilde{s}) - k_s \tilde{s}_{in} + k_s \Delta s_{in2}$. Therefore, if one gets $\tilde{s}(2k_s + \tilde{s}) - k_s \tilde{s}_{in} + k_s \Delta s_{in2} < 0$ then it follows $\lambda_c < \lambda_u$. From (7), we know that $\tilde{s} \geq 0$, hence the resolution of the above inequality implies to choose $\tilde{s}$ such that $0 \leq \tilde{s} \leq s_{max} = -k_s + k_s \sqrt{1 + \frac{\tilde{s}_{in} - \Delta s_{in2}}{k_s}}$. Thus from (7) this condition on $\tilde{s}$ will always be satisfied if and only if condition (14) holds. $\square$

In the sequel, we suppose that s_{max} has been chosen to respect Lemma 1. It is coherent since in practice $-\tilde{\mu}$ corresponds to the slow mode.

Furthermore, the pair (A, B) is stabilizable for all admissible perturbations Δs_{in}. The uncontrollable mode is $-\tilde{\mu}$ which is asymptotically stable.

3 Control Design Objectives

For a given output matrix C relying the state x_e to an available output y_e, one considers a saturated output feedback law:

$$u_e = sat(KCx_e + v) \; ; \; K \in \Re \; ; \; v \in \Re \tag{15}$$

defined as:

$$sat(KCx_e + v) = \begin{cases} d_{max} - \tilde{\mu} & \text{if } KCx_e + v > d_{max} - \tilde{\mu} \\ KCx_e + v & \text{if } -\tilde{\mu} \leq KCx_e + v \leq d_{max} - \tilde{\mu} \\ -\tilde{\mu} & \text{if } KCx_e + v < -\tilde{\mu} \end{cases}$$

The closed-loop nonlinear system is then given by:

$$\dot{x}_e = Ax_e + Bsat(KCx_e + v) - A \begin{bmatrix} Y \Delta s_{in} \\ 0 \end{bmatrix} \tag{16}$$

If the control does not saturate, that is, if $KCx_e + v \in \mathcal{D}(u_e)$ or, equivalently, if $x_e \in \mathcal{D}(K)$, where domain $\mathcal{D}(K)$ is defined as:

$$\mathcal{D}(K) = \{x_e \in \Re^2 ; -\tilde{\mu} \leq KCx_e + v \leq d_{max} - \tilde{\mu}\} \tag{17}$$

the closed-loop system (16) admits the following linear model:

$$\begin{aligned}
\dot{x}_e &= (A + BKC)x_e + Bv - A\begin{bmatrix} Y\Delta s_{in} \\ 0 \end{bmatrix} \\
&= A_0 x_e + Bv - A\begin{bmatrix} Y\Delta s_{in} \\ 0 \end{bmatrix}
\end{aligned} \tag{18}$$

The problems related both to perturbations Δs_{in} and to the state and input constraints, impose the study of the local stability of closed-loop system (16). Hence, a classical approach to find a local domain of invariance and stability respecting all the constraints is to determine a positively invariant set $\mathcal{D}_{inv} \supset \{0\}$ with respect to system (18) such that:

$$\mathcal{D}_{inv} \subseteq (\mathcal{D}(x_e) \cap \mathcal{D}(K)) \tag{19}$$

In this case, the control is never saturated. Condition (19) expresses that the set $(\mathcal{D}(x_e) \cap \mathcal{D}(K))$ is potentially the maximal set of invariance and stability, allowing to respect the constraints both on state and input [7], [24], [25]. In general, if no additional condition is added, this set is however not a positively invariant set for system (18).

Finally the input constrained control problem may be stated as follows:

Problem 1. According to the closed-loop saturated system (16), determine:

1. An output feedback matrix K such that the controllable eigenvalue is faster in closed-loop than in open-loop, for all admissible perturbations Δs_{in};
2. A reference input v such that the equilibrium points of the closed-loop system (18) are the same as those of the open-loop system ($u_e = 0$);
3. A local polyhedral domain $\mathcal{D}_{inv}$, of invariance and stability satisfying (19), with respect to system (18), for all admissible perturbations Δs_{in}.

The resolution of this design problem involves the notion of polyhedral stabilizability which can be defined as the classical quadratic stability [15] by considering a polyhedral Lyapunov function. Thus the determination of a polyhedral domain $\mathcal{D}_{inv}$ implies the existence of a dissymmetrical or symmetrical polyhedral Lyapunov function valid for all admissible perturbations.

Let us turn now to the output used in the control law. Different cases may be considered according to the possibility to measure the cell concentration or not. Throughout the chapter, two cases are considered. The first case is based on the measure on the substrate concentration, which turns out to be the more realistic case and corresponds to $C = \begin{bmatrix} 0 & 1 \end{bmatrix}$. Then, it follows:

$$KC = \begin{bmatrix} 0 & f_s \end{bmatrix} \tag{20}$$

and matrix $A_0 = A + BKC$ is defined by

$$A_0 = A_{0s} = \begin{bmatrix} 0 & (\tilde{c} + Y\Delta s_{in})((\frac{\tilde{\partial\mu}}{\partial s}) - f_s) \\ -\frac{\tilde{\mu}}{Y} & -\tilde{\mu} - (\frac{\tilde{c}}{Y} + \Delta s_{in})((\frac{\tilde{\partial\mu}}{\partial s}) - f_s) \end{bmatrix} \tag{21}$$

The other option is to consider that the cell concentration is available, which corresponds to $C = \begin{bmatrix} 1 & 0 \end{bmatrix}$. In such a case, one gets:

$$KC = \begin{bmatrix} f_c & 0 \end{bmatrix} \tag{22}$$

and matrix A_0 is defined by

$$A_0 = A_{0c} = \begin{bmatrix} -(\tilde{c} + Y\Delta s_{in})f_c & (\tilde{c} + Y\Delta s_{in})(\frac{\partial \tilde{\mu}}{\partial s}) \\ -\frac{\tilde{\mu}}{Y} + (\frac{\tilde{c}}{Y} + \Delta s_{in})f_c & -\tilde{\mu} - (\frac{\tilde{c}}{Y} + \Delta s_{in})(\frac{\partial \tilde{\mu}}{\partial s}) \end{bmatrix} \tag{23}$$

Even if this second case may be considered as implausible from a practical operation point of view, both cases are considered in the following since they allow to illustrate that the properties of the closed-loop system (16) strongly depends on the available measurements. Moreover, the complete case where both the substrate concentration and the cell are available will not be addressed in the chapter, but may be directly extended from any case with a single measured concentration.

4   Basic Mathematical Tools

The solution of Problem 1 is based on the notion of M-matrices, the positive invariance property and the definition of some polyhedral cones. Such tools are briefly recalled in this section.

Z-matrices and M-matrices [11], [4] are defined as follows.

Definition 1. *[11] A Z-matrix is a matrix $A \in \Re^{n \times n}$, for which all off-diagonal terms are negative or null.*

Definition 2. *A Z-matrix A is an M-matrix if and only if one of the following equivalent properties holds:*

(i) every eigenvalue of A is with positive real-part.
(ii) all principal minors of A are nonnegative.
(iii) A^{-1} exists and is a nonnegative matrix, i.e., all its elements are positive or null.
(iv) there exists a strictly positive vector x such that Ax is also strictly positive.

The positive invariance property is recalled in the following definition.

Definition 3. *A set $\mathcal{D}_{inv} \subset \Re^n$ is positively invariant with respect to a linear system $\dot{x} = Ax$ if and only if $\forall x_0 \in \mathcal{D}_{inv}$, the trajectory $x(t; x_0)$ initialized from x_0 belongs to $\mathcal{D}_{inv}$, $\forall t \geq 0$, i.e., if and only if $\mathcal{D}_{inv}$ is such that $e^{A_0 t}\mathcal{D}_{inv} \subseteq \mathcal{D}_{inv}, \forall t \geq 0$.*

This definition is general, and the set $\mathcal{D}_{inv}$ can be, for example, a bounded polyhedron, a cone or a vectorial subspace. We are more particularly concerned with constructing polyhedral domains of positive invariance.

To express some invariance conditions, cones properties are very useful. We only give some definitions on the polyhedral cones [17]. The reader may also refer to Berman, Neumann and Stern [4] and the references herein.

Definition 4. *A polyhedral cone $\mathcal{K}_+ \subseteq \Re^n$ is defined by:*

$$\mathcal{K}_+ = N\Re_+^k \tag{24}$$

where column vectors of matrix $N \in \Re^{n \times k}$ are the finitely extremal rays, i.e., $\mathcal{K}_+$ has one-dimensional faces. $\mathcal{K}_+$ is also defined by:

$$\mathcal{K}_+ = \{x \in \Re^n \; ; \; Gx \succeq 0\} \; , \quad G \in \Re^{k \times n} \tag{25}$$

Definition 5. *A polyhedral cone $\mathcal{K} \subseteq \Re^n$ is said to be simplicial if $\mathcal{K}_+ = N\Re_+^n$ where $N \in \Re^{n \times n}$ is a nonsingular matrix, that is, if it possesses n extremal rays.*

Let us consider the convex dissymmetrical polyhedron $\mathcal{D}(K; a, b)$ defined by:

$$\mathcal{D}(G; a, b) = \{x \in \Re^n; -b \preceq Gx \preceq a\} \tag{26}$$

with $G \in \Re^{g \times n}$, $rank(G) = g, a \in \Re_+^g$ and $b \in \Re_+^g$. If we denote a proper cone $\mathcal{K} = \mathcal{K}_+ = -\mathcal{K}_-$ given by (25), the sets $\mathcal{D}(G; a, b)$ considered in this chapter are given by:

$$\mathcal{D}(G; a, b) = (\mathcal{K}_+ - b) \cap (\mathcal{K}_- + a) \tag{27}$$

Proposition 1. *[23] A nonempty set $\mathcal{D}(G; a, b)$ is positively invariant with respect to system $\dot{x} = A_0 x + r$, with $r \in \Omega \subset \Re^n$, Ω being a compact set, with property for matrix e^{tA_0} of leaving a proper cone $\mathcal{K}$ positively invariant ($e^{tA_0}\mathcal{K} \subseteq \mathcal{K}$) if and only if:*

$$\Omega \subseteq \Omega_0 \tag{28}$$

where

$$\Omega_0 = (\mathcal{K}_+ + A_0 b) \cap (\mathcal{K}_- - A_0 a) \tag{29}$$

Some properties concerning matrices A_0 such that $e^{tA_0}\mathcal{K} \subseteq \mathcal{K}$ can be consulted for instance in [4], [22].

5   Main Results

5.1   Substrate Measurement Based Strategy

Let us first consider the case where the substrate concentration only is being available. Addressing the first objective of Problem 1 related to the asymptotic stability of the closed-loop system (18), we get the following lemma.

Lemma 2. *The output feedback matrix K defined in (20) assigns to the closed-loop system an asymptotically stable eigenvalue faster than those of the open-loop system, for all admissible perturbations Δs_{in}, if and only if*

$$f_s < 0 \tag{30}$$

Proof: The eigenvalues of matrix A_{0s} defined in (21) are $\lambda_1 = -\tilde{\mu}$ and $\lambda_2 = -(\frac{\tilde{c}}{Y} + \Delta s_{in})((\frac{\partial \tilde{\mu}}{\partial s}) - f_s)$. Hence, matrix A_{0s} will be asymptotically stable for all admissible disturbances Δs_{in} if and only if $\lambda_2 < 0$, that is, from (7) one gets $0 < (\frac{\partial \tilde{\mu}}{\partial s}) - f_s$. Moreover from Lemma 1, s_{max} is chosen such that $\frac{-(\tilde{c}+Y\Delta s_{in})}{Y}(\frac{\partial \tilde{\mu}}{\partial s}) < -\tilde{\mu}$, for all admissible Δs_{in}. Hence, the closed-loop eigenvalue λ_2 will be faster than those of the open-loop if and only if $-(\frac{\tilde{c}}{Y} + \Delta s_{in})((\frac{\partial \tilde{\mu}}{\partial s}) - f_s) < -\tilde{\mu}$, for all admissible perturbations Δs_{in}, that is, if and only if condition (30) is satisfied. Furthermore, note that condition (30) contains the above condition of asymptotic stability since $(\frac{\partial \tilde{\mu}}{\partial s}) > 0$. $\qquad\square$

Addressing the second objective of Problem 1, one gets the following.

Lemma 3. *The open-loop system (13) and the linear closed-loop system (18) have the same equilibrium points if and only if, in the control law (15), scalar v is chosen such that:*

$$v = -KC \begin{bmatrix} Y\Delta s_{in} \\ 0 \end{bmatrix} \tag{31}$$

Proof:
(If). If v is chosen as in (31) then the linear closed-loop system (18) becomes

$$\dot{x}_e = A_{0s} x_e - A_{0s} \begin{bmatrix} Y\Delta s_{in} \\ 0 \end{bmatrix} \tag{32}$$

Hence, since A_{0s} is invertible, the equilibrium points $(\dot{x}_e = 0)$ are such that $x_e = \begin{bmatrix} Y\Delta s_{in} \\ 0 \end{bmatrix}$ which are the same as in open-loop.

(Only if). Assume that the equilibrium points in closed-loop are the same than those of the open-loop system. It follows: $\dot{x}_e = 0 = A_{0s} x_e + Bv - A \begin{bmatrix} Y\Delta s_{in} \\ 0 \end{bmatrix}$ with $x_e = \begin{bmatrix} Y\Delta s_{in} \\ 0 \end{bmatrix}$. Thus, one gets : $B(KC \begin{bmatrix} Y\Delta s_{in} \\ 0 \end{bmatrix} + v) = 0$ then condition (31) follows. $\qquad\square$

According to the control law (15), (20) and by considering that Lemmas 2 and 3 hold, the set $\mathcal{D}(K)$ previously defined in (17), can be described as follows:

$$\mathcal{D}(K) = \left\{ x_e = \begin{bmatrix} x_e^1 \\ x_e^2 \end{bmatrix} \in \Re^2 ; \frac{d_{max} - \tilde{\mu}}{f_s} \leq x_e^2 \leq -\frac{\tilde{\mu}}{f_s} \right\} \tag{33}$$

From (21) and Lemma 2, matrix A_{0s} cannot simultaneously be asymptotically stable and have the structure of an $-M$-matrix [11], [23]. Nevertheless, its spectrum is that of an $-M$-matrix. Hence matrix $e^{tA_{0s}}$ has the property of leaving a polyhedral cone positively invariant [4], [22]. Moreover, there exists some matrices G such that $GA_{0s} = HG$ where H is an $-M$-matrix. Thus, all a family of polyhedral cones may allow to construct some polyhedral domains obtained from the intersection of shifted cones defined by:

$$\mathcal{D}(G; x_{max}, x_{min}) = \{x_e \in \Re^2; -Gx_{min} \preceq Gx_e \preceq Gx_{max}\} \\ = (L_+ - x_{min}) \cap (L_- + x_{max}) \tag{34}$$

where L is a polyhedral cone [17] defined by $L_+ = -L_- = \{x \in \Re^2; \ Gx \succeq 0\}$. The potential maximal invariant set is the domain $\mathcal{D}(K) \cap \mathcal{D}(x_e)$. However, according to the results in [5], [8], it can be proven that this domain cannot be positively invariant for the closed-loop system (32). Hence, in order to address the third objective of Problem 1, we will use Proposition 1 to compute $\mathcal{D}_{inv}$ as defined in (34). In order to maximize some directions of this set, we want to consider that $\mathcal{D}_{inv}$ possesses some common boundaries with $\mathcal{D}(K)$. By this way, we have to verify that $\mathcal{D}(K) \cap \mathcal{D}(x_e) \subseteq \mathcal{D}(x_e)$. Note that some interesting solutions to enlarge the domain of indexpositive invariancepositive invariance are also proposed in [9] in the context of discrete-time systems.

In the case of system (32) the condition $\mathcal{D}(K) \cap \mathcal{D}(x_e) \subseteq \mathcal{D}(x_e)$ is guaranteed if and only if:

$$f_s \leq -\max\left(\frac{d_{max} - \tilde{\mu}}{\tilde{s}}, \ \frac{\tilde{\mu}}{s_{max} - \tilde{s}}\right) \tag{35}$$

which directly follows from the definition of domains $\mathcal{D}(K)$, in (33), and $\mathcal{D}(x_e)$, in (12), recalling that from Lemma 2 one gets $f_s < 0$.

Consider that the matrix G defining a simplicial cone L is chosen as:

$$G = \begin{bmatrix} KC \\ G_2 \end{bmatrix} = \begin{bmatrix} 0 & f_s \\ g_1 & g_2 \end{bmatrix} \tag{36}$$

The following lemma deals with some structural property of the domain $\mathcal{D}(G; x_{max}, x_{min})$, defined from (36), with respect to the autonomous system:

$$\dot{x}_e = A_{0s} \ x_e \tag{37}$$

Lemma 4. *The matrix G, defining the set $\mathcal{D}(G; x_{max}, x_{min})$ as in (34), satisfies the equality $GA_{0s} = HG$ with H an $-M$-matrix, if the following conditions hold:*

$$\begin{aligned} g_1 &> 0 \\ g_2 &\geq 0 \\ Y \leq \frac{g_2}{g_1} &\leq \frac{(\frac{\tilde{\partial}\mu}{\partial s} - f_s)(\tilde{c} + Y\Delta s_{in})}{\tilde{\mu}} \end{aligned} \tag{38}$$

Proof: First, matrix G is a nonsingular one if and only if $g_1 \neq 0$, since from Lemma 2 one gets $f_2 < 0$. According to the structure of G it follows that the matrix H solution to $GA_{0s} = HG$, or equivalently, $H = GA_{0s}G^{-1}$, is an $-M$-matrix if the set of relations (38) hold [4], [11]. $\qquad\square$

By considering the structure of matrix G given in (36), and from the definition of the polyhedral set $\mathcal{D}(G; x_{max}, x_{min})$ given in (34), we can express Gx_{max} and Gx_{min} as follows:

$$Gx_{max} = \begin{bmatrix} d_{max} - \tilde{\mu} \\ \alpha \end{bmatrix} \text{ and } Gx_{min} = \begin{bmatrix} \tilde{\mu} \\ \beta \end{bmatrix} \tag{39}$$

where α and β have to satisfy some conditions in order for the set $\mathcal{D}(G; x_{max}, x_{min})$ to be positively invariant with respect to the closed-loop system (32). The following proposition exhibits such suitable conditions.

Proposition 2. *The polyhedral domain* $\mathcal{D}(G; x_{max}, x_{min})$ *defined in (34), with (36) and (39), is a positively invariant set for system (32) and satisfies (19), for all admissible perturbations* Δs_{in} *given by (4), if and only if:*

$$Y g_1 \Delta s_{in1} \le \alpha \le Y g_1 (\Delta s_{in1} + \tilde{s}) + \frac{g_2}{f_s}(d_{max} - \tilde{\mu})$$
$$Y g_1 \Delta s_{in2} \le \beta \le Y g_1 (\Delta s_{in2} + s_{max} - \tilde{s}) + \frac{g_2}{f_s}\tilde{\mu} \tag{40}$$

$$\frac{g_2}{g_1} \le min\left(-Y\frac{\tilde{s} - f_s}{d_{max} - \mu} \; ; \; -Y\frac{(s_{max} - \tilde{s})f_s}{\tilde{\mu}}\right) \tag{41}$$

Proof:

(Left-hand inequality (40)). According to Proposition 1, with $r = -A_{0s}\begin{bmatrix} Y\Delta s_{in} \\ 0 \end{bmatrix}$, the condition of positive invariance of $\mathcal{D}(G; x_{max}, x_{min})$ is expressed as:

$$GA_{0s}x_{min} \preceq -GA_{0s}\begin{bmatrix} Y\Delta s_{in} \\ 0 \end{bmatrix} \preceq -GA_{0s}x_{max}$$
$$\Leftrightarrow HGx_{min} \preceq -HG\begin{bmatrix} Y\Delta s_{in} \\ 0 \end{bmatrix} \preceq -HGx_{max}$$

Since under the conditions of Lemma 4, H is an $-M$-matrix, $(-H)^{-1}$ exists and is a nonnegative matrix. Thus, it follows:

$$-Gx_{min} \preceq G\begin{bmatrix} Y\Delta s_{in} \\ 0 \end{bmatrix} \preceq Gx_{max}$$

From the expressions of Gx_{max} and Gx_{min} given in (39), some conditions are derived on α and β which must be satisfied for every Δs_{in} satisfying (4). Hence left-hand inequalities of condition (40) are obtained.

(Right-hand inequality (40)). It remains to show that x_{max} and $-x_{min}$ belong to $\mathcal{D}(x_e)$, described in (12). Indeed, according to the shapes of $\mathcal{D}(K)$ and $\mathcal{D}(x_e)$, the inclusion $\mathcal{D}(K) \cap \mathcal{D}(x_e) \subseteq \mathcal{D}(x_e)$ only concerns the state x_e^2. Hence, (19) will be satisfied if:

$$Y(\tilde{s} - s_{max} - \Delta s_{in2}) \le x_{max}^1 \le Y(\tilde{s} + \Delta s_{in1})$$
$$Y(\tilde{s} - s_{max} - \Delta s_{in2}) \le -x_{min}^1 \le Y(\tilde{s} + \Delta s_{in1})$$

Since under the conditions of Lemma 4, matrix G is nonsingular, we can write:

$$x_{max} = G^{-1}\begin{bmatrix} d_{max} - \tilde{\mu} \\ \alpha \end{bmatrix} \text{ and } -x_{min} = -G^{-1}\begin{bmatrix} \tilde{\mu} \\ \beta \end{bmatrix}$$

and hence we must satisfy:

$$Y(\tilde{s} - s_{max} - \Delta s_{in2}) \le -\frac{g_2}{g_1 f_s}(d_{max} - \tilde{\mu}) + \frac{\alpha}{g_1} \le Y(\tilde{s} + \Delta s_{in1})$$
$$Y(\tilde{s} - s_{max} - \Delta s_{in2}) \le \quad \frac{g_2}{g_1 f_s}\tilde{\mu} - \frac{\beta}{g_1} \quad \le Y(\tilde{s} + \Delta s_{in1})$$

As $g_1 > 0$, $g_2 \ge 0$ and $f_s < 0$, we finally obtain the right-hand conditions of (40).

Finally, the sets of suitable α and β defined by (40) are not empty if and only if condition (41) is satisfied.   □

Remark 1. Conditions (38) and (41) on $\frac{g_2}{g_1}$ may be associated and therefore one obtains:

$$Y \le \frac{g_2}{g_1} \le \min\left(-Y\frac{\tilde{s}f_s}{d_{max} - \tilde{\mu}} \;\; ; \;\; -Y\frac{(s_{max} - \tilde{s})f_{2s}}{\tilde{\mu}} \;\; ; \;\; \frac{(\frac{\tilde{\partial \mu}}{\partial s} - f_s)(\tilde{c} + Y\Delta s_{in})}{\tilde{\mu}} \right)$$

$$(42)$$

which is a non-empty set in terms of g_1 and g_2 if and only if condition (35) on the output feedback f_s is satisfied.

5.2 Cell Measurement Based Strategy

Let us turn now to the case where the cell concentration is being available. Addressing the first objective of Problem 1 related to the asymptotic stability of the linear closed-loop system (18), we get the following lemmas.

Lemma 5. *The output feedback matrix K defined in (22) assigns to the closed-loop system an asymptotically stable eigenvalue faster than those of the open-loop system, for all admissible perturbations Δs_{in}, if and only if*

$$0 < f_c \tag{43}$$

Proof: The proof mimics that one of Lemma 2 considering now the spectrum of matrix A_{0c} defined in (23).   □

As in the previous case, the second objective of Problem 1 is given by Lemma 3.

As in the previous case where the cell concentration was not available, the spectrum of matrix A_{0c} is that of an $-M$-matrix. Hence matrix $e^{tA_{0c}}$ has the property of leaving a polyhedral cone positively invariant, and the same strategy as in the previous case may be applied to build some polyhedral domains obtained from the intersection of shifted cones. Nevertheless, contrarily to the previous case, it is also possible to determine conditions on matrix K such that A_{0c} is an $-M$-matrix. With these conditions, $e^{tA_{0c}}$ is a nonnegative matrix, which has the property of leaving the positive orthant $\Re_+^n$ positively invariant, that is, $e^{tA_{0c}}\Re_+^n \subset \Re_+^n$. This may be used to determine more easily domains of positive invariance $\mathcal{D}_{inv}$, with $G = I_2$, for the linear system (18) under constraints (12) and (17).

Lemma 6. *Matrix K defined in (22) is such that matrix A_{0c}, defined in (23), is an $-M$-matrix for all admissible perturbations Δs_{in} if and only if*

$$f_c \geq \frac{\tilde{\mu}}{Y(s_{in} - \tilde{s} - \Delta s_{in2})} \qquad (44)$$

Proof: From Definition 1, matrix A_{0c} is an $-Z$-matrix if all off-diagonal elements are positive or null, that is, if $(\tilde{c} + Y\Delta s_{in})(\frac{\partial\mu}{\partial s}) \leq 0$ and $-\frac{\mu}{Y} + (\frac{\tilde{c}}{Y} + \Delta s_{in})f_c \leq 0$ for all admissible perturbations Δs_{in}. Those two conditions imply $0 \leq (\frac{\partial\mu}{\partial s})$, which is true by hypothesis, and $f_c \geq \frac{\tilde{\mu}}{(\tilde{c} + Y\Delta s_{in})}$ for all admissible perturbations Δs_{in}. Hence from (4), condition on f_c reads as $f_c \geq \frac{\tilde{\mu}}{(\tilde{c} - Y\Delta s_{in2})}$. From (ii) of Definition 2, matrix A_{0c} is an $-M$-matrix, or equivalently $-A_{0c}$ is an M-matrix, if and only if $(\tilde{c} + Y\Delta s_{in})f_c \geq 0$, $\tilde{\mu} + (\frac{\tilde{c}}{Y} + \Delta s_{in})(\frac{\partial\mu}{\partial s}) \geq 0$ and $det(-A_{0c}) = \tilde{\mu}(\tilde{c} + Y\Delta s_{in})(\frac{\partial\mu}{\partial s}) \geq 0$, for all admissible Δs_{in}. Then recalling (3), condition (44) follows. $\square$

It must be noted that condition (44) which gives the $-M$-matrix property of A_{0c} implies asymptotic stability of the closed-loop system (18). Furthermore, this condition also implies that the $-M$-matrix A_{0c} has its assigned controllable eigenvalue faster than that of the open-loop system. Considering the control law (15), admissible disturbances belonging to (4) and that Lemmas 3 and 6 hold, the set $\mathcal{D}(K)$ previously defined in (17) can be described as follows:

$$\mathcal{D}(K) = \left\{ x_e = \begin{bmatrix} x_e^1 \\ x_e^2 \end{bmatrix} \in \Re^2; -\tilde{\mu} + f_c Y \Delta s_{in1} \leq f_c x_e^1 \leq d_{max} - \tilde{\mu} - f_c Y \Delta s_{in2} \right\}$$
$$(45)$$

Remark 2. It can be noticed that the set $\mathcal{D}(K)$, given in (45), is a non-empty set if and only if:

$$f_c \leq \frac{d_{max}}{Y(\Delta s_{in1} + \Delta s_{in2})} \qquad (46)$$

According both to the shape of domains $\mathcal{D}(x_e)$ and $\mathcal{D}(K)$ and to the fact that there exists an output feedback K satisfying (44), we are especially interested in determining a positively set as:

$$\mathcal{D}_{inv} = \mathcal{D}(I_2; x_{max}, x_{min}) = \{x_e \in \Re^2 \ ; \ -x_{min} \preceq x_e \preceq x_{max}\} \qquad (47)$$

in order to solve the third objective of Problem 1. The following propositions present conditions of positive invariance relative to this set $\mathcal{D}_{inv}$.

Proposition 3. *The polyhedral domain $\mathcal{D}_{inv}$ defined in (47) is a positively invariant set for system (18), for all admissible perturbations Δs_{in}, if and only if:*

$$\begin{aligned} x_{max}^1 &\geq Y\Delta s_{in1} \\ x_{min}^1 &\geq Y\Delta s_{in2} \end{aligned} \qquad (48)$$

Proof: According to Proposition 1 with $r = -A_{0c} \begin{bmatrix} Y \Delta s_{in} \\ 0 \end{bmatrix}$, condition of positive invariance of $\mathcal{D}_{inv}$ is expressed as: $A_{0c} x_{min} \preceq -A_{0c} \begin{bmatrix} Y \Delta s_{in} \\ 0 \end{bmatrix} \preceq -A_{0c} x_{max}$. As A_{0c} is an $-M$-matrix, $(-A_{0c})^{-1}$ exists and is a non-negative matrix. The condition of positive invariance of $\mathcal{D}_{inv}$ becomes: $-x_{min} \preceq \begin{bmatrix} Y \Delta s_{in} \\ 0 \end{bmatrix} \preceq x_{max}$, which must be satisfied for every Δs_{in} satisfying (4). Hence condition (48) is obtained. $\qquad\square$

Proposition 4. *The polyhedral domain $\mathcal{D}_{inv}$ defined in (47) is included in $\mathcal{D}(K) \cap \mathcal{D}(x_e)$ if and only if:*

$$
\begin{aligned}
0 \preceq x_{max} \preceq &\begin{bmatrix} \min\left(\frac{d_{max} - \tilde{\mu}}{f_c} - Y \Delta s_{in2} \; , \; Y(\tilde{s} + \Delta s_{in1}) \right) \\ s_{max} - \tilde{s} \end{bmatrix} \\
0 \preceq x_{min} \preceq &\begin{bmatrix} \min\left(\frac{\tilde{\mu}}{f_c} - Y \Delta s_{in1} \; , \; Y(s_{max} - \tilde{s} + \Delta s_{in2}) \right) \\ \tilde{s} \end{bmatrix}
\end{aligned}
\tag{49}
$$

Proof: $\mathcal{D}_{inv} \subseteq \mathcal{D}(K) \cap \mathcal{D}(x_e)$ if and only if vertices of $\mathcal{D}(I_2; x_{max}, x_{min})$ belong to $\mathcal{D}(K) \cap \mathcal{D}(x_e)$, that is

$$
\left\{ \begin{bmatrix} x^1_{max} \\ x^2_{max} \end{bmatrix}, \; \begin{bmatrix} -x^1_{min} \\ x^2_{max} \end{bmatrix}, \; \begin{bmatrix} -x^1_{min} \\ -x^2_{min} \end{bmatrix}, \; \begin{bmatrix} x^1_{max} \\ -x^2_{min} \end{bmatrix} \right\} \in \mathcal{D}(K) \cap \mathcal{D}(x_e)
$$

According to definition (45) of domain $\mathcal{D}(K)$, one can write each vertex of $\mathcal{D}_{inv}$ as follows:

$$
\begin{aligned}
-\frac{\tilde{\mu}}{f_c} + Y \Delta s_{in1} \leq \; & x^1_{max} \; \leq \frac{d_{max} - \tilde{\mu}}{f_c} - Y \Delta s_{in2} \\
-\frac{\tilde{\mu}}{f_c} + Y \Delta s_{in1} \leq \; & -x^1_{min} \; \leq \frac{d_{max} - \tilde{\mu}}{f_c} - Y \Delta s_{in2}
\end{aligned}
\tag{50}
$$

As x_{max} and x_{min} belong to the positive orthant $\Re^2_+$, one obtains:

$$
\begin{aligned}
0 \leq x^1_{max} &\leq \frac{d_{max} - \tilde{\mu}}{f_c} - Y \Delta s_{in2} \\
0 \leq x^1_{min} &\leq \frac{\tilde{\mu}}{f_c} - Y \Delta s_{in1}
\end{aligned}
\tag{51}
$$

In the same way, according to definition (12) of domain $\mathcal{D}(x_e)$, every vertices of $\mathcal{D}_{inv}$ can be written as follows:

$$
\begin{aligned}
Y(\tilde{s} - s_{max} - \Delta s_{in2}) \leq \; & x^1_{max} \; \leq Y(\tilde{s} + \Delta s_{in1}) \\
-\tilde{s} \leq \; & x^2_{max} \; \leq s_{max} - \tilde{s} \\
Y(\tilde{s} - s_{max} - \Delta s_{in2}) \leq \; & -x^1_{min} \; \leq Y(\tilde{s} + \Delta s_{in1}) \\
-\tilde{s} \leq \; & -x^2_{min} \; \leq s_{max} - \tilde{s}
\end{aligned}
\tag{52}
$$

As x_{max} and x_{min} belong to the positive orthant $\Re^2_+$, one obtains:

$$
\begin{aligned}
0 \leq x^1_{max} &\leq Y(\tilde{s} + \Delta s_{in1}) \\
0 \leq x^2_{max} &\leq s_{max} - \tilde{s} \\
0 \leq x^1_{min} &\leq Y(s_{max} - \tilde{s} + \Delta s_{in2}) \\
o \leq x^2_{min} &\leq \tilde{s}
\end{aligned}
\tag{53}
$$

By associating (51) and (53), condition (49) such that $\mathcal{D}(I_2; x_{max}, x_{min}) \subseteq \mathcal{D}(K) \cap \mathcal{D}(x_e)$ follows. $\qquad\square$

The objective 3 of Problem 1 is then addressed through the following proposition.

Proposition 5. *The polyhedral domain defined in (47) satisfies inclusion (19) and is positively invariant with respect to system (18), where A_{0c} is an $-M$-matrix, if and only if:*

$$\begin{bmatrix} Y\Delta s_{in1} \\ 0 \end{bmatrix} \preceq x_{max} \preceq \begin{bmatrix} min\left(\frac{d_{max}-\tilde{\mu}}{f_c} - Y\Delta s_{in2} \; ; \; Y(\tilde{s} + \Delta s_{in1})\right) \\ s_{max} - \tilde{s} \end{bmatrix} \tag{54}$$

$$\begin{bmatrix} Y\Delta s_{in2} \\ 0 \end{bmatrix} \preceq x_{min} \preceq \begin{bmatrix} min\left(\frac{\tilde{\mu}}{f_c} - Y\Delta s_{in1} \; ; \; Y(s_{max} - \tilde{s} + \Delta s_{in2})\right) \\ \tilde{s} \end{bmatrix} \tag{55}$$

$$\frac{\tilde{\mu}}{Y(s_{in} - \tilde{s} - \Delta s_{in2})} \leq f_c \leq min\left(\frac{d_{max} - \tilde{\mu}}{Y(\Delta s_{in1} + \Delta s_{in2})} \; ; \; \frac{\tilde{\mu}}{Y(\Delta s_{in1} + \Delta s_{in2})}\right) \tag{56}$$

Proof: Proposition 5 gathers the results of Propositions 3 and 4. The low bound of inequality (56) consists in satisfying (44), for which A_{0c} is an $-M$-matrix. The upper bound of (56) corresponds to the following set of conditions on f_c:

- Condition (46) to ensure $\mathcal{D}(K) \neq \emptyset$.
- $f_c \leq \frac{d_{max}-\tilde{\mu}}{Y(\Delta s_{in1}+\Delta s_{in2})}$ in order that there exists x_{max} satisfying (54).
- $f_c \leq \frac{\tilde{\mu}}{Y(\Delta s_{in1}+\Delta s_{in2})}$ in order that there exists x_{min} satisfying (55). $\qquad\square$

Remark 3. The existence conditions of x_{max} and x_{min} satisfying (54) and (55), respectively, are implicitly given in the bounds of f_c, in equation (56). Moreover, the existence condition of f_c satisfying inequality (56) are given by:

$$\tilde{\mu}\Delta s_{in1} + d_{max}\Delta s_{in2} \leq (d_{max} - \tilde{\mu})(s_{in} - \tilde{s}) \tag{57}$$

$$\Delta s_{in1} + 2\Delta s_{in2} \leq s_{in} - \tilde{s} \tag{58}$$

5.3   Optimization Issues

Propositions stated in the previous subsections 5.1 and 5.2 establish conditions of existence to solve Problem 1. An optimization related problem may then be to maximize the size of the domain of positive invariance. This issue has been partly addressed in [13] in the case of a symmetrical normalized domain of positive invariance or in [9] in the context of discrete-time systems. A solution to maximize the size of the domain in the directions which are the most sensitive to disturbances has been proposed.

In the case of the dissymmetrical domain $\mathcal{D}(G; x_{max}, x_{min})$ not only the matrix G defining the polyhedral cone L has to be judiciously chosen but also the vertices x_{max} and x_{min}. The optimization of the volume of this polyhedral domain can be a way to study it.

In the considered two-dimensional case, the maximization of the size of $\mathcal{D}(G; x_{max}, x_{min})$ defined in Proposition 2 under the existence constraints (35), (38), (40) and (42) corresponds to the maximization of its surface defined by:

$$S = -\frac{1}{g_1 f_s} \, (\alpha + \beta) \, (u_{min} + u_{max})$$

that is, by the product of the determinant of G^{-1} and the distances between x_{max} and x_{min} projected on the plane axis. Finally the optimization problem can be expressed as follows:

$$\max_{g_1, g_2, f_s, \alpha, \beta} \left\{ -\frac{1}{g_1 f_s} \, (\alpha + \beta) \, (u_{min} + u_{max}) \right\} \tag{59}$$

subject to

$$\begin{cases} f_s \leq -\max \left(\frac{d_{max} - \tilde{\mu}}{\tilde{s}} \, , \, \frac{\tilde{\mu}}{s_{max} - \tilde{s}} \right) \\ g_1 > 0 \\ g_2 \geq 0 \\ Y \leq \frac{g_2}{g_1} \leq \min \left(-Y \frac{\tilde{s} f_s}{d_{max} - \tilde{\mu}} \; ; \; -Y \frac{(s_{max} - \tilde{s}) f_s}{\tilde{\mu}} \; ; \; \frac{(\frac{\partial \mu}{\partial s} - f_s)(\tilde{c} + Y \Delta s_{in})}{\tilde{\mu}} \right) \\ Y g_1 \Delta s_{a1} \leq \alpha \leq Y g_1 (\Delta s_{in1} + \tilde{s}) + \frac{g_2}{f_s} (d_{max} - \tilde{\mu}) \\ Y g_1 \Delta s_{in2} \leq \beta \leq Y g_1 (\Delta s_{in2} + s_{max} - \tilde{s}) + \frac{g_2}{f_s} \tilde{\mu} \end{cases}$$

In the cell measurement based approach, a similar optimization problem may be described in the same way.

6   Numerical Issues

Simulations have been carried out by numerically integrating the nonlinear model of the bio-reactor described by (1). The specific growth rate parameters of the Monod law (2) correspond to physical values of the micro-organism. In the case of alcohol production by *Saccharomyces cerevisiae*, $\mu_{max} = 0.3h^{-1}$ and $k_s = 5g/l$ represent realistic values of Monod parameters. The yield of cell mass is $Y = 0.07$.

The maximum dilution rate is $d_{max} = 0.3h^{-1}$. The influent substrate concentration is $s_{in} = 105g/l$. Perturbations Δs_{in} on the influent substrate concentration around the nominal value s_{in} given by (4) are $\Delta s_{in1} = 45g/l$ and $\Delta s_{in2} = 25g/l$. The disturbance associated to the influent substrate concentration s_{in} has been set as a step response between $s_{in} - \Delta s_{in2}$ and $s_{in} + \Delta s_{in1}$ during 250 hours then as a sinusoidal signal with 50 hours period (cf Fig. 1).

The control objective is to maintain the substrate concentration at its nominal value $\tilde{s} = 5g/l$. According to the growth model (2) and to the steady state (3), one has, for the nominal feeding concentration $s_{in} = 105g/l$, the following nominal values: $\tilde{\mu} = \tilde{d} = 0.15h^{-1}$ and $\tilde{c} = 7g/l$.

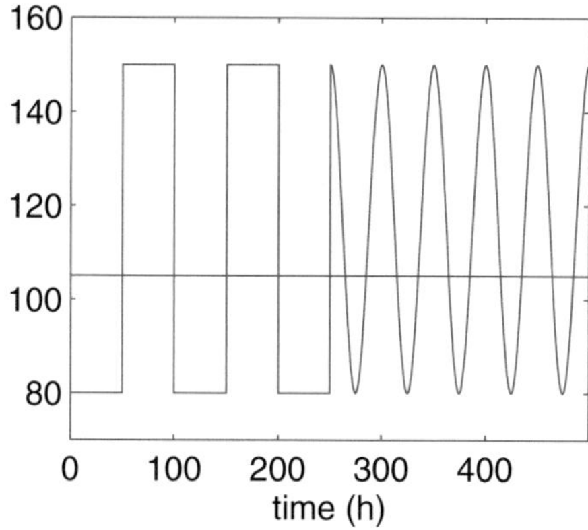

Fig. 1. Time-evolution of the disturbance $s_{in} + \Delta s_{in}$

Let us first consider the substrate measurement based control strategy. A solution to Problem 1 is given by:

$$f_s \leq -0.03, \quad G = \begin{bmatrix} 0 & -0.03 \\ 1 & 0.07 \end{bmatrix}, \quad x_{max} = \begin{bmatrix} 3.5 \\ -5 \end{bmatrix} \text{ and } x_{min} = \begin{bmatrix} 2.1 \\ -5 \end{bmatrix}$$

Simulations are carried out by numerically integrating via Matlab ODE tool the continuous-flow fermentation model given by (1), where the dilution rate d (i.e. the control variable u) is calculated according to the following expression: $d = \tilde{\mu} + f_s * x_e^2$. The initial value of the state vector is set to $x_0 = \begin{bmatrix} 9 \\ 8 \end{bmatrix}$. Hence $x_{e0} = x_0 - \tilde{x} = \begin{bmatrix} 2 \\ 3 \end{bmatrix}$ belongs to $\mathcal{D}_{inv}$. The time evolution of the states x^1 and x^2 are shown in Fig. 2 and 3 respectively, both in open-loop and closed-loop.

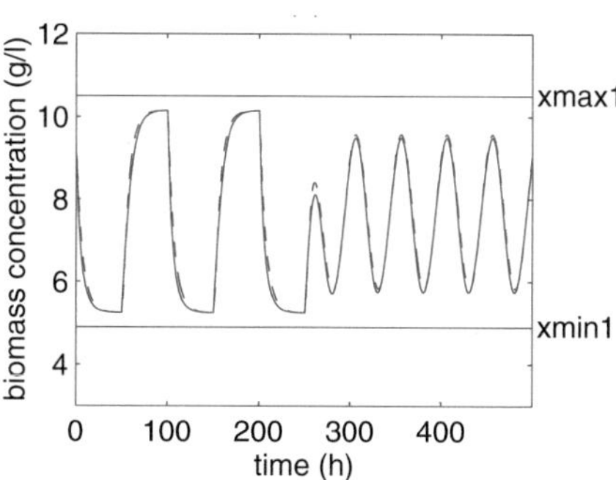

Fig. 2. Substrate measurement based strategy - Open-loop (dashed line) and closed-loop (solid line) evolution of the biomass concentration $x^1 = x_e^1 + \tilde{x}^1$

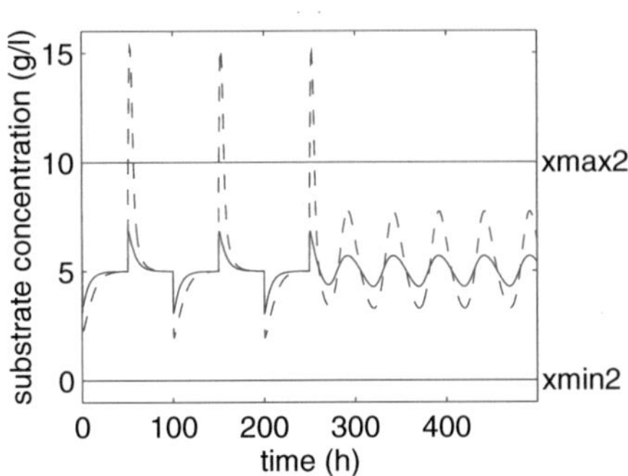

Fig. 3. Substrate measurement based strategy - Open-loop (dashed line) and closed-loop (solid line) evolution of the substrate concentration $x^2 = x_e^2 + \tilde{x}^2$

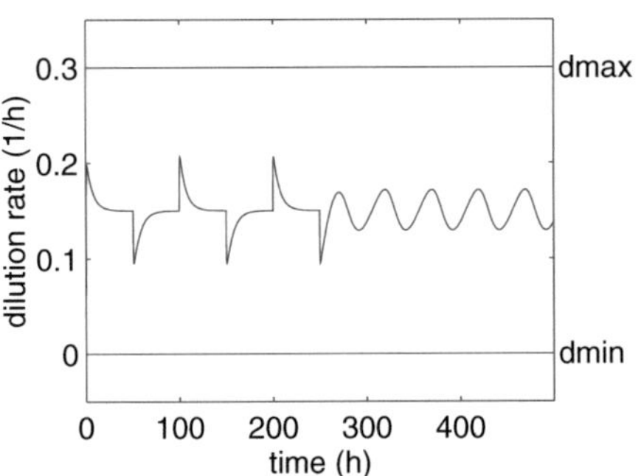

Fig. 4. Substrate measurement based strategy - Closed-loop evolution of the dilution rate d (control variable $u = u_e + \tilde{u}$)

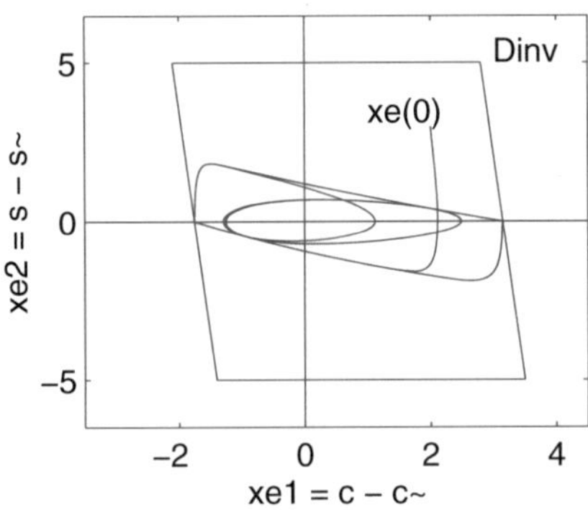

Fig. 5. Substrate measurement based strategy - $\mathcal{D}(G; x_{max}, x_{min})$ and phase plane evolution

The dilution rate d (i.e. the control variable $u = u_e + \tilde{u}$) is shown in Fig. 4. It can be seen on Fig. 5 that the closed-loop response (solid line) does not violate the constraints, i.e. remains inside $\mathcal{D}(G; x_{max}, x_{min})$.

Let us turn now to the cell measurement based control strategy. With those numerical values (57) and (58) are satisfied and Proposition 5 can be applied. The domain of admissible f_c described in (56) numerically becomes: $0.029 \leq f_c \leq 0.032$. Moreover $\mathcal{D}(x_e)$ is included in $\mathcal{D}(K)$ if $f_c \geq 0.029$. A solution to the Problem 1 is given by:

$$f_c = 0.029, \quad x_{max} = \begin{bmatrix} 3.5 \\ 5 \end{bmatrix} \text{ and } x_{min} = \begin{bmatrix} 2.1 \\ 5 \end{bmatrix}$$

The control action, i.e., the dilution rate d is computed according to the following expression:

$$d = \tilde{\mu} + f_c * (x_e^1 - Y \Delta s_{in}) \tag{60}$$

The initial value of the state vector is set to $x = \begin{bmatrix} 9 \\ 8 \end{bmatrix}$, that is, $x_e = x - \tilde{x}$ belongs to $\mathcal{D}_{inv}$. The time evolution of the states x^1 and x^2 are shown in Figs. 6 and 6 respectively, both in open-loop and closed-loop.

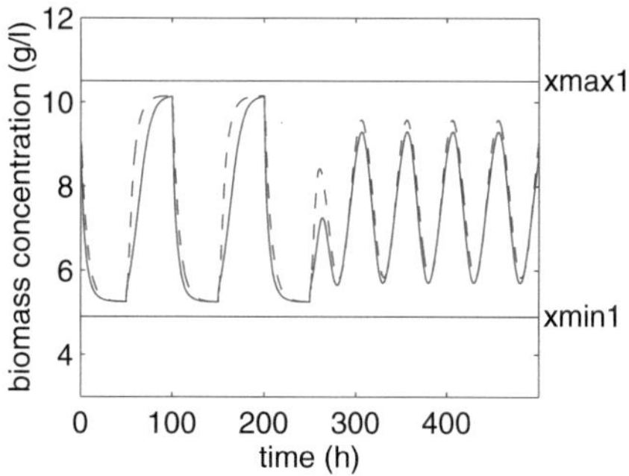

Fig. 6. Cell measurement based strategy - Open-loop (dashed line) and closed-loop (solid line) evolution of the biomass concentration $x^1 = x_e^1 + \tilde{x}^1$

It can be seen on Fig. 7 that the closed-loop response (solid line) does not violate the constraints x_{max} and x_{min}, contrary to the open-loop response (dashed line). The dilution rate d is shown in Fig. 8.

Note that the control objective expressed as a constant setpoint for the substrate concentration imposes to know the time-evolution of the additive disturbance Δs_{in}, according to (60. In the case where only the bounds on Δs_{in} are known, the second objective of Problem 1 cannot be verified, but the control strategy $d = \tilde{\mu} + f_c * x_e^1$ satisfies the objectives 1 and 3 of the Problem.

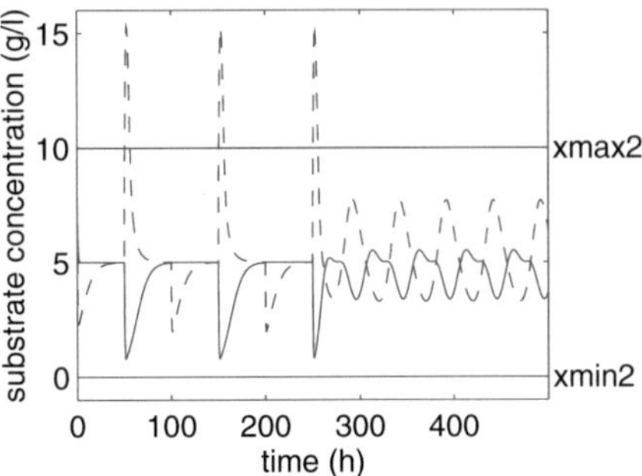

Fig. 7. Cell measurement based strategy - Open-loop (dashed line) and closed-loop (solid line) evolution of the substrate concentration $x^2 = x_e^2 + \tilde{x}^2$

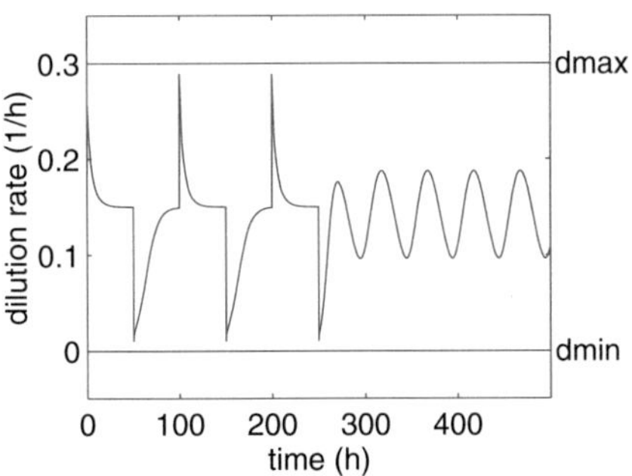

Fig. 8. Cell measurement based strategy - Closed-loop evolution of the dilution rate d (control variable $u = u_e + \tilde{u}$)

7 Conclusion

The aim of the work was to analyze and control a class of biological systems at the level of cells. More precisely, we have considered some tools like positive invariance and polyhedral Lyapunov functions to characterize some properties of biological systems involving amplitude limitations both on output and input variables. The main results are given in terms of constructive conditions allowing to build invariant sets in which all the constraints of the system under study are satisfied.

When dealing with such systems, there are still some open questions. The first one resides in the way to consider the limitations on the process, namely by taking into account the effective saturation of the variables rather than their linear satisfaction as pursued in the current chapter. In this case, some recent results as those developed in [1], [25] could be considered by taking care of the resulting domains in which the properties wished are satisfied. Moreover, another interesting objective should to consider the synthesis problem by using a recirculation loop and a by-pass with the aim to complete the results of [12] in taking into account both output and input limitations.

References

1. T. Alamo, A. Cepeda, D. Limon. Improved computation of ellipsoidal invariant sets for saturated control systems, *Proc. of the 44rd IEEE Conf. on Decision and Control (CDC)*, pp.6216-6221, Sevilla, Spain, December 2005.
2. B.R. Barmish, M. Fu, S. Saleh. Stability of a polytope of matrices: counterexamples, *IEEE Trans. Aut. Contr.*, vol.33, no.6, pp.569-572, 1988.
3. G. Bastin and D. Dochain. *Online estimation and adaptive control of bioreactors*, Elsevier, Amsterdam, 1990.
4. A. Berman, M. Neumann and R.J. Stern. *Nonnegative matrices in dynamic systems*, John Wiley and Sons, Inc., New York, 1989.
5. G. Bitsoris: Existence of positively invariant polyhedral sets for continuous-time linear systems, *Control-Theory and Advanced Technology*, vol.7, no.3, pp.407-427, 1991.
6. F. Blanchini. Feedback control of linear time-invariant systems with state and control bounds in the presence of disturbances, *IEEE Trans. Aut. Control*, vol.35, no.11, pp.1231-1234, 1990.
7. F. Blanchini. Set invariance in control, *Automatica*, vol.35, no.11, pp.1747-1767, 1999.
8. E.B. Castelan, J.M. Gomes da Silva Jr., J.E.R. Cury. A reduced-order framework applied to linear systems with constrained controls, *IEEE Trans. Aut. Control*, vol.41, no.2, pp.249-255, 1996.
9. A. Cepeda, D. Limon, T. Alamo, E.F. Camacho. Computation of polyhedral *H*-invariant sets for saturated systems, *Proc. of the 43rd IEEE Conf. on Decision and Control (CDC)*, pp.1176-1181, Bahamas, December 2004.
10. V. Chellaboina, W.M. Haddad, J.M. Bailey and J. Ramakrishnan. On nonoscillation and monotonicity of solutions of nonnegative and comportmental dynamical systems, *IEEE Trans. Biomedical Engineering* vol.51, no.3, pp:408-414, 2004.
11. M. Fiedler and V. Ptak. On matrices with nonpositive off-diagonal elements and positive principal minors, *Czech. Math. J.*, vol.12, pp:382-400, 1962.
12. J. Harmand, A. Rapaport, F. Mazenc. Output tracking of continuous bioreactors through recirculation and by-pass, *Automatica*, vol.42, pp.1025-1032, 2006.
13. J.C. Hennet and E.B. Castelan. Constrained control of unstable multivariable linear systems, *Proc. of 2nd ECC*, vol.4, pp. 2039-2043, Groningen (The Netherlands), 1993.
14. L. Mailleret, O. Bernard, J.P. Steyer. Nonlinear adaptive control for bioreactors with unknown kinetics. *Automatica*, vol.40, no.8, pp.1379-1385, 2004.
15. P.L.D. Peres, J.C. Geromel, J. Bernussou. Quadratic stabilizability of linear uncertain systems in convex-bounded domains, *Syst. and Contr. Letters*, vol.29, no.2, pp.491-493, 1993.
16. S. Roy, S. Alam and J. Chattopadhyay. Role of nutrient bound of prey on the dynamics of predator-mediated competitive-coexistence, *Biosystems*, vol.82, pp.143-153, 2005.
17. A. Schrijver. *Theory of linear and integer programming*, John Wiley and Sons, 1987.

18. L.A. Segel. *Modeling dynamic phenomena in molecular and cellular biology*, Cambridge University Press, Cambridge, 1984.
19. G.N. Stephanopoulos, A.A. Aristidou and J. Nielsen. *Metabolic Engineering: principles and methodologies*, Academic Press, San Diego, 1998.
20. E.D. Sontag. Molecular systems biology and control. Fundamental issues in control, *European J. of Control*, vol.11, no.4-5, pp.396-435, 2005.
21. M. Sznaier. A set induced norm approach to the robust control of constrained systems, *SIAM J. Contr. and Optim.*, vol.31, no.3, pp.733-746, 1993.
22. S. Tarbouriech and C. Burgat. Positively invariant sets for continuous-time systems with the cone-preserving property, *Int. J. of Systems Sciences*, vol.24, no.6, pp:1037-1047, 1993.
23. S. Tarbouriech and C. Burgat. Positively invariant sets of constrained continuous-time systems with cone properties, *IEEE Trans. Aut. Control*, vol.39, no.2, pp.401-405, 1994.
24. S. Tarbouriech and G. Garcia. Control of uncertain systems with bounded inputs, *Lecture Notes in Control and Information Science*, vol.227, Springer-Verlag, 1997.
25. S. Tarbouriech, G. Garcia, A.H. Glattfelder (Eds.). Advanced strategies in control systems with input and output constraints *Lecture Notes in Control and Information Science*, vol.346, Springer Verlag, 2007.
26. H.L. Smith and P. Waltman. *The theory of the chemostat*, Cambridge University Press, Cambridge, 1995.
27. E. Zafiriou and H.W. Chiou. Output constraint softening for SISO model predictive control, *Proc. of ACC*, pp. 372-376, 1993.

Index

Lecture Notes in Control and Information Sciences

Edited by M. Thoma, M. Morari

Further volumes of this series can be found on our homepage:
springer.com

Vol. 357: Queinnec I.; Tarbouriech S.; Garcia G.;
Niculescu S.-I. (Eds.):
Biology and Control Theory: Current Challenges
340 p. 2007 [978-3-540-71987-8]

Vol. 356: Karatkevich A.:
Dynamic Analysis of Petri Net-Based Discrete
Systems
166 p. 2007 [978-3-540-71464-4]

Vol. 355: Zhang H.; Xie L.:
Control and Estimation of Systems with
Input/Output Delays
213 p. 2007 [978-3-540-71118-6]

Vol. 354: Witczak M.:
Modelling and Estimation Strategies for Fault
Diagnosis of Non-Linear Systems
215 p. 2007 [978-3-540-71114-8]

Vol. 353: Bonivento C.; Isidori A.; Marconi L.;
Rossi C. (Eds.)
Advances in Control Theory and Applications
305 p. 2007 [978-3-540-70700-4]

Vol. 352: Chiasson, J.; Loiseau, J.J. (Eds.)
Applications of Time Delay Systems
358 p. 2007 [978-3-540-49555-0]

Vol. 351: Lin, C.; Wang, Q.-G.; Lee, T.H., He, Y.
LMI Approach to Analysis and Control of
Takagi-Sugeno Fuzzy Systems with Time Delay
204 p. 2007 [978-3-540-49552-9]

Vol. 350: Bandyopadhyay, B.; Manjunath, T.C.;
Umapathy, M.
Modeling, Control and Implementation of Smart
Structures 250 p. 2007 [978-3-540-48393-9]

Vol. 349: Rogers, E.T.A.; Galkowski, K.;
Owens, D.H.
Control Systems Theory
and Applications for Linear
Repetitive Processes 482 p. 2007 [978-3-540-
42663-9]

Vol. 347: Assawinchaichote, W.; Nguang, K.S.;
Shi P.
Fuzzy Control and Filter Design
for Uncertain Fuzzy Systems
188 p. 2006 [978-3-540-37011-6]

Vol. 346: Tarbouriech, S.; Garcia, G.; Glattfelder,
A.H. (Eds.)
Advanced Strategies in Control Systems
with Input and Output Constraints
480 p. 2006 [978-3-540-37009-3]

Vol. 345: Huang, D.-S.; Li, K.; Irwin, G.W. (Eds.)
Intelligent Computing in Signal Processing
and Pattern Recognition
1179 p. 2006 [978-3-540-37257-8]

Vol. 344: Huang, D.-S.; Li, K.; Irwin, G.W. (Eds.)
Intelligent Control and Automation
1121 p. 2006 [978-3-540-37255-4]

Vol. 341: Commault, C.; Marchand, N. (Eds.)
Positive Systems
448 p. 2006 [978-3-540-34771-2]

Vol. 340: Diehl, M.; Mombaur, K. (Eds.)
Fast Motions in Biomechanics and Robotics
500 p. 2006 [978-3-540-36118-3]

Vol. 339: Alamir, M.
Stabilization of Nonlinear Systems Using
Receding-horizon Control Schemes
325 p. 2006 [978-1-84628-470-0]

Vol. 338: Tokarzewski, J.
Finite Zeros in Discrete Time Control Systems
325 p. 2006 [978-3-540-33464-4]

Vol. 337: Blom, H.; Lygeros, J. (Eds.)
Stochastic Hybrid Systems
395 p. 2006 [978-3-540-33466-8]

Vol. 336: Pettersen, K.Y.; Gravdahl, J.T.;
Nijmeijer, H. (Eds.)
Group Coordination and Cooperative Control
310 p. 2006 [978-3-540-33468-2]

Vol. 335: Kozłowski, K. (Ed.)
Robot Motion and Control
424 p. 2006 [978-1-84628-404-5]

Vol. 334: Edwards, C.; Fossas Colet, E.;
Fridman, L. (Eds.)
Advances in Variable Structure and Sliding Mode
Control
504 p. 2006 [978-3-540-32800-1]

Vol. 333: Banavar, R.N.; Sankaranarayanan, V.
Switched Finite Time Control of a Class of
Underactuated Systems
99 p. 2006 [978-3-540-32799-8]

Vol. 332: Xu, S.; Lam, J.
Robust Control and Filtering of Singular Systems
234 p. 2006 [978-3-540-32797-4]

Vol. 331: Antsaklis, P.J.; Tabuada, P. (Eds.)
Networked Embedded Sensing and Control
367 p. 2006 [978-3-540-32794-3]

Vol. 330: Koumoutsakos, P.; Mezic, I. (Eds.)
Control of Fluid Flow
200 p. 2006 [978-3-540-25140-8]

Vol. 329: Francis, B.A.; Smith, M.C.; Willems, J.C. (Eds.)
Control of Uncertain Systems: Modelling, Approximation, and Design
429 p. 2006 [978-3-540-31754-8]

Vol. 328: Loría, A.; Lamnabhi-Lagarrigue, F.; Panteley, E. (Eds.)
Advanced Topics in Control Systems Theory
305 p. 2006 [978-1-84628-313-0]

Vol. 327: Fournier, J.-D.; Grimm, J.; Leblond, J.; Partington, J.R. (Eds.)
Harmonic Analysis and Rational Approximation
301 p. 2006 [978-3-540-30922-2]

Vol. 326: Wang, H.-S.; Yung, C.-F.; Chang, F.-R.
H_∞ Control for Nonlinear Descriptor Systems
164 p. 2006 [978-1-84628-289-8]

Vol. 325: Amato, F.
Robust Control of Linear Systems Subject to Uncertain
Time-Varying Parameters
180 p. 2006 [978-3-540-23950-5]

Vol. 324: Christofides, P.; El-Farra, N.
Control of Nonlinear and Hybrid Process Systems
446 p. 2005 [978-3-540-28456-7]

Vol. 323: Bandyopadhyay, B.; Janardhanan, S.
Discrete-time Sliding Mode Control
147 p. 2005 [978-3-540-28140-5]

Vol. 322: Meurer, T.; Graichen, K.; Gilles, E.D. (Eds.)
Control and Observer Design for Nonlinear Finite and Infinite Dimensional Systems
422 p. 2005 [978-3-540-27938-9]

Vol. 321: Dayawansa, W.P.; Lindquist, A.; Zhou, Y. (Eds.)
New Directions and Applications in Control Theory
400 p. 2005 [978-3-540-23953-6]

Vol. 320: Steffen, T.
Control Reconfiguration of Dynamical Systems
290 p. 2005 [978-3-540-25730-1]

Vol. 319: Hofbaur, M.W.
Hybrid Estimation of Complex Systems
148 p. 2005 [978-3-540-25727-1]

Vol. 318: Gershon, E.; Shaked, U.; Yaesh, I.
H_∞ Control and Estimation of State-multiplicative Linear Systems
256 p. 2005 [978-1-85233-997-5]

Vol. 317: Ma, C.; Wonham, M.
Nonblocking Supervisory Control of State Tree Structures
208 p. 2005 [978-3-540-25069-2]

Vol. 316: Patel, R.V.; Shadpey, F.
Control of Redundant Robot Manipulators
224 p. 2005 [978-3-540-25071-5]

Vol. 315: Herbordt, W.
Sound Capture for Human/Machine Interfaces: Practical Aspects of Microphone Array Signal Processing
286 p. 2005 [978-3-540-23954-3]

Vol. 314: Gil', M.I.
Explicit Stability Conditions for Continuous Systems
193 p. 2005 [978-3-540-23984-0]

Vol. 313: Li, Z.; Soh, Y.; Wen, C.
Switched and Impulsive Systems
277 p. 2005 [978-3-540-23952-9]

Vol. 312: Henrion, D.; Garulli, A. (Eds.)
Positive Polynomials in Control
313 p. 2005 [978-3-540-23948-2]

Vol. 311: Lamnabhi-Lagarrigue, F.; Loría, A.; Panteley, E. (Eds.)
Advanced Topics in Control Systems Theory
294 p. 2005 [978-1-85233-923-4]

Vol. 310: Janczak, A.
Identification of Nonlinear Systems Using Neural Networks and Polynomial Models
197 p. 2005 [978-3-540-23185-1]

Vol. 309: Kumar, V.; Leonard, N.; Morse, A.S. (Eds.)
Cooperative Control
301 p. 2005 [978-3-540-22861-5]

Vol. 308: Tarbouriech, S.; Abdallah, C.T.; Chiasson, J. (Eds.)
Advances in Communication Control Networks
358 p. 2005 [978-3-540-22819-6]

Vol. 307: Kwon, S.J.; Chung, W.K.
Perturbation Compensator based Robust Tracking Control and State Estimation of Mechanical Systems
158 p. 2004 [978-3-540-22077-0]

Vol. 306: Bien, Z.Z.; Stefanov, D. (Eds.)
Advances in Rehabilitation
472 p. 2004 [978-3-540-21986-6]

Vol. 305: Nebylov, A.
Ensuring Control Accuracy
256 p. 2004 [978-3-540-21876-0]

Vol. 304: Margaris, N.I.
Theory of the Non-linear Analog Phase Locked Loop
303 p. 2004 [978-3-540-21339-0]

Vol. 303: Mahmoud, M.S.
Resilient Control of Uncertain Dynamical Systems
278 p. 2004 [978-3-540-21351-2]

Vol. 302: Filatov, N.M.; Unbehauen, H.
Adaptive Dual Control: Theory and Applications
237 p. 2004 [978-3-540-21373-4]

Made in the USA
Monee, IL
07 July 2026